Mobile Computing and Technology Applications in Tourism and Hospitality

Agustinus Borgy Waluyo
La Trobe University, Australia

Ling Tan
Australian Council for Educational Research, Australia

A volume in the Advances in Hospitality, Tourism, and the Services Industry (AHTSI) Book Series

IGI Global
PUBLISHER of TIMELY KNOWLEDGE

Published in the United States of America by
IGI Global
Business Science Reference (an imprint of IGI Global)
701 E. Chocolate Avenue
Hershey PA, USA 17033
Tel: 717-533-8845
Fax: 717-533-8661
E-mail: cust@igi-global.com
Web site: http://www.igi-global.com

Library of Congress Cataloging-in-Publication Data

Names: Waluyo, Agustinus Borgy, 1973- editor. | Tan, Ling, 1973- editor.
Title: Mobile computing and technology applications in tourism and hospitality / Agustinus Waluyo, and Ling Tan, editor.
Description: Hershey PA : Business Science Reference, [2021] | Includes bibliographical references and index. | Summary: "This book presents the latest research and development in mobile and intelligence computing with a focus on tourism and hospitality sectors and how during this current pandemic crisis, the role of research on innovative technologies and applications will offer timely help to bring the tourism and hospitality industry back to its normal state"-- Provided by publisher.
Identifiers: LCCN 2020057661 (print) | LCCN 2020057662 (ebook) | ISBN 9781799869047 (hardcover) | ISBN 9781799869054 (paperback) | ISBN 9781799869061 (ebook)
Subjects: LCSH: Tourism--Information technology. | Hospitality industry--Information technology. | Mobile computing--Economic aspects. | Wireless communication systems--Economic aspects. | COVID-19 (Disease)--Economic aspects.
Classification: LCC G156.5.I5 M634 2021 (print) | LCC G156.5.I5 (ebook) | DDC 910.285--dc23
LC record available at https://lccn.loc.gov/2020057661
LC ebook record available at https://lccn.loc.gov/2020057662

This book is published in the IGI Global book series Advances in Hospitality, Tourism, and the Services Industry (AHTSI) (ISSN: 2475-6547; eISSN: 2475-6555)

British Cataloguing in Publication Data
A Cataloguing in Publication record for this book is available from the British Library.

All work contributed to this book is new, previously-unpublished material.
The views expressed in this book are those of the authors, but not necessarily of the publisher.

For electronic access to this publication, please contact: eresources@igi-global.com.

Advances in Hospitality, Tourism, and the Services Industry (AHTSI) Book Series

ISSN:2475-6547
EISSN:2475-6555

Editor-in-Chief: Maximiliano Korstanje University of Palermo, Argentina

MISSION

Globally, the hospitality, travel, tourism, and services industries generate a significant percentage of revenue and represent a large portion of the business world. Even in tough economic times, these industries thrive as individuals continue to spend on leisure and recreation activities as well as services.

The Advances in Hospitality, Tourism, and the Services Industry (AHTSI) book series offers diverse publications relating to the management, promotion, and profitability of the leisure, recreation, and services industries. Highlighting current research pertaining to various topics within the realm of hospitality, travel, tourism, and services management, the titles found within the AHTSI book series are pertinent to the research and professional needs of managers, business practitioners, researchers, and upper-level students studying in the field.

COVERAGE

- Service Design
- Destination Marketing and Management
- Health and Wellness Tourism
- International Tourism
- Food and Beverage Management
- Leisure & Business Travel
- Customer Service Issues
- Casino Management
- Hotel Management
- Travel Agency Management

IGI Global is currently accepting manuscripts for publication within this series. To submit a proposal for a volume in this series, please contact our Acquisition Editors at Acquisitions@igi-global.com or visit: http://www.igi-global.com/publish/.

The Advances in Hospitality, Tourism, and the Services Industry (AHTSI) Book Series (ISSN 2475-6547) is published by IGI Global, 701 E. Chocolate Avenue, Hershey, PA 17033-1240, USA, www.igi-global.com. This series is composed of titles available for purchase individually; each title is edited to be contextually exclusive from any other title within the series. For pricing and ordering information please visit http://www.igi-global.com/book-series/advances-hospitality-tourism-services-industry/121014. Postmaster: Send all address changes to above address.

Titles in this Series

For a list of additional titles in this series, please visit:
http://www.igi-global.com/book-series/advances-hospitality-tourism-services-industry/121014

Handbook of Research on Digital Communications, Internet of Things, and the Future of Cultural Tourism
Lídia Oliveira (University of Aveiro, Portugal)
Business Science Reference • © 2022 • 587pp • H/C (ISBN: 9781799885283) • US $325.00

New Governance and Management in Touristic Destinations
Marco Valeri (Niccolò Cusano University, Italy)
Business Science Reference • © 2022 • 357pp • H/C (ISBN: 9781668438893) • US $225.00

Global Perspectives on Literary Tourism and Film-Induced Tourism
Rita Baleiro (University of the Algarve, Portugal & CiTUR, Portugal) and Rosária Pereira (University of the Algarve, Portugal & CinTurs, Portugal)
Business Science Reference • © 2022 • 371pp • H/C (ISBN: 9781799882626) • US $215.00

Planning and Managing the Experience Economy in Tourism
Rui Augusto Costa (GOVCOPP, University of Aveiro, Portugal) Filipa Brandão (GOVCOPP, University of Aveiro, Portugal) Zelia Breda (GOVCOPP, University of Aveiro, Portugal) and Carlos Costa (GOVCOPP, University of Aveiro, Portugal)
Business Science Reference • © 2022 • 407pp • H/C (ISBN: 9781799887751) • US $195.00

Food Safety Practices in the Restaurant Industry
Siti Nurhayati Khairatun (Universiti Putra Malaysia, Malaysia) Ainul Zakiah Abu Bakar (Universiti Putra Malaysia, Malaysia) Noor Azira Abdul Mutalib (Universiti Putra Malaysia, Malaysia) and Ungku Fatimah Ungku Zainal Abidin (Universiti Putra Malaysia, Malaysia)
Business Science Reference • © 2022 • 334pp • H/C (ISBN: 9781799874157) • US $215.00

Prospects and Challenges of Community-Based Tourism and Changing Demographics
Ishmael Mensah (University of Cape Coast, Ghana) and Ewoenam Afenyo-Agbe (University of Cape Coast, Ghana)
Business Science Reference • © 2022 • 300pp • H/C (ISBN: 9781799873358) • US $195.00

For an entire list of titles in this series, please visit:
http://www.igi-global.com/book-series/advances-hospitality-tourism-services-industry/121014

IGI Global
PUBLISHER of TIMELY KNOWLEDGE

701 East Chocolate Avenue, Hershey, PA 17033, USA
Tel: 717-533-8845 x100 • Fax: 717-533-8661
E-Mail: cust@igi-global.com • www.igi-global.com

Table of Contents

Detailed Table of Contents

The COVID-19 pandemic process, which has affected the whole world, has accelerated technological developments by laying the groundwork for digital transformation in many fields. In this context, various technological innovations have been introduced, such as a social distance tracking system that warns when the appropriate social distance rules are violated between employees, robot dogs on patrol, the use of drones for delivery, and robots that can do many jobs that humans can do. The restaurant industry has also tried to integrate these rapid technological developments as much as possible. In this study, which is about how the recent advances in technology have or can be affected restaurants, examples from the world are given for each technological development mentioned in order to better understand the subject. It is considered that the study will contribute to the literature, future studies, sector representatives who want to apply technological developments in their restaurants, and investors who want to open a new restaurant with technological infrastructure.

In today's scenario, there has been a dramatic transformation in the way services are being provided and taken by the consumers. With the changing preferences of consumers, and businesses, vendors across the regions are emphasizing towards the introduction of touchless solutions, which are likely to change the way people interact and use the services. Incorporation of smart technologies enables the operators to create a positive and seamless experience for the users. Convergence of technologies with the current market trends is expected to bring the concept of "internet of things" nearer to reality. Embracing IoT solutions transform the overall hospitality business scenario and pose several opportunities for the smart hotel solutions. In this chapter, various IoT solutions used in the hospitality and leisure sector have been discussed. This chapter further emphasizes how efficiently the touchless technology products with minimal contact enhance the operations in hospitality sector and how the hotel industry is focusing towards nurturing the dreams of travellers.

Smart technologies are becoming rapidly used in various industries successfully. The tourism industry stands as one of the evolving industries, benefiting from these smart technological developments such as virtual and augmented reality, robotics, and internet of things. Anticipatory, experiential, and reflective are the three main phases of the consumer behavior process in tourism, which are regarded as pre-travel, during travel, and post-travel phases in tourism. Smart tourism technologies are being implemented to enhance the tourist experience in these phases of their journey. This chapter aims to highlight the smart tourism technology applications in every phase of consumer experience by presenting examples from the tourism industry.

The hospitality and tourism industry is one of the most diversified and biggest industries in the world. Increasing trends of technology and innovation are expected to favorably influence the future scenario of the hospitality and tourism industry. Growing adoption of new technologies in these industries over the past few years has significantly reformed the services provided by the same. One of the major goals of the upgradation of services by the hotels or leisure segments is digitalization of services and personalization of customer experiences. In this chapter, state-of-the-art technologies currently being used in the hospitality and tourism sector have been discussed. This chapter further emphasizes how information technology is proving significant in keeping pace with the competitiveness in terms of services in the hospitality and leisure segments. In addition to all of the above, the chapter also focuses on the trends, impacts, and opportunities of information technology applications in the industry.

Tourism has been one of the engines powering the thriving economies, leveraging the quality of lives and transforming the people's sociocultural status worldwide. There is no doubt that innovative technologies are critical and significantly efficient for the economies during a hard time and in blossoming times. Moreover, it is also vital to understand and present some essential issues that harm tourism development. Thus, this chapter strives to offer innovative technology-oriented alternative strategies and practical implications to benefit all the stakeholders at the destination and the consumers by focusing on the long-term benefits. This chapter also aims to understand how technology can take a role to stimulate visitors' motivation and behaviour towards a touristic product or a destination.

Chapter 6
Museum Experience and Digital Consumption: The Role of Mobile Augmented Reality in Tunisia's Cultural Heritage 120
Khouloud Mouihbi, IHEC of Carthage, Tunisia
Nedra Bahri-Ammari, IHEC of Carthage, Tunisia
Daouda Coulibaly, ISTEC Paris, France

This study explores the role of mobile augmented reality (MAR) technology in enhancing user experience (UX) in the context of cultural heritage. Data were collected via individual semi-structured interviews, according to a theoretical model of UX. The findings show that MAR can enhance UX in cultural heritage museums when the product's features and characteristics match users' needs. A contemporary challenge in the cultural heritage industry is to find creative ways to attract new visitors and enhance their museum experience. Therefore, this study was conducted to improve museum curators' understanding about the UX of MAR apps to satisfy their visitors' expectations.

Chapter 7
Blockchain Technology: A Revolution in the Hospitality and Tourism Industry 145
Anusha Thakur, University of Petroleum and Energy Studies, India

Travel and technology tend to possess seamless amalgamation. Advancements in technologies are significantly changing the way people travel, with exciting and promising experiences. Incorporation of smart devices and solutions in businesses enables easier ways to analyze, collect, as well as share the real-time data with vendors. One such concept includes blockchain technology, inclusion of which brings about an extreme change in the way the data are being accessed and stored. This technology enables the businesses to fortify operations, transactions, as well as processes, thereby bolstering productivity and growth and discovering newer profit models. This chapter discusses the different competitive advantages offered by the tourism businesses on inclusion of blockchain technology in their strategic-decisions

Preface

COVID-19 AND IMPACT TO THE TOURISM AND HOSPITALITY SECTOR

In December 2019, the first Coronavirus (COVID-19) case was reported in Wuhan, China (World Health Organization, 2020). The virus was spiralling out to the whole world, leading to uncertain circumstances which warranted calls for stringent measure including various stages of lockdown. Only essential sectors were allowed to open their business during lockdown, and the tourism and hospitality sector was not exempted from this regulation. Though many sectors have been drastically affected, aviation, tourism and hospitality were brought to a complete halt with a real threat of significant declines in insolvencies, revenue and job losses (United Nations World Tourism Organization (UNWTO), 2020). The current pandemic has resulted in a global economic and health crisis and created unprecedented disruptions to the world economy's leading sectors.

To put into context, in 2019 travel and tourism was considered a significant contributor to the world economy when it directly contributed USD 2.9 trillion to the overall growth of the global economy of USD 9.3 trillion (Abbas et al, 2021). During the pandemic, the travel and tourism industry has been hit hard by the travel restrictions worldwide, and a 98% sharp decline was identified in May 2020. More specifically, based to the available data, it was estimated the arrivals of global tourists decreased about 300 million from January to May 2020 compared to 2019 for the same period (Abbas et al, 2021). Further, it reported a 320 billion dollars loss in relation to international tourism receipts based on export revenue. Asia and the Pacific region were the first to suffer and recorded a steep 60% drop in arrivals from January to May 2020.

This occurrence is considered as the first in the history, where businesses in almost 200 countries have been paralyzed due to a common factor. Only time will truly tell the full impact of COVID-19 on the global business scenario. The landscape of hospitality could possibly change forever, and in order to stay relevant, we must create new and creative ways to rejuvenate the industry.

BEHAVIOUR AND SOCIETAL CHANGE

While the hospitality industry appears to be slowly recovering, COVID-19 continues to exert profound impacts on how hospitality and tourism businesses operate. Substantial changes are required to sustain their operations in the COVID-19 environment in order to ensure employees' and customers' health and safety, which would hopefully increase the confidence level of the customers to patronize their business (Gössling et al., 2020). With unprecedented challenges faced by the hospitality industry in the COVID-10 era, it also presents research opportunities for hospitality scholars to shift their research focus to develop solutions for the industry.

The impact of COVID-19 and the associated measures and restrictions in place across the globe has triggered some behaviour and societal change. Preliminary observation of a longitudinal study suggests that reopening the sit-down restaurants and easing travel restrictions will not bring customers back immediately (Gursoy et al., 2020). A large portion of individuals (over 50%) are not willing to dine in at a restaurant immediately. The same applies to staying at hotels. Most customers (over 50%) are not willing to travel to a destination and stay at a hotel any time soon. Only around a quarter of the customers have already dined in a restaurant and only around one-third are willing to travel to a destination and stay at a hotel in the foreseeable future (Gursoy et al., 2020). These findings suggest that customers in general still do not feel comfortable to dine in at a sit-down restaurant, travel to a destination and stay at a hotel. Since the breakeven point in the hospitality industry is relatively high due to high operating costs, the survival of many hospitality businesses heavily depends on increasing the demand for their services and products.

The magnitude of this crisis and its devastating effects on operations, employees, and customers are unparalleled compared to previous crises. Thus, identifying what will make customers return is essential and this requires intensive research efforts. The industry and the academia are in urgent need of behavioral and operational hospitality marketing, management and technological research to guide the hospitality operations in the time of COVID-19 pandemic.

MOVING FORWARD

One of the most significant findings indicate that a large proportion of restaurant customers (64.71%) and the majority of hotel customers (70.42%) believe that the use of various technologies in service delivery will be necessary in the COVID-19 environment in order to minimize human-to-human contact (e.g. service robots, contactless payment, digital menus that can be viewed on personal mobile devices via QR codes, contactless digital payments, keyless entry, touchless elevators, etc.)

(Gursoy et al., 2020). These findings strongly suggest that technology integration and adoption into hospitality operations will likely be integral in the near future.

THIS BOOK

While hospitality researchers have been studying the use of various technologies in hospitality over the years, most of those studies have focused on unintelligent technology adoption. Therefore, it is becoming essential to explore and leverage the power of mobile and wireless intelligence to assist with the recovery of the tourism and hospitality sector during and after the pandemic, which constitutes the theme of this book.

More specifically, this book aims to present the latest research and development in mobile and intelligence computing to support tourism and hospitality sectors in adapting to the behaviour and societal change of the community in general and the ongoing impact of the COVID-19 pandemic. The role of research on innovative technologies and applications will never be as significance to timely help the tourism and hospitality industry to prevail in the next normal.

The COVID-19 pandemic has incurred a substantial economic and social impact around the world. Many sectors have been drastically affected, and the tourism and hospitality industry were one of the hardest hits by the pandemic globally.

Strategies to manage the COVID-19 outbreaks such as community lockdowns, stay-at-home orders, travel and mobility restrictions have resulted in temporary closure of many hospitality businesses and significantly decreased or ceased the demand for tourism (Bartik et al., 2020) with a sharp decline in hotel occupancies and revenues. Paradigm-shifting is now taking place such as social distancing, limiting the number of customers served, more rigorous and frequent cleaning of common areas are part of the most important safety precautions customers expect from a restaurant and a hotel (Gursoy et al., 2020).

The role of technology to the tourism and hospitality's revitalization has never been more critical. Mobile computing and wireless technology have grown at a phenomenal rate in the recent years, and so has artificial intelligence. The most advanced software applications that the world ever witnesses can be realised through a combined aspect of these disciplines.

This book attempts to dedicate to the research on advanced solutions, applications and services in tourism and hospitality with the purpose to disseminate the most recent research under this theme and how technological innovations can change and contribute to the recovery and sustainability of the sectors going forward. We have received 28 chapter proposals from different parts of the world, and have conducted a double-blind review process for each chapter. Based on the reviews, 9 chapters are selected for inclusion in this book.

The chapters in this book cover four most significant areas of research and how each could contribute to the future of tourism and hospitality sectors, namely: (1) Mobile Technology, (2) Blockchain, (3) Virtual and Augmented Reality, and (4) Artificial Intelligence as outlined below.

Chapter 1 presents the state-of-the-art review of digital transformation in restaurants including the evolution of traditional restaurants from labour-intensive services to the food and beverage automation systems, QR menus, and robotic chefs to name a few. The digitalized restaurants with smart technologies would not only offer personalized service, but also enable innovations into restaurants as a one step ahead in supporting the safety measures introduced across jurisdictions and mitigating the impact of COVID-19 pandemic.

Chapter 2 is dedicated to the assessment of various Internet-of-Things (IoT) solutions used in the hospitality and tourism sector. More specifically, it discusses touchless and sensing technology solutions and how their role in the operations in hospitality sector including the impacts, challenges and opportunities, offered by IoT technology in the industry.

Chapter 3 focuses on the technological trends in the hospitality and tourism industry, along with the future aspects of next generation technologies, such as digital solutions for classifying the touchpoints and confirming minimum physical contact among the employees and guests at every point of contact which is expected to be significant. In addition to these trends, some challenges, and future-proof solutions with enhanced technological aspects to the hospitality industry are also discussed in this chapter.

Chapter 4 addresses some emerging technologies that can assist the hospitality industry and achieve more sustainable models of tourism during and post-COVID-19 pandemic. Part of these technologies is the use of augmented and virtual reality in the tourism and hospitality sector, which is discussed in detail in Chapters 5 and 6. Other relates to the blockchain technology, and how this technology could contribute to the hospitality and tourism industry is described in Chapter 7.

Chapter 8 presents artificial intelligence integrated mobile technologies in tourism in terms of technological, marketing, and managerial perspectives. A framework is introduced for artificial intelligence integration to mobile tourism services and theoretical-managerial implications. The chapter argues to succeed, hospitality businesses should integrate AI applications into their mobile applications, enriching tourist's continuous, well-predicted and instant experiences.

Chapter 9 studies a multi-generational population (Gen X, Y and Z) including what each generation prefers and what they value by gathering the different reactions needed to succeed. This study aims to inform tourism industries especially mobile app developers, smart device manufacturing companies, travelling organisation and tourism stakeholders to incorporate suitable or new features into smart devices and introducing new services to the apps for promoting services and products in more effective and sustainable methods.

We would like to thank all the authors for their valuable contributions to the book.

Agustinus Borgy Waluyo
La Trobe University, Australia

Ling Tan
Australian Council for Educational Research, Australia

REFERENCES

Bartik, A. W., Bertrand, M., Cullen, Z. B., Glaeser, E. L., Luca, M., & Stanton, C. T. (2020). *How are small businesses adjusting to COVID-19? Early evidence from a survey (No. w26989)*. National Bureau of Economic Research. doi:10.3386/w26989

World Health Organization. (n.d.). *Listings of WHO's response to COVID-19*. https://www.who.int/news/item/29-06-2020-covidtimeline

Gursoy, D., Chi, C. G., & Chi, O. H. (2020). *COVID-19 Study 2 Report: Restaurant and Hotel Industry: Restaurant and hotel customers' sentiment analysis. Would they come back? If they would, WHEN?* (Report No. 2), Carson College of Business, Washington State University.

Gössling, S., Scott, D., & Hall, C. M. (2020). Pandemics, tourism and global change: A rapid assessment of COVID-19. *Journal of Sustainable Tourism*, 1–20.

United Nations World Tourism Organization. (2020). UNWTO *World Tourism Barometer, 18*(2).

Abbas, J., Mubeen, R., Iorember, P., T., Razad, S., Mamirkulova, G. (2021). Exploring the impact of COVID-19 on tourism: transformational potential and implications for a sustainable recovery of the travel and leisure industry. *Current Research in Behavioral Sciences, 2*.

Chapter 1
Digital Transformation in Restaurants

Aysu Altaş
Aksaray University, Turkey

ABSTRACT

The COVID-19 pandemic process, which has affected the whole world, has accelerated technological developments by laying the groundwork for digital transformation in many fields. In this context, various technological innovations have been introduced, such as a social distance tracking system that warns when the appropriate social distance rules are violated between employees, robot dogs on patrol, the use of drones for delivery, and robots that can do many jobs that humans can do. The restaurant industry has also tried to integrate these rapid technological developments as much as possible. In this study, which is about how the recent advances in technology have or can be affected restaurants, examples from the world are given for each technological development mentioned in order to better understand the subject. It is considered that the study will contribute to the literature, future studies, sector representatives who want to apply technological developments in their restaurants, and investors who want to open a new restaurant with technological infrastructure.

INTRODUCTION

The industrial revolution, which started after the 1760s, continued gradually in parallel with technological developments. The period between 1760 and 1830 is called the 1st Industrial Revolution by Industrial historians. In this period, the production sector, which was monopolized by the craftsmen, shifted to factories with the development of watermills and then steam engines. Thus, with their abilities far

DOI: 10.4018/978-1-7998-6904-7.ch001

above human power, the machines both increased the production speed and ensured standardization. The period between 1830 and 1945 is called the second Industrial Revolution. This period is divided into two as the age of coal and the age of oil. In the first stage, the railways increased the speed of transportation and the transportation of goods and people became cheaper, comfortable, and faster. Interchangeable parts laid the foundations for industry standards. This early period of the 2nd Industrial Revolution can be called the age of steam engines. Coal is the energy source of this period. Developments in electricity generation and internal combustion engines are at the center of the second period of the 2nd Industrial Revolution. Oil, another energy source, has replaced coal. The challenging competitive environment created by the two world wars in the first half of the 20th century has advanced the pace of development for decades. After 1945, in the Third Industrial Revolution era, programmable logic circuits that enable machines to do complex tasks provided new opportunities to the industry. Coordinated programmable robots that can perform tasks that were previously considered to be impractical by machines have increased industrial production. Materials science is an important pillar of the 3rd Industrial Revolution. Polymers and nano-technological materials have made possible new applications that have not been seen before. The greatest advances of the period were nuclear technologies, spacecraft, and computers (Proente Automation, n.d.).

Today, Industry 4.0, which is the latest stage of the industrial revolution, is being experienced. With Industry 4.0, concepts such as Smartphone technology, mobile internet, social media usage, cloud technology, virtual reality, artificial intelligence, and the Internet of Things have become an integral part of today's people. Although these concepts have a very recent history, they have become indispensable in people's lives in a very short time. For example, considering that the first Smartphone was invented by IBM Company in August 1994 (Cumhuriyet Newspaper, 2015); it becomes clear once again how quickly people adapt to these technologies. All these innovations have been integrated into each sector in different ways. The first examples of digital transformations in the tourism industry, where restaurants are also a pillar, were seen in hotels. For example, in the 290-room FlyZoo Future hotel of Chinese electronic e-commerce giant Alibaba Group in Hangzhou, China, many services such as registration, control of lights, and room service are performed by artificial intelligence robots. Hilton hotels, on the other hand, have started using a real robot called "Connie" who assists guests at check-in and advises on famous local sights and things to do, in Virginia, US. Connie, a small robot standing at the hotel reception desk, learns and develops itself as it interacts with visitors. It can also access WayBlazer's travel information to help customers get the information they need (New Software Solutions, 2020). Aiming to provide "concierge" service to hotel customers, Connie is named after Conrad Hilton, one of the founders of the Hilton hotel chain, and was developed upon the cooperation of IBM, one of the world's largest information technology companies, with Hilton Hotels (Pr Outlet Plus, 2016).

Apart from these, the following examples can be given regarding the digitalization of the tourism sector: Atlanta-based InterContinental Hotels Group (IHG) has announced the launch of a new delivery robot, named Dash, at the Crowne Plaza San Jose-Silicon Valley hotel in 2015. The Starwood Aloft Hotel has introduced the robot butler Boltr to provide comfort and help to the hotel guests. Singapore has tested a robotic virtual assistant, SARA (Singapore Auto-Response Assistant) offering touristic assistance for visitors coming to Singapore. SARA is one of the robotics that able to detect the user's location on a map by using a GPS integrated module and accordingly can provide real-time orientation and direction help. Sanbot provides flight information to tourists at Chinese airports and guides passengers to their destinations. Hen-na Hotel, which includes robot employees, started its operations in Nagasaki, Japan in 2015. Hen-na Hotel greets guests at the reception with a lifelike woman Android service robot, answers simple questions, and explains how to check in to the hotel. This robot receptionist, who has a smile on her face, is practically no different from any human employee. Except for the security personnel, all the employees of the hotel are robots. In this hotel where robots are employees, people are in the position of managers who control errors. Junko Chihira, a realistic android robot produced by Toshiba, works full-time at a tourist information center in Tokyo. Junko Chihira can greet customers, inform visitors about current events, and speak Japanese, Chinese, English, German, and even sign language (Library of the Hidden Sciences, n.d.).

Restaurant businesses, which are part of a labor-intensive industry, did not remain uninvolved in these developments, and digital transformation has started in restaurants. It can be said that the first step of digital transformation in restaurants is taken with the use of wireless internet (Wi-Fi). Providing unlimited internet to the restaurant customers during the period they receive food and beverage service was considered as a very important development at the beginning. Another significant development regarding restaurants is the creation of Smartphone applications that are used to evaluate all restaurants around the world, and these applications become a part of people's daily life. It is known that today's consumers tend to get service by researching before choosing a restaurant, and they usually used these applications installed on Smartphone for researching restaurants. Consumers who have a restaurant experience can share their opinions as comments with these applications and can evaluate restaurants according to many criteria such as food, taste, service speed, cleanliness, atmosphere, and price. In addition, the fact that many of these applications allow country and language selection, makes them available worldwide.

The food and beverage automation systems used in cafes and restaurants, QR menus, the use of robots as service personnel, the work of robotic chefs in food preparation units, food production with 3D printers, and many innovations designed within the scope of the internet of things is considered as the latest developments in digital transformation in restaurants for now, and studies in the literature generally follow this axis. While these developments in technology bring a new perspective to the economy and form the basis of future industrial developments, they also raise serious concerns about privacy, security, and protection of personal data. It also brings the question to mind "Is it a danger for the future of human beings to be taken out of the system in working life?" Rapid progress in technology will inevitably lead to different improvements and accompanying concerns to the agenda in the near future.

LITERATURE REVIEW

There are numerous studies in the literature related to digitalization in the food and beverage sector, although not all of them are directly related to the digital transformation in restaurants. In their study titled "Robotics and Food Technology: A Mini-Review" Nayik, Muzaffar and Gull (2015) pointed out that there are some difficulties related to the use of robots in the food industries. They emphasized that the first of these difficulties is that the objects dealt with are variable in size, shape, weight, and position, the second is that the objects to be processed are usually coated with sensitive, slippery or sticky substances, and the third is the hygiene, quality, and consumer safety concerns. Since the study was published in 2015, it can be stated that it has a general opinion that it would be more appropriate to use robots in food packaging. However, the point reached today is much further than that.

Sun et al. (2015) conducted a study titled "An Overview of 3D Printing Technologies for Food Fabrication" to gather, analyze, classify and summarize published articles on 3D food printing and its impact on food processing, as well as to give an idea of the direction of its future development. The study concluded that 3D food printing integrates 3D printing and digital gastronomy to revolutionize food manufacturing, and food products could be designed and produced by controlling the amount of printing material and nutrient content to meet individual needs (Sun et al., 2015a). The same year, Sun et al. carried out another study titled "A Review on 3D Printing for Customized Food Fabrication" that introduced the concept designs and viable prototypes of the first generation food printers aiming to revolutionize the customized food production with 3D printing (Sun et al., 2015b).

Mogali (2015), in her study titled "Internet of Things and its Role in Smart Kitchen", focused on the internet of things and presented information about smart kitchen equipment developed in this context. Özgüneş et al. (2020) examined the reports of the IFR International Robotics Federation and discussed the future changes in the food and beverage industry and the possible effects of these changes on the individual and society in their study titled "Towards Advanced Technology and Pandemic Order in Food and Beverage Sector: Is It Possible to Say, 'Bless Your Hands!' to a Robot in the Near Future?" In their study titled "Use of Artificial Intelligence in Food and Beverage Business" Çerkez & Kızıldemir (2020) aimed to reveal the current status of consumers on technology usage and expectations and the differences between age groups in a restaurant where artificial intelligence and robotic technology applications are involved. Hazarhun & Yılmaz (2020) evaluated the data they obtained as a result of interviews with the owner of a restaurant in Istanbul, where robot waiters work and smart tables are used in their study titled "Digital Transformation in Restaurants: The Touch Restaurant Example."

Another study on digitalization in restaurants is the study titled "On the Road from Tourism 4.0 to Gastronomy 4.0: Restaurants and Management of the Future" by Yıldız & Davutoğlu (2020). Yıldız & Davutoğlu (2020) in their study, in which they conducted a literature review and document analysis, argued that harmonizing digital technologies with each other and offering the consumer the digital experience they seek with a comprehensive service model will be the most important task of the restaurant managers of the future. Asif et al. (2015) mentioned the design and development process of a waiter robot and the fact that robot technology creates a rapid shift from manual labor to robotic labor in their study titled "Waiter Robot–Solution to Restaurant Automation", which they presented in a congress. They also mentioned that customers in classical cafes, restaurants, and hotels face many problems due to the insufficient number of waiters and manual order taking, especially during rush hours, and emphasized that these deficiencies can be overcome by a restaurant automation system with 'Robot Waiters'. Considering that today, there are great developments in the direction stated by the authors in that study, it can be stated that that study, which belongs to the year 2015, is highly predictive. On the other hand, considering that those who conducted the study worked in the electrical engineering department, it was once again revealed how multidisciplinary the digital transformation in restaurants is.

Kim et al. (2021) examined robot service in restaurants and expectations between customers and employees with a survey study titled "On the Common and Different Expectations on Robot Service in Restaurant between Customers and Employees" and presented in a congress. They concluded that depending on the service elements that the restaurant industry generally adopts, customers and employees share many common expectations and that personalized service is important for customer-oriented robot service in restaurants. Seo & Lee (2021), in their study titled "The Emergence of Service Robots at Restaurants: Integrating Trust, Perceived Risk, and Satisfaction," conducted a scenario-based online survey on 338 participants. Seo & Lee (2021) concluded that perceived risk decreases satisfaction, increased trust in robot service increases satisfaction, and perceived ease of use has an indirect effect on consumers' intention to revisit robot restaurants.

In his study titled "Robot Chefs in Gastronomy Tourism: What's on the Menu?" Fuste-Forne (2021) conducted semi-structured interviews with 12 tourists about the impressions created by robot chefs on tourists with the robotization of gastronomic tourism experiences. To reveal perceptions of tourists about a dining experience where service robots are involved and the reasons why tourists visit or do not visit restaurants where robots cook, the questions "How do tourists perceive a dining experience where service robots participate?", and "What are the reasons why tourists would or wouldn't visit restaurants where robots cook?" were asked to participants consisting of ten female and two male tourists. Kwak et al. (2021), in their study titled "Senior Consumer Motivations and Perceived Value of Robot Service Restaurants in Korea", analyzed 243 questionnaires to reveal the reasons why older customers prefer robot service restaurants in South Korea. It was revealed that older customers have a positive attitude towards robot service restaurants, but customer attitudes differ according to their income levels.

In their study titled "Robotic Applications and Artificial Intelligence in Gastronomy", Bozkurt Uzan & Sevimli (2020) conducted a literature review to discuss the intersection of robotic field and gastronomy field and robot chefs used in the gastronomy sector. They concluded that the use of robots in kitchens saves space and personnel and increases the attractiveness of the business. In their study titled "Serving Robots: Management and Applications for Restaurant Business Sustainability", Jang and Lee (2020) focused on the "anthropomorphism", "animacy", "likeability", "intelligence", and "safety" characteristics of robots serving in restaurants and their effects on restaurant customers. In this context, 294 online surveys were conducted with customers aged eighteen and older at restaurants using service robots. Researchers aimed to provide a sustainable development model for restaurants with manpower shortages and concluded that customers have a positive perception about service robots and that the perceived value positively affects their satisfaction and revisit intention.

Examples of Digital Transformation in Restaurants

Examples of digital transformation in restaurants can be gathered under the headings of food and beverage automation systems, QR menus, use of humanoid service robots in restaurants, use of robotic chefs in restaurants, internet of objects technologies in restaurants, and 3D food printers.

Food and Beverage Automation Systems

The design of early examples of these devices almost look like a bank ATM, and later resized according to the counters in restaurants/bars, have been redesigned in parallel with the rapid developments in Smartphone technology. Today, "handheld terminals", that look like a mobile phone, have become widely used. In addition to basic functions such as taking orders and reservations, these systems have capabilities that extend to much more complex operations such as customer relationship management, stock tracking, loyalty programs, creating promotions, and targeting customers by segmenting. These systems are recommended to be used especially by medium and large-sized businesses. Considering that there is restaurant personnel who earn a commission at a certain rate over the sales price of the food and beverage they sell, instead of getting a salary; the automation system can issue separate sales documents for each personnel on a daily basis, and the process of determining the wages for each personnel is managed transparently. Since these systems accelerate the coordination between the kitchen and service personnel, shorten the service times, save time and increase customer satisfaction, they are very effective in minimizing the problems experienced in cafes and restaurants. Food ordering applications, which have become widespread with the development of mobile applications for mobile phones and tablets, can also be integrated into automation systems. Thus, orders given by customers from their homes or workplaces are automatically transmitted to the kitchen and bar section of the relevant restaurant.

QR menus

Another development related to digital transformation in restaurants has been experienced in menus. In the beginning, the menus were classified as classic and modern menus according to the ages, table d'hote, a la carte, cycle, California, and carte du jour menus according to the prices, breakfast, lunch, brunch, dinner, supper menus according to the time of the day, and menus according to the seasons. Afterward, the restaurants presented "Menus Prepared in Line with Special Needs", taking into account the specific demands of the customers. For example, the demands of customers who are allergic to various foods such as peanuts and exotic fruits, who have celiac disease, who have religious dietary restrictions and ethical concerns, who are on a diet due to a disease, or just because of their worries about healthy eating have been taken into account and special menus have been designed by the restaurants for these customers. Vegetarian and vegan menus are also among the menu types that can be included in the menus group prepared in line with special needs.

Apart from all these menus, it is now possible to classify the menus in terms of printing techniques and technology. In this context, in terms of printing technique and technology, the menus are divided into three as Braille menus, audio menus, and electronic menus. Braille menus are created by printing the à la carte menus on special papers (the weight of these papers is 160 gr/m2) with the 'Braille alphabet' that the visually impaired can read. Although it appeals to a limited number of consumers, in terms of "accessible tourism" and "accessible restaurants", Braille menus are a significant development. The first emergence of electronic menus was the use of tablets belonging to the restaurant instead of paper-printed menu cards. In this system, when customers want to see the menu, they are given tablets with the restaurant menu installed and customers can choose their food and drink from the photos on the tablet. However, in these menus, the orders still taking by service personnel. It has been observed that this development, which excited people when it first came out, has some disadvantages. These are:

- Although prices vary according to the preferred brand, tablets are expensive products.
- Since they are electronic goods, there is a possibility of breakdown. In case of breakdown, it is necessary to send them to authorized services for repair, which takes a few days.
- Just like a Smartphone, daily charge and periodic updates are required.
- Although daily cleaning is done by the service personnel, it does not seem possible to clean after each guest, especially during rush hours. Therefore, it is not hygienic.

Today, all the disadvantages mentioned here and the Covid-19 measures have been taken into account, and the menus that can be used with QR codes have started to become widespread. These menus, which are given different names such as QR menus or Digital QR menus, are used as follows (Protel, n.d.):

- The customer at the table scans the QR code using his/her mobile phone's camera.
- The customer reaches the most recent menu on the QR Menu Order page integrated with the POS system and places an order through this same page.
- The order reaches the kitchen and the POS system both at the same time.
- The prepared order is delivered to the customer's table by the service personnel.
- The online payment takes place via credit card registered on the QR Menu Order page.
- The customer can also prefer to register an email address in order to receive an e-invoice relating to the transaction.

Figure 1. The digital QR menu
Source. (Pill Digital, n.d.).

Figure 1 shows an example of the digital QR menu.

The advantages of QR menus for restaurants can be listed as follows (Pardon, n.d .; Unilever Food Solutions, n.d.):

- *The workload reduces:* The workload is one of the biggest burdens in the restaurant industry. Thanks to digital QR menus, the need for labor will be relatively reduced. If a restaurant's order-taking process is digitized, the number of required personnel to run that restaurant will also be significantly reduced. However, this does not mean that employment has to be reduced. The digital QR menu complements the waiters in most restaurants and does not replace them. It should not be forgotten that happy personnel who are less tired lead to happy customers.
- *It is cheaper:* Digital QR menus are cheaper than written menus that require reprinting every time a change is required. Also, since they can be updated instantly instead of reprinting the menu, they eliminate the need for cellulose to produce printed menus and the cutting of trees.
- *Productivity Increases:* The digital QR menu increases the productivity of the workforce. Restaurants that use digital QR menus minimize errors while receiving orders from customers. In addition, in the digitalizing world, customers now prefer to order from digital menus instead of old menus. Thus, there is no communication problem between the customer and the restaurant.
- *Revenue Increase:* The biggest benefit of having a digital menu is increased revenues. Restaurants with digitized menus record higher sales.
- *Other Advantages:* Eliminates clutter caused by using printed menus in general, especially in restaurants that use multiple menus. Thanks to digital screens, it is possible to display high-resolution images of foods. Digital menus are much more advantageous in terms of hygiene and health, as they are not passed from hand to hand and are not touched by different people.

Restaurants can place their QR codes wherever potential customers can access them. Restaurant doors, brochures sent in the takeaway, and restaurant tables can be given as examples. Although digital menus are more advantageous than printed menus in many aspects, it is still possible guests who are not good with technology may want to see the menu printed on paper. It is possible that especially tourists over the age of 65 may not very familiar with the new technology. In such cases, the service personnel who take the orders have a great responsibility and are expected to manage the process well. On the other hand, considering that Y and Z generations prefer technological interaction rather than human interaction, QR menus are very practical for this age group.

Use of Humanoid Robots in the Service Department of Restaurants

Another development related to digital transformation in restaurants is the use of humanoid robots. At first, some "themed restaurants" used robots as a theme to attract customers. Later on, the use of humanoid robots became increasingly widespread and took the place of the human, which is considered an indispensable part of the service sector. Considering that there is a group of people who have not yet get used a food and beverage vending machine, which can be considered the ancestor of robotization in the food and beverage industry, serving by robots instead of humans is revolutionary. In the next five years, robots are expected to enter into every section of the restaurant system, including robots that cook, clean floors, clean kitchens, serve customers, and collect plates (Service Robots, 2020). Examples of the use of humanoid robots in restaurants are as follows:

Royal Palace Restaurant-Renesse/Netherlands

Two red-and-white robot waiters are working in this restaurant to greet customers, serve food and pick up dishes from tables. The robot waiters who greet the guests say "hello" and "welcome." It is also noteworthy that the waiters wear little scarves to look more human. The fact that robot waiters can't get the Corona virus and the lack of direct human contact is a great advantage for people who are afraid of being infected. However, although robots help restaurant owners, the fact that real people are needed does not change. Leah Hu, whose family owns the Royal Palace, stated the following in an interview:

They help us with the work we do. We are often busy and cleaning tables and the robots give us an extra hand. We are not disappearing. We are still here. They will always need people in this industry. (Küstür, 2020)

Figure 2. Robot waiters
Source. (Küstür, 2020).

Touch Restaurant-Istanbul/Turkey

The technological restaurant, which started to serve in Ataköy-Istanbul in 2019, serves with tablet tables and robot waiters. Turkey's first technological restaurant was inspired by similar restaurants abroad. Customers who come to the restaurant choose the food they want from the digital menu on the tablet tables with internet infrastructure. The menu selected by the customer is displayed on the computer screen in the kitchen, and the guests can watch the preparation of their meals live from the dining table. The meals prepared by the cook are placed on the serving tray of the robot waiters. Three robot waiters named Rozzy, Robi, and Robo serve at the restaurant. The robots can also engage in dialogue with customers. Robots say "*I am Rozz, this is your plate. Can you touch my arm so I can turn?*" After the robot waiters bring the order to the customers' table, they return to the kitchen for a new service task. While eating their meals, the customers can browse the internet through the applications on the tablet tables. In an interview on the subject, the manager of the restaurant said:

We saw restaurants that used robots in serving in Germany, China, and Japan, and we started working to bring this project to life in Turkey. We received very positive feedback about the restaurant. Especially the children loved the restaurant. Robots attract the attention of people passing by the restaurant. Since technology is now at the forefront in our country too, we got very good reactions. Although robots serve, we always need humans. After all, we are the ones who developed the technology, but also we are the ones who should manage it. We are happy to offer different services and especially to make children happy. (Independent Turkish, 2019)

Robotazia/England

Four robot waiters serve in the country's first sci-fi-themed restaurant, Robotazia, in Milton Keynes, England. In addition to ten human employees, the robot waiters working in the restaurant, which opened at the end of August 2020. The robot waiters can bring drinks and meals to the customers' tables. However, customers still have to take their plates and glasses from the trays of robot waiters. Owners of Robotzia argue that operating robots is an advantage during the pandemic period (Euronews, 2020).

Using Robotic Chef in the Restaurants Kitchen

Another development related to digital transformation in restaurants is the use of robotic chefs in kitchens. Some examples of restaurants that use robotic chefs in their kitchens are as follows:

Robotic Chef Pazzi

A French company EKIM has worked with Universal Robots on the pizza-making robot called 'Pazzi.' PAZZI, which is capable of making up to 10 pizzas simultaneously from nearly 5 million recipe combinations on a 24-hour basis, created using Universal Robots' UR10 and UR5 arms. EKIM planned to use this project to provide an optimal and quality catering solution in high traffic areas, such as railways, airports, main boulevards, shopping centers, and campuses (Universal Robots, n.d.). The robot was created by the works of two engineers, Cyrill and Sébastien. While developing the robot, engineers worked with three-time world pizza champion Thierry Graffagnino. This robot is capable of spreading tomato sauce on pizza dough, putting it in the oven, putting it in a cardboard box, and cutting it. While a chef can cook an average of 40 pizzas per hour, Pazzi can reach up to 120 pizzas by making a pizza every thirty seconds (Bozkurt Uzan & Sevimli, 2020).

Robotic Hamburger Chef Flippy

In 2017, John Miller, the owner and operator of 36-restaurant Cali Burger chain, cooperated with Buck Jordan, the investor and founder of Miso Robotics, and started automation studies to keep the employees safer and increase their productivity. With

Figure 3. Robotic chef making ramen
Source. (Vuuren, 2009).

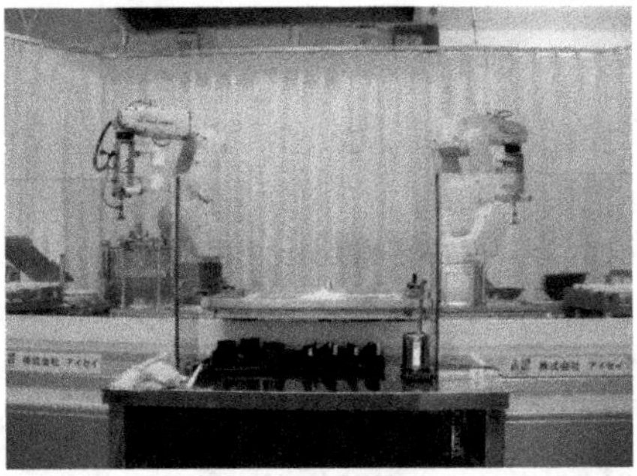

the support of the engineering team, robotic arms called Flippy were produced (Miso Robotics, n.d.). In 2017, Flippy started working as an AI-powered kitchen assistant at the Cali Burger fast food restaurant in Pasadena. Flippy can work at the grill or fryer station in the kitchen. Flippy, which is programmed to work at the grill station, can automatically detect when the burgers are placed on the grill, monitor the burgers in real-time during cooking, turn the cooked burgers with a spatula and clean the grill after cooking. Flippy, which is programmed to work in the fryer station, can place the fryer baskets in the fryer, gently shakes the baskets while frying the food, monitors the cooking times of all the baskets, hangs the baskets to remove excess oil from the cooked foods, and takes the products to a hot station (Bozkurt Uzan & Sevimli, 2020).

Other Robotic Chefs

At Fuamen, a ramen restaurant in Nagoya, Japan, a two-armed robotic chef can prepare 800 bowls of ramen on a busy day. Machines are pre-programmed to boil noodles, pour soup into bowls and sprinkle ingredients on top. Ramen making process takes 1 minute and 40 seconds per bowl, and customers can watch the entire process (ABS-CBN News, 2009). On the other hand, the prices of the meals prepared by robotic chefs in the restaurant are different from the meals prepared by human chefs.

Another example of restaurants using robotic chefs is Zume Pizza. Zume Pizza was founded in 2015 in Mountain View, California, and uses a combination of robots, predictive analytics, and mobile ovens. There are multiple robots in Zume Pizza. These robots share work, and each has a name: Pepe and Giorgio pours the dough sauces, Marta spreads the sauce on the dough, and Bruno puts the pizza into the oven (Bozkurt Uzan & Sevimli, 2020). Another robot named Vincenzo takes on the task of taking out pizzas from hot ovens to reduce human injuries. Vincenzo is a robot assigned to perform this dangerous task in case people in the pizza cooking business stick their arms in and out of the 800-degree oven hundreds of times a day (Shaw, 2018).

Use of Internet of Things Technology in Restaurants

The Internet of Things concept was first introduced by Kevin Ashton in 1991. The Internet of Things is defined as all systems that can transfer data over the network without the need for human beings, provided by interrelated computing devices, digital machines, mechanical objects, or unique identifiers. The concept of things has a very broad meaning in terms of the Internet of Things. All kinds of monitoring devices, sensors, biochips, or access devices are considered as "things" in this system. In order for a device to be considered "smart" and qualify as "things", it must have a unique id, be connectable, and have a sensor. In this way, "smart things" become accessible and controllable from anywhere in the world (Wikipedia, n.d.). Radio Frequency Identification (RFID), Wireless Sensor Networks (WSN), Cloud Computing, Networking Technologies, and Nano Technologies are among the technologies used by the Internet of Things technology (Sundarapandiyan et al., 2019). Numerous digital kitchen products have been developed by taking advantage of the Internet of Things technology. With this technology, the equipment acts like an artificial intelligence connected to the internet. Thanks to the Internet of Things, it is predicted that the Things interacting with the internet will work faster, more productively, and with a saving mechanism (Taş & Olum, 2020). Although some of these digital kitchen products are designed only for homes, some products can also be used in restaurants. Examples of digital kitchen products designed with the Internet of Things and used in restaurants are given below.

Smart Pans

Smart pans, with a temperature sensor mounted on the bottom, connect with the user's Smartphone via Bluetooth. The application instructs the user for mixing, turning, adding ingredients, or changing the heat setting. Since it is possible to predict how long the desired product should be cooked at which temperature, smart pans ensure that the product is cooked at the desired time and setting. Thanks to these pans, energy loss due to overcooking are prevented (Taş & Olum, 2020). Smart pans can also compute the calorie value of every dish cooked, syncs seamlessly with fitness tracking assistants to help people to meet their fitness goals, and allow the recipes saved in memory to be shared on social media platforms (Smartypans, n.d.). Smart pans are 10 inches in size and suitable for use in both restaurants and homes. These pans, which are called "pantelligent" due to their artificial intelligence (Mogali,

2015), can also measure the weight and moisture level of the food. Smart pans are also easy to clean and highly durable (Decortips, 2021).

Smart Chopsticks

Smart chopsticks designed by one of the major technology companies in China can measure the temperature of various foods precisely and show the expiration date of the foods. The chopsticks, which also include a sodium analyzer to help users regulate their daily salt intake, can connect to any computer via Wi-Fi and Bluetooth (BBC, 2014). Chopsticks have sensors that light up red when they detect an ingredient or substance that could be dangerous to human health (such as unsanitary cooking oil), and blue if the food is safe. Able to track the chemical make-up of oil and water present within foodstuffs, the smart chopsticks can also monitor a dish's PH levels, temperature, and calorie content. Users can sync them with their Smartphone to provide further insight into the composition of their food (Stylus, 2014).

Egg Minder

The Egg Minder is a wireless egg tray that helps keep track of how long each egg is inside. When the tray is opened, a blue LED indicates the oldest egg that has been inside the smart device for the longest time. Based on this, the decision to break a particular egg can be made (Mogoli, 2015). The Egg Minder has been developed based on the idea that the deterioration in the structure of the egg cannot be seen with the naked eye. Each time the oldest egg is removed from the Egg Minder, the light next to the next oldest egg continues to flash (Taş & Olum, 2020).

B4RM4N Smart Shaker

The B4RM4N is a product designed for users to make a perfect cocktail every time. It has been announced that the app will be available on both iOS and Android. This shaker uses a sensitive weight sensor to measure how much alcohol or juice has been poured into it. B4RM4N gives instant feedback from the row of LEDs on the side, which changes color according to the material used, as well as sound indicators. The progress also can be watched on the phone's screen. When pouring of an ingredient is done, the phone tells what to do next (Dillet, 2014).

3D Food Printers

3D printing can be defined as "a method that includes techniques applied to create physical parts or structures by depositing materials in layers" (Aksan Aldanmaz &

Figure 4. B4RM4N Smart Shaker
Source. (Dillet, 2014).

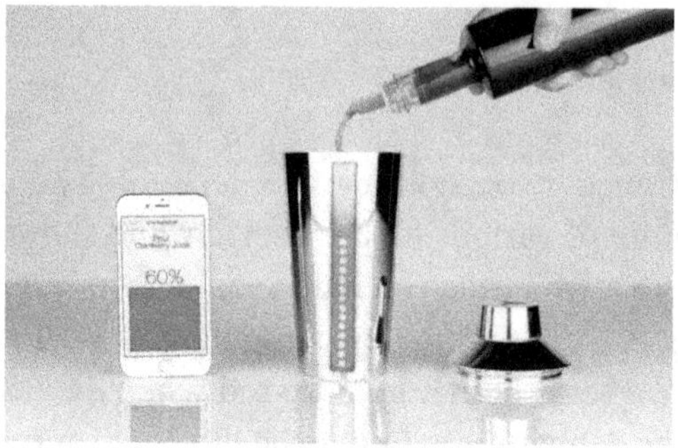

Sever, 2017). The first 3D printer was produced in 1984 by Chuck Hull of 3D Systems. However, it started to be heard more frequently after 2010 and has become much more widely used today. The fact that 3D printing is getting more media coverage, many entrepreneurial companies invest in this technology, the increasing number of academic studies on the subject, the conveniences and advantages of technology in many areas, and the decrease in production costs can be cited among the reasons for this (Yaprak Aydın et al., 2019). While some 3D printers are not produced to print food but can be made to print food with an extruder (printhead) change, some devices are designed only for 3D food printing. Different production techniques are available for 3D printers to obtain healthy, personalized, and restructured products (rich in fiber, protein, and starch) and print according to the desired product recipe. Selective Laser Sintering (SLS), Fused Deposition Modeling (FDM), Binder Jetting (BJ), and Ink Jet Printing Systems have widely used food industry. The main distinction between the techniques used is the method used to deposit layers. While in some techniques, layers are produced by melting and softening the material, in other methods, the liquid materials are stacked directly and hardened immediately, and the product takes its final form (Yaprak Aydın et al., 2019).

In food printing, the process starts with making the food desired to be printed into a puree. This puree is filled into the printing capsules, and the desired geometric shape is defined to the device; the 3D printer makes the defined shape with the puree. The latest developed 3D devices can now cook as well. In other words, the printed product is cooked without removing it from the device and is completely ready to eat (Arıkan, n.d.). 3D food printers save a lot of time compared to traditional production processes. It also provides producers with advantages such as creativity and personalized food production. However, at today's technology level, 3D food printers are not suitable for all types of food and cannot be used in all food production

Figure 5. 3D food printer
Source: (Evli, 2020)

processes. Also, these 3D printers are not suitable for mass production yet (Özer, 2019).

Foodini

Foodini is a 3D printer for foodstuffs. Its Florida-based makers are hoping their time-saving device becomes as ubiquitous on kitchen countertops as the microwave oven has become. The Foodini 3D printer automates some of the cooking preparation process, by for instance, printing out individual ravioli instead of the cook having to make the dough and fill and assemble each individual piece themselves. It's not replacing all the preparation required to make the meal, but it's likely going to speed up aspects of food preparation (without necessarily sacrificing the freshness of the ingredients being used). The machine can also print both savory and sweet dishes, so isn't limited to just printing sugary candies as some 3D printers are (Lomas, 2014). Print times depend on the ingredients, the recipe, and quantity you are printing. Some things print very fast in a matter of minutes (e.g., flatter type foods like crackers, simple plate decorations), while other things take longer times (e.g., intricate chocolate sculptures can take 2 hours to print.) Quantities also have to be taken into consideration. For example, a small, single serving of ravioli will print faster than quantity to feed a family of 4 (Natural Machines, n.d.). Paco Perez, a revered chef who has won Michelin stars for several of his restaurants including La Enoteca at the Hotel of Arts in Barcelona, has been using the Foodini 3D printer to enhance his food presentation. Perez has used the food 3D printer to create a stunning flowery design on one of his plates out of a seafood puree, which will complement a range of other tastes, from caviar, to fresh sea-urchin, to a carrot foam. The dish, called Sea Coral, is meant to recall the sea floor and its rich colors and textures (Tess, 2016).

CONCLUSION

The tourism industry is one of the industries that benefit from positive developments thanks to technological advances, offers different experiences to the customers in parallel with these developments, and offers practical options. In this context, many innovative products have been developed for restaurants, which are indispensable for the tourism industry in the digitalized world. While some of these products are used during the preparation of the meal, some are used while serving the meal, and some are used during the delivery of the meal. Various new technology products can now be paired with Smartphone, which have become an indispensable part of our daily lives. These new technology products are generally designed to produce foods that seem complicated to prepare, to produce in accordance with hygiene standards without human touch, to prevent unnecessary material use or waste, to save time, labor, and materials, and to contribute to sustainability.

There is intense competition all over the world between restaurants, which are the first businesses that come to mind in the food and beverage sector. Today, the survival of a business in this intensely competitive sector depends on its ability to be up-to-date and making improvements and innovations accordingly. During the Covid-19 pandemic, which has affected the whole world, it has been seen that even traditional restaurants need digitalization and somehow integrate innovations into their restaurants. Today's restaurant consumers, especially the Y and Z generations, are no longer in the profile of customers who go to a restaurant that is mentioned in a printed map or tourist guidebook or that they see by chance. Today's' travelers, share restaurant and food photos on Instagram, make restaurant comments on Twitter and Facebook, reserve a table in the restaurant with the mobile application, and want to have different experiences. Therefore, digitalized restaurants that use smart technologies and provide special or, in other words, personalized service are one step ahead as they also arouse a sense of curiosity in the customers.

While the tourism industry uses technology to transform itself, it also causes concerns such as removing people from the system and being unemployed in the future. However, the general opinion is that the use of automation in restaurants exists not to replace human workers but to improve the customer experience and provide more job satisfaction and training opportunities. Although both the human workforce and consumers, in general, find robot workers exciting and entertaining, for now, it does not seem possible that robots completely replace humans. Robot workers come to the fore as a solution for businesses that have difficulties in finding a human workforce.

This study has concluded that the restaurants equipped with technology, which has become more popular have many advantages. These advantages are:

- Reducing increased labor costs,
- To improve food quality by cooking exactly according to its characteristics,
- To increase beverage quality by preparing drinks, especially cocktails, according to their specifications,
- Possibility of more hygienic production by reducing the risks of cross-contamination,
- To eliminate the risks of work accidents such as burning and slipping by undertaking dangerous works such as cooking in deep fat, removing food from high-temperature ovens,
- To provide uninterrupted working opportunities without a shortage of personnel in cases within the scope of personal rights of human employees such as break, sick leave, maternity and death leave,
- To contribute to the promotion of the business.

REFERENCES

ABS-CBN News. (2009). *Japan launches 'ramen robot chef'*. https://news.abs-cbn.com/lifestyle/08/05/09/japan-launches-ramen-robot-chef

Aksan Aldanmaz, E., & Sever, R. (2017). Gıdaların dizaynında 3 boyutlu yazıcı teknolojisi uygulamaları. AB2017, XIX. In Akademik Bilişim Konferansı, Aksaray.

Arıkan, I. (n.d.). *Gastronomide 3D gıda yazıcısı uygulamaları*. https://www.fiyatimbu.com/blog/gastronomide-3d-gida-yazicisi-uygulamalari_213

Asif, M., Sabeel, M., Rahman, M. U., & Khan, Z. (2015). Waiter robot–solution to restaurant automation. *MDSRC Proceedings*.

BBC. (2014). *'Smart' chopsticks unveiled in China*. https://www.bbc.com/news/technology-29062509

Bozkurt Uzan, Ş., & Sevimli, Y. (2020). Gastronomideki robotik uygulamalar ve yapay zeka. *Tourism and Recreation*, *2*(2), 46–58.

Çerkez, M., & Kızıldemir, Ö. (2020). Yiyecek–içecek işletmelerinde yapay zekâ kullanımı. *Türk Turizm Araştırmaları Dergisi*, *4*(2), 1264–1278. doi:10.26677/TR1010.2020.394

Decortips. (2021). *Mutfakta yüksek teknoloji: Akıllı tava*. https://decortips.com/tr/mutfak/mutfakta-teknoloji-akilli-tava/

Dillet, R. (2014). *B4RM4N is a smart shaker that turns the art of cocktail mixing into a science*. https://techcrunch.com/2014/10/28/b4rm4n-is-a-smart-shaker-that-turns-the-art-of-cocktail-mixing-into-a-science/

Euronews. (2020). *No social distancing worries at 'UK's first' sci-fi robot-themed diner*. https://www.euronews.com/travel/2020/09/24/no-social-distancing-worries-at-uk-s-first-sci-fi-robot-themed-diner

Evli, A. (2020). *3D gıda baskısı tüketimde yeni bir çağ açıyor*. https://teknoloji.org/3d-gida-baskisi-tuketimde-yeni-bir-cag-aciyor/

Fuste-Forne, F. (2021). Robot chefs in gastronomy tourism: What's on the menu? *Tourism Management Perspectives*, *37*, 37. doi:10.1016/j.tmp.2020.100774

Gizli İlimler Kütüphanesi. (n.d.). *Robotik kültür ve turizm*. https://www.gizliilimler.org/Dijital-Sehirler-Cagi-ve-Robotlar-Sehri-Neom.htm

Hazarhun, E., & Yılmaz, Ö. (2020). Restoranlarda dijital dönüşüm: Touch restoran örneği. *Gastroia: Journal of Gastronomy and Travel Research*, *4*(3), 384–399.

Independent Türkçe. (2019). *Teknoloji hizmet sektöründe: Robot garsonlar işbaşı yaptı*. https://www.indyturk.com/node/89406/haber/teknoloji-hizmet-sekt%C3%B6r%C3%BCnde-robot-garsonlar-i%C5%9Fba%C5%9F%C4%B1-yapt%C4%B1

Kim, M., Park, M., Kim, J., Kwon, Y., Sohn, D., Yoon, H., & Seo, K. (2021). On the common and different expectations on robot service in restaurant between customers and employees. *ACM/IEEE International Conference on Human-Robot Interaction*, 262–265. 10.1145/3434074.3447172

Küstür, S. (2020). *Hollanda'da bir restoran sosyal mesafe için robot garsonlar kullanıyor*. https://www.teknoblog.com/sosyal-mesafe-icin-robot-garson-cozumu/

Kwak, M. K., Lee, J., & Cha, S. S. (2021). Senior consumer motivations and perceived value of robot service restaurants in Korea. *Sustainability*, *13*(5), 13. doi:10.3390u13052755

Lomas, N. (2014). *Foodini ıs a 3D printer that lets you print dishes with fresh ingredients*. Academic Press.

Miso Robotics. (n.d.). *About us*. https://misorobotics.com/about-us/

Mogali, S. (2015). Internet of things and its role in smart kitchen. *4th National Conference on Scientometrics*, 1-11.

Natural Machines. (n.d.). *How long does it take to print something? What is the print speed?* https://www.naturalmachines.com/faq

Nayik, G., Muzaffar, K., & Gull, A. (2015). Robotics and food technology: A mini review. *Journal of Nutrition & Food Sciences*, *5*(4), 1–5.

New Software Solutions. (2020). *Akıllı turizm otelcilik ve yapay zeka uygulamaları*. https://www.new.com.tr/blog/akilli-turizm-otelcilik-ve-yapay-zeka-uygulamalari/

Özer, B. (2019). *Gıda işlemede yeni bir teknolojik yaklaşım: 3-boyutlu gıda yazıcıları*. https://sutdunyasi.com/gida-islemede-yeni-bir-teknolojik-yaklasim-3-boyutlu-gida-yazicilari/

Özgüneş, R., Bozok, D., & Küçükaltan, D. (2020). Yiyecek içecek sektöründe ileri teknoloji ve pandemik düzene doğru: Yakın gelecekte bir robota 'eline sağlık!' diyebilir miyiz? *Afyon Kocatepe Üniversitesi Sosyal Bilimler Dergisi*, *22*(4), 1124–1139. doi:10.32709/akusosbil.797343

Pardon. (n.d.). *Dijital QR Menü*. https://pardon-app.com/dijital-qr-menu/

Pill Digital. (n.d.). *Restoran ve cafeler için QR kod ile menu*. https://www.pill.com.tr/icerik/100/restoran-ve-cafeler-icin-qr-kod-ile-menu

Pr Outlet Plus. (2016). *IBM watson ve hilton iş birliği*. https://www.proutletplus.com/2016/09/03/ibm-watson-ve-hilton-is-birligi/

Proente Otomasyon. (n.d.). *Endüstri alanında yaşanan gelişmelerin sınıflandırılması*. https://proente.com/endustri-alaninda-yasanan-gelismelerin-siniflandirilmasi/

Protel. (n.d.). *QR menu order ile temassız sipariş & ödeme*. https://www.protel.com.tr/restoran-otomasyonu/qr-menu-order-temassiz-siparis-odeme-sistemi/

Seo, K. H., & Lee, J. H. (2021). The emergence of service robots at restaurants: Integrating trust, perceived risk and satisfaction. *Sustainability*, *13*(8), 4431. doi:10.3390u13084431

Service Robots. (2020). *Robot technology in the restaurant industry*. https://www.servicerobots.com/blog/robot-technology-in-the-restaurant-industry/

Shaw, K. (2018). *Zume pizza names new president, adds new robot to pizza-making staff.* https://www.roboticsbusinessreview.com/news/zume-pizza-adds-robot-pizza-making-staff/

Smartypans. (n.d.). https://smartypans.io/

Stylus. (2014). *Smart chopsticks.* https://www.stylus.com/wvmyhn

Sun, J., Peng, Z., Zhou, W., Fuh, J., Hong, G., & Chiu, A. (2015b). A review on 3D printing for customized food fabrication. *Procedia Manufacturing*, *1*, 308–319. doi:10.1016/j.promfg.2015.09.057

Sun, J., Zhou, W., Huang, D., Fuh, J. Y. H., & Hong, G. S. (2015a). An overview of 3D printing technologies for food fabrication. *Food and Bioprocess Technology*, *8*(8), 1605–1615. doi:10.100711947-015-1528-6

Sundarapandiyan, M., Karthik, S., & Daniel, J. A. (2019). IOT based smart kitchen. *International Journal of Computer Science Trends and Technology*, *7*(2), 13–16.

Taş, D., & Olum, E. (2020). Yiyecek-içecek sektöründe sürdürülebilirlik ve yenilikçi yaklaşımlar. *Türk Turizm Araştırmaları Dergisi*, *4*(3), 3082–3098. doi:10.26677/TR1010.2020.527

Tess. (2016). *Cooking revolution: Are food 3D printers changing the face of high end cuisine?* http://www.3ders.org/articles/20160301-cooking-revolution-are-food-3d-printers-changing-the-face-of-high-end-cuisine.html

Türkiye İş Bankası. (2020). *Nesnelerin interneti nedir ve hayatımızı nasıl değiştirecek?* https://www.isbank.com.tr/blog/nesnelerin-interneti-nedir

Unilever Food Solutions. (n.d.). *Normalleşme döneminde dijital menüler.* https://www.unileverfoodsolutions.com.tr/konsept-uygulamalarimiz/normallesme-doneminde-restoranlar-icin-10-altin-kural/normallesme-doneminde-dijital-menuler.html

Universal Robots. (n.d.). *Universal robots' cobots prepare delicious pizzas.* https://www.universal-robots.com/about-universal-robots/news-centre/universal-robots-cobots-prepare-delicious-pizzas/

Vuuren, M. V. (2009). *Eating robot ramen in nagoya.* https://www.reformatt.com/blog/nagoya

Wikipedia. (n.d.). *Nesnelerin interneti.* https://tr.wikipedia.org/wiki/Nesnelerin_interneti

Yaprak Aydın, H., Kılıç, A., & Tekin, A. R. (2019). Geleneksel Türk gıdalarının 3B yazıcı ile yazdırılması. *International Journal of 3D Printing Technologies and Digital Industry, 3*(1), 1-10.

Yıldız, E., & Davutoğlu, N. (2020). Turizm 4.0‘dan turizm gastronomi 4.0’a giden yolda: Geleceğin restoranları ve yönetimi. *Akademik Sosyal Araştırmalar Dergisi, 8*(109), 301–318.

Chapter 2
Sensor-Based Technology in the Hospitality Industry

Anusha Thakur
https://orcid.org/0000-0001-8761-2250
University of Petroleum and Energy Studies, India

ABSTRACT

In today's scenario, there has been a dramatic transformation in the way services are being provided and taken by the consumers. With the changing preferences of consumers, and businesses, vendors across the regions are emphasizing towards the introduction of touchless solutions, which are likely to change the way people interact and use the services. Incorporation of smart technologies enables the operators to create a positive and seamless experience for the users. Convergence of technologies with the current market trends is expected to bring the concept of "internet of things" nearer to reality. Embracing IoT solutions transform the overall hospitality business scenario and pose several opportunities for the smart hotel solutions. In this chapter, various IoT solutions used in the hospitality and leisure sector have been discussed. This chapter further emphasizes how efficiently the touchless technology products with minimal contact enhance the operations in hospitality sector and how the hotel industry is focusing towards nurturing the dreams of travellers.

INTRODUCTION

In today's scenario, the hospitality industry is majorly being shaped by the current surge in the Internet of Things technology. The concept of "IoT" outspreads internet connectivity outside conventional devices such as laptop & desktop computers,

DOI: 10.4018/978-1-7998-6904-7.ch002

tablets, and smartphones to varied range of devices, which significantly exploits embedded technology for enhanced interaction and communication via internet (Aluri, 2016). Implementation of proper sensor-activated technological solutions poses to be an essential aspect in any hospitality or hotel businesses. IoT is paving the way, various products and services in hotel businesses deliver value and redefine user experiences in the forthcoming future by solving real-time queries and interpreting hyper-personalized recommendations for the guests (Team Trilyo, 2019). The businesses are on the leading edge of IoT technology in order, to maintain a competitive edge in the market. The key vendors are developing unique versatility in terms of functionality, design, and quality standard of products and solutions. This includes the interconnection of physical devices such as identification tags, actuators, sensors, as well as mobile devices, which enables them to communicate directly or indirectly with each other.

One of the key strategies adopted by leisure businesses against the influx of new players includes technological developments, product innovations, brand development, and brand acquisition. Technology is significantly the major factor expected to propel the shifting preferences of consumers, whether it is, in terms of seeking entertainment, dining out, leisure activities, stay at hotels, and other activities. User-experience plays a vital role among all the hoteliers and hospitality providers, thus enabling the businesses to acquire services, and strategies which are competitive and significant in the changing market place.

With the increasing need of minimal contact nowadays, touchless technology poses to be one of the solutions gaining surge in popularity. These include the well-assimilated technologies such as sensor-based and voice-activated technologies which offers safer environment, and seamless experience to the users. For instance, incorporation of touchless technology helps in preventing the risks of contamination among the users in hotel kitchens, pubs, restaurants, rooms, and others, hence, making it safe for them. In addition to the hotels, in case of high-traffic areas, such as airport lounges, cruises, and a few others, there is high risk of transmission of the germs. Introduction of smart sensor-based technologies, thereby, poses to be advantageous to the customers in terms of their health and well-being as well. With the changing market scenario, it is hence, necessary for the businesses, to emphasize on the overall safety, and excellence in terms of guest experience. The innovative and advanced techniques introduced by various vendors are expected to enable the end-users to cope up with the changing time and consumer preferences, and cater to the concerns of health and safety as well.

LITERATURE REVIEW

The hospitality and leisure industries are one of the most diversified and biggest industries in the various regions of the world. Economic indicators, such as leaning on low unemployment, continued GDP growth, as well as recent tax cuts are expected to impact the growth in the hospitality industry. Increasing trends of technology and innovation are expected to favorably influence the future scenario of the hospitality industry. The innovative strategies in the hospitality sector are anticipated to cater to the changing needs and expectations of the consumers across the different regions in the world. The travelers from various parts of the country, mainly emphasize much on their health and well-being.

Trends Shaping the Hospitality Industry

- **Digital Technologies:** Digital technologies are increasingly engaged in shaping the overall consumer experiences. In today's scenario, the concept of artificial intelligence, and automated products and solutions are offering enriching experience to the customers.
- **Sustainability:** Hospitality businesses are emphasizing towards the concept of sustainability, because it poses to be an essential factor in today's changing competitive world (Booking global, 2018).
- **Evolution of New Business Models:** In today's scenario, with the advent of technological developments, hotel businesses are nowadays focusing on technology-focused and innovative design amenities. In addition to these the businesses are also focusing on implementing smart technology in order to, cater to the convenience of the guests. Implementation of smart room functions appeals the hoteliers, because of the fact, that these can easily be integrated into both retrofits and new builds (Imbardelli, 2019).

With the incorporation of latest technologies, the hospitality sector focuses on offering its guests a seamless experience, with customization and convenience. Incorporation of smart room technology is expected to cater to the changing customer expectations and wants. Most of the hotels have already shifted their focus towards smart technology, thereby offering sense of control over their stay, hassle-free and excellent solutions to the customers. The hotel industries are mainly emphasizing on partnerships, acquisitions, as well as programming to cater to the various opportunities in the market. In today's scenario, the hotel technology trends are continuously evolving in terms of automation and simplification, thereby offering convenience to both professionals as well as customers. For instance, technological trends such as big data, cloud technology, and artificial intelligence, have significantly contributed to the growth of the overall hospitality businesses.

Over the past few years, sensing technology has witnessed significant rise in developments in order to cater to the changing market scenario. Sensors designed for smart hotels offers benefits such as cost savings, and revenue opportunities, thereby improving the overall guest experiences. For example: occupancy sensors, and thermostats permits the concept of smart energy management which offers light settings, automated temperature controls, and utilization of energy. Few benefits offered by sensors in hospitality industry includes:

- Providing personalized experience to the guests
- Automated room entry with the self-check-out and check-in feature
- Touchless products in restroom areas, thereby maintaining hygiene and sanitation facilities

The goal of smart technology (IoT) hotels is to offer its guests with much better control and the convenience for controlling thermostats, lights, door locks, and few more. With the advent of technology, the innovators are continuously emphasizing towards the development and improvement of IoT technology in businesses. As smart products are making their way to consumers' mainstream lifestyle, the vendors are investing in this arena to get better returns. The emergence of smart solutions across industries is already having a profound impact on product design, development, manufacturing, marketing, sales, aftersales, and offering customized consumer experience. If properly instrumented, these solutions can offer reams of insights on product usage which can further help in continuous product improvement. These insights can also influence the company's strategic moves into connected markets. Hence, the opportunities offered by smart products are seemingly game-changing for companies seeking new level of efficiencies and product expansions.

Internet of Things: A Smart Solution for the Hospitality Industry

Leisure in the hospitality sector poses to be a vital factor in terms of gaining profitability and expanding the customer base. In today's scenario, a huge section of the hospitality industry is witnessing unparalleled mobilization across different regions, in order to mitigate the adverse impacts on environment and confront significant challenges in future as well (Trivago, 2022). The hotels are increasingly being engaged in redesigning the user experiences by offering them simplified and enhanced services.

With the advent of the "Internet of Things" (IoT), hotels can leverage technology that tracks temperature, room occupancy and more from any web-based device and adjust as needed in real time. This translates into a huge win for properties

Incorporation of IoT connectivity in the hospitality sector, offers its guests the potential to solve various problems which have been faced by the industry since long. Advancements in the concept of IoT enables the hotel businesses to leverage the technology which efficiently tracks the occupancy of rooms, and temperature from any web-based devices and accordingly adjust on real time basis. These technologies offer endless avenues and possibilities for increasing returns as well as reducing the expenditures. Further, integration of these solutions is expected to impact the overall guest journey with smart and differentiated value propositions, and flawless experience, thereby also contributing to the overall profitability of the businesses.

Energy management

The hotel businesses use substantial amount of energy for various recreational activities as well as daily operational purposes. The guests making their stay at the hotels are expected to have higher expectations, wherein energy management, "smart home" technologies, poses to be few of them. One of the latest trends in the tourism industry includes the fact that, nearly 87% of the travelers nowadays, want to travel sustainably (Meisrow and Blumenthal, 2019). In today's scenario, the consumers are increasingly supporting the brands which represent their values, along with Corporate Social Responsibility (CSR) initiatives, such as energy efficiency management and solutions. Without proper planning, energy management variables for the hospitality industry are considerably difficult to control. The hotel businesses witness variation in terms of the room rate, energy usage per room (on an average basis), different occupancy rates, and few more. Implementation of IoT solutions is expected to support the CSR programs, thereby backing the sustainability initiatives as well.

IoT in the hospitality businesses are expected to add "smartness" required to deliver enhanced value to the consumers, thereby further improving the overall operational processes of the businesses as well (Rangnekar, 2021).

In real-time, it is at times difficult to get electricity bills. Hence, implementation of specialized cameras helps in reading the display on a utility meter, thereby offering updates related to the energy usage immediately. Incorporation of the IoT solutions, thereby enables the hotel businesses to save nearly 25% of the utility costs. These solutions also help in improving the asset utilization with connected luggage carts and room service trays, thereby reducing the energy costs across connected buildings. In addition to these, preventative maintenance IoT systems helps in identifying the equipment or appliance issues. The sensors can easily monitor HVAC systems, and alert for maintenance in case, if required.

Key goals of the IoT solutions offers the guests with better control over the environment. Smart technology in the hotel rooms offers the customers with the ease and convenience of automating and controlling the different settings for thermostats, lights, and door locks (Igor, 2020).

With the usage of connected thermostats, it becomes convenient for guests to adjust the room temperatures during the check-out or check-in process, thereby eliminating the costs of heating or cooling the rooms when vacant. In addition to these, thermostats linked with sensors enables the air conditioner to automatically turn off as soon as the guests open a balcony or the window (Alcatel Lucent Enterprise, 2020).

Implementation of smart energy management solutions helps the hotelier to know when a room is occupied or unoccupied, and accordingly adjust the temperature to minimize the energy consumption by nearly 20 to 45 percent (Potter, 2019). Hence, improved energy management solutions are expected to save considerable amount of money, and propel the guest satisfaction levels by enabling the users to effortlessly adjust themselves to the resources in the hotel room.

Virtual and Augmented Reality

Over the past few years, the hospitality industry has prospered by adding value to the user experience and continually focusing on expanding the consumer base. From methods of arrangement, food & drinks, to accommodation, the hospitality industry has been witnessing betterment in terms of customer engagement. The industry witnessed significant transformation in terms of the way the services and products are being consumed and purchased. Incorporation of digital reality in the hospitality businesses, has substantially led to the dramatic growth of the market with an increased revenue and share of the market players. The concept of Virtual Reality (VR) and Augmented Reality (AR) play a significant role in the hospitality industry in serving the guests and keeping them entertained. The real-world experiences are expected not to be enough for the customers habituated to supplementing their experiences with social networks and digital facts. Integration of digital reality enables the businesses to cater to the changing customer expectations and market scenario. These concepts enable the consumers to recognize the environment near them, thereby transforming into a range of application segments.

AR has significantly developed in the hospitality industry over the past few years since it enables the industries to enhance their physical environment along with the experience of the customers. It creates a virtual picture of the hotel rooms for the customers before their check-in process, as an alternative of trusting on the brochures or on website. In addition to these, the concept of AR also offers its guests with 3D entertainment via the AR hotspots, which are easily accessible through the smartphones. The technology offers two-fold benefits in the hospitality industry: as a back and front of house application incorporated with varied other technologies such as Computerized Maintenance Management System (CMMS), or Enterprise Asset Management (EAM) mobile application. CMMS have considerable usage for maintenance management for inventory management, preventive management, and scheduling the inspections (Simulation in hospitality: Sparking travelers' imagination, 2021)

https://www.ediiie.com/hospitality.html .

Few examples of AR in the hospitality industry includes:

- **Gamification:** In today's scenario, the hotel businesses are significantly encouraged by the success of "Pokémon Go", thereby hoping to launch the AR-based games which offers some surprises and rewards to the guests by using their smartphones for exploring the amenities, thereby keeping them engaged.
- **Augmented Environments in Hotels:** The restaurants are engaged in using the "AR" concept in creating the theatre like experience, thereby making it more attractive for the customers. Partnering with the different travel & tour companies are expected to help the hotels to showcase various activities via virtual reality, thereby enhancing their customer experiences.

In addition to all these, it is expected that virtual concierge concept will be standardized by 2025. Synchronizing guests with the previous and current stay records enables the hotels to provide virtual assistants such as, delivering breakfasts to the rooms of the guests who need to leave the hotel premises early morning (Friedman and Sen, 2017). The autonomous indoor delivery robots are expected to help in automating the hotel room services. These automated solutions include embedding the concept of artificial intelligence, machine learning and sensors, which enables the robots to navigate crowds and operate elevators, while making room deliveries.

Hence, there are countless opportunities in terms of digital reality, which is expected to expedite personalization for the users, thereby making tailored experiences a reality for them. Hence, these advanced technological concepts offer better and enhanced opportunities to the businesses to cater to their needs and requirements.

Role of biometrics

Various public events and gatherings such as conferences, workshops, marathons, weddings, and few more, require badge verification and lengthy registration processes, thereby leading to longer waiting time. Few solutions, such as biometric tokens help in backing up queue management algorithms and facial recognition, thereby protecting the personal information of the consumers and streamlining the entire process.

In the hotel segments, the guests are emphasizing on the frictionless and easy check-in procedures. With the increasing technological advancements, entering the room straight away, will be hassle-free, with the check-in lines being reduced to the maximum over the forecast period. Fingerprint scans provides safe and secured access to the rooms and facial recognition helps in easy identification of the guests. Hence, growing demand for the programmed and easily installed biometric readers, is also expected to gain traction among the hotel properties, which needs to update the access daily to the restricted areas (Short, 2015). Few other factors expected to impact the adoption of biometrics in the hotel industry includes the increasing number of the biometrics-enabled mobile devices. According to the study conducted, it is estimated that nearly all smartphones are expected to have varied embedded biometric features (Muelen, 2015). The other factor impacting the demand for biometric solutions includes the acceptance of the biometric technology among the customers. One of the challenges in the biometric technology implementation is convincing the users that their sensitive information is secure. Facial recognition products pose to be difficult to sell among that of the privacy-conscious consumers.

Implementation of biometric solutions helps in reducing the "card clutter", thereby enhancing the existing card systems, hence, making it customer centric. One of the most essential factors includes the comfort with technology and convenient stay, thereby giving excellent experience to the guests at the hotel. Hence, in today's scenario incorporating these solutions in the businesses offers secure, personalized, and enhanced consumer experiences. Implementation of these technological solutions is expected to enhance the competitiveness among different business segments in the industry.

Data-Enabled Hyper-Personalization

Hospitality industry is considered, to be enormously data-intensive, owing to the fact, that huge amount of data is being generated, right from making enquiries, to virtual tours to reserving the tickets, to making payments. Generating valuable insights from the guest's data is expected to enhance the user experience. It is very essential for the businesses to influence smart data to offer real-time developments in the guest experience, according to the changing customer needs and requirements. In order to cater to these changing needs, the businesses need to shift towards the usage of AI-driven assistants, from personalized booking to chatbots.

In today's scenario, the hospitality businesses, need smart technology such as self-learning machines which embraces the communication potential between travelers and businesses. Technologies such as smart bots are expected to reduce friction at the time of maintenance, occupancy, and check-in, thereby freeing the staffs from routine tasks.

The concepts of Artificial Intelligence, is expected to enable the tools to craft precise response in the appropriate situation, thereby freeing the time for assistance and interaction as an when required (Davis and Bosma, 2020). Hence, the workers can offer customized offerings to the guests, thereby making their experience a unique and special one.

How IoT Technology is Impacting the Businesses

Automation has proven to be one of the leading factors for businesses to succeed. In today's scenario, group of hotels are increasingly focusing on the implementation of Hospitality-IoT, as a technological development for their businesses. The Internet of Things concept has been witnessing exponential growth in demand particularly among the restaurant and hotel chain sectors. The improvisation of IoT technology along with cloud computing concept in the hospitality businesses has largely impacted the way businesses operate. IoT solutions for the hospitality industry enables the businesses to understand the behavior and patterns of consumers, and accordingly recommend customization of services, thereby catering to their needs (Coppol, 2019). The businesses offer an integrated approach with the help of web applications, which enables them to easily access both big data as well as the controllers/sensors.

The concept of "Smart Tourism Network" comprises of IoT connected with sensors, controller interfaces, and devices, which owes the capability to easily assess and sense the information. Most of the top-notch businesses are recognizing the need for embracing these advanced solutions (Biz4Intellia, 2022). IoT solutions converts the collected data into business strategies in order to, cater to the increasing competitiveness among the players in market. IoT in the hospitality businesses has significantly broadened their approach, by the incorporation of smart logistics management, ease-of-access, and automated solutions, thereby offering enhanced consumer experiences. The technology owes the potential to convert the hospitality industry by largely altering the way resorts, cruise, hotels, casinos, restaurants along with other leisure service businesses, automates the processes, interface with users, as well as gathers data. The concept of IoT discusses the networking of physical objects via the usage of actuators, and embedded sensors, which can easily transmit relevant information about real-time activities. Combination of IoT with technologies such as data analytics and user mobility, has led to a paradigm shift in hospitality businesses. The Internet of Things technology offers the following benefits to the businesses:

- **Enhanced Guest Experiences:** With unique digital solutions as well as technologically advanced services, and amenities, the businesses are engaged in offering enhanced and differentiated experience to its customers.
- **Optimized Processes:** Development of new and latest solutions, along with minimization in operating costs, is expected to enhance the productivity and offer optimized processes.
- **Preferences and Needs of the Guests:** Implementation of IoT technology enables the businesses to learn about the preferences and needs of the guests, thereby enabling businesses to deliver more personalized services and products.
- **Enhanced Efficiency:** These advanced technology makes the businesses much more efficient and smarter by proactively creating efficient processes as well as monitoring critical infrastructures.

How IoT Technology is Impacting the Guest Experience

The hotel industry has always been at the front-line of an economy which is continuously evolving to cater to the changing upgradations and innovativeness in terms of technology and customer requirements. The hotel industry poses a unique opportunity to explain the changing customer needs and accordingly make the physical assets well-advanced with at par technological upgradation (Meisrow and Blumenthal, 2019).

Nearly 42.6 billion of IoT enabled devices are expected to be linked to the internet by the year 2022 (Chauhan and Saxena, 2021). Embracing smart technology by the hoteliers is one of the best strategies to enhance the overall customer experience by offering them unparalleled services.

There are numerous ways in which smart technology is expected to impact the hotel experience among the consumers, by offering them convenience and a better environment to stay at. With the help of smart controls and voice commands, the guests have easier accessibility to multiple devices. Further, the guests also owe the ability to set up their room as per their liking and conveniently address their amenity requirements. Incorporation of IoT in the hotel rooms offers its guests with accessibility to air conditioning and lighting in the room. Few hotels are also engaged in implementing technologies such as Alexa and Amazon Echo into their designs, thereby enabling its guests to offer the same device-setting abilities via voice activation processes.

Another technology expected to gain traction in the hospitality businesses includes interactive maps, which enables the guests the ability to download an app, which will offer advanced concierge services from anywhere. The guests nowadays, are also looking forward for personalized services such as setting lighting at personal or setting the temperature, to having a favorite TV show or radio station playing. Personalization generates individualized guest experiences by incorporating customizability and flexibility to the accessible service packages (Kansakar, 2019). Smart technology is expected to help in the reduction of costs for the hotel owners. In addition to these, these technological solutions also enable the staffs to remotely monitor the repair needs, thereby offering an efficient and seamless repair processes, which are convenient both for the guests as well as for the maintenance staffs.

Challenges in the Positioning of IoT Technology

The concept of the "Internet of Things" leads to unparalleled flow of data, along with the operational, and management issues as well as security risks for the network infrastructure. In order to, cope up with these challenges, the hospitality segment need to acclimatize with the traditional network designs, thereby offering enhanced level of automation, security and intelligence in the business environment.

Hence, the businesses require a cost-effective network infrastructure which is easy to operate, manage and can securely handle enormous flow of data. The structure must:

- **Offer Automated and Simple Method for IoT Device Onboarding:** IoT systems comprises of sensors or devices. Managing and physically provisioning all the endpoints has proven to be error-prone and complex. With the incorporation of automated onboarding concept, the network infrastructure can easily identify devices and accordingly allocate them to the proper secured network.
- **Offer a Safe Environment against Data Loss and Cyberattacks:** Security poses to be a challenging factor for mitigating the risks of cybercrime, owing to the fact that the networked sensors and devices in the IoT technology led the way to a plenty of potential attack vectors. Hence, security poses to be an essential factor at multiple levels, which includes the control of IoT networks.
- **Efficient and Proper Working of the Network Resources for IoT Systems:** The devices in IoT system provides information which requires a specific level of Quality of Service (QoS). For instance, Proper bandwidth reservations are essential for Smart TV's (IPTV) on an efficient network infrastructure, thereby offering service delivery excellence.

Ways to use Smart technology and IoT Applications

The businesses are engaged in the applications of IoT solutions across different segments. Every application area focuses on enhancing the customer experience and increase in the overall revenue and sales of the business segments. These connected solutions provide various benefits which supports seamless hotel interactions and experiences. Few of the benefits offered includes:

- **Improved Efficiency with In-Room Automation Tools:** Incorporation of automated products and solutions leads to reduction in costs at the same time enhancing the customer experience. Occupancy sensors helps in tracking the room activity, thereby shutting down unnecessary features, when the room is unoccupied. These technologies help in controlling:
- Thermostats and HVAC systems
- Hotel-wide and in-room lighting
- **Offering Voice-Controlled Services:** IoT solutions such as Amazon Alexa enables the users to perform varied activities instead of them picking up the phone and giving instructions. These in-room voice-controlled services offer its guests with same experience as they are at home. These services enable the guests to conveniently schedule the appointments for any service, avail housekeeping services, and few more.
- **Supplement a Data-Driven Culture:** Data derived from different Internet of Thing devices back predictive analytics. Usage of external and internal factors offers a holistic view of the consumer experience.
- **Enhanced Wi-Fi Experience:** Offering an excellent network tends to act as a differentiating factor between the guests having “satisfactory” and “excellent” experiences. Incorporating a capable managed network offers the guests with enhanced experiences.

The Concept of “Smart Washroom”

Facility management is a vital service and component for any business and its buildings, with personnel routine ensuring areas, such as washrooms, wherein, higher standard of hygiene needs to be maintained. It requires regular checking of the filled levels of paper towels, dispenser units, and bins to maintain the hygiene facilities in such places. Washrooms in the areas such as large airports, hotels, and transport hubs, need to cope up with increasing demand for the visitors on a regular basis. With too many numbers of users using the restroom facilities, the risk of bad odours, infections, and maintenance problems are gaining significant traction. Deployment of an efficient touchless system by the hotels, airports, cruises, restaurants and few more, are gaining significant traction, to prevent the risks of cross contamination which are otherwise difficult to identify and even difficult to stop.

Usage of touchless technologies helps in safeguarding the health of individuals, thereby reducing health risks among the same. Sensor-operated and touchless products such as dispensers, sanitizers, and dryers contribute to a higher level of handwashing hygiene. The advantages offered by sensor-operated products extend beyond public restrooms to any other type of handwashing station, particularly in the applications, wherein, high level of cleanliness is required.

Hence, in order, to cope up with the changing needs and requirements, cleaning and sanitation facilities in the various end-user applications are becoming more modernized. In addition to this, the washroom solutions are becoming more robust, as well as resource and cost efficient, in today's scenario.

Enhanced connectivity solutions and sensors are gaining traction owing to their smaller size and cheaper costs. The cleaning facilities are thereby emphasizing on the adoption of a number, of connected solutions such as Internet of Things (IoT). Adoption of the smart washroom solutions poses to be an ideal solution for the hospitality businesses, wherein, larger footfall is common, and they can easily identify any operational or hygienic issues, breakdowns or leakage, and accordingly send SMS alerts to address the issue. These technologies help in monitoring the washroom resources and usage, with the help of a 3D IoT smart technology platform.

For instance, in case of automatic soap dispensers, the smart washroom solutions developed by the leading vendors in today's scenario, enables cleaners to make real-time adjustments to regular schedules as per the sensor data or information collected from the dispensers. This information collected are then clearly displayed in a data dashboard. With the combination of revolutionary software and Internet of Things (IoT), the levels in dispensers can easily be measured with the help of sensors, which further transmits the data to a centralized platform. Cloud-based and connected technology, offers more real-time trend data insights, thereby creating more hygienic and engaging washrooms. With the incorporation of smart technologies, the operators are creating a positive experience for the users in washrooms.

Few of the smart solutions for the dispenser products are illustrated in the figure mentioned below:

- 24/7 filled-level monitoring with smart dispensers
- Automated notification of re-ordering and new cleaning
- Predictive maintenance with big data analysis and IOT

Adoption of latest technologies, enables the high-traffic facilities to run at an optimum efficiency, thereby maintaining cleaner washroom environment.

The guests are nowadays, seeking hotels which are strictly following the social distancing measures and ultra-enhanced safety in place. Along with comfort and seamless experience, the guests are also focusing on the safety and hygiene standards. This includes the near-contactless services or the touchless products and solutions. Growing concerns over hygiene and increasing disease outbreaks across the world is expected to propel the demand for touch-free sensing technological products. Usage of 2D images for identifying gesture less interactions are being used to track the motion of hands. In addition to all of the above, the outbreak of COVID-19 pandemic has exponentially fuelled the demand for the touchless technological products and solutions to protect the individuals.

Growing Demand for the Touchless Products

Touching a manual faucet, sanitizer, or dispenser, are expected to result in contact with bloodborne pathogens and hazardous biological waste. Inadequate hygiene strategies are thereby expected to lead to the growth of harmful pathogens, particularly in the restrooms, wherein, the water droplets allow the bacteria to grow and spread easily.

The health-conscious consumers are increasingly focusing on the usage of touchless sanitary equipment in the restrooms. These products include touchless dispensers, flushes, faucets, and trash bins. These products reduce the risk of cross-contamination of germs by removing the need to touch cranks or paper, levers, bars, or faucets. Incorporation of touchless solutions in the restrooms offers enhanced efficiency savings, and also caters to the sustainability goals. Few products with their market analysis are mentioned below:

Touchless dispensers

Automatic soap dispensers also known as the smart/touchless/no-touch dispensers are the devices which dispenses the soap or sanitizers. These products enable the control of consumption of soap or sanitizer which are being used in the hospitals, residential buildings, and public restrooms. These products work in conjunction with the electronic faucets. The touchless products comprise of electrically connected sensing device structure which helps in detecting the presence of an external object, as an when it comes in the proximity of the same.

Sensor-operated and touchless dispenser contributes to a higher level of handwashing hygiene. The advantages offered by sensor-operated dispensers extend beyond public restrooms to any other type of handwashing station, particularly in the applications, wherein, high level of cleanliness is required. For instance, these products are gaining traction at the airports, hotels, cruises, and other highly-touched areas, to prevent the risks of cross contamination which are otherwise difficult to identify and even difficult to stop.

Over the past few years, the automatic dispenser products have witnessed significant rise in demand. This can majorly be contributed to the growing trends of the smart restroom technologies. The dispenser products play a vital role in maintaining the hygiene standards of the highly touched areas or surfaces. The market for these products tends to get influenced by a number, of, factors from direct impact on manufacturers or a transferred impact on the same. Over the years, the automatic dispenser market has undergone several changes in terms of quality standards and other crucial innovative advancements.

Preference for automated products or technically advanced products is expected to open various avenues for the manufacturers. Hence, vendors across the regions are engaged in emphasizing towards the introduction of no-touch solutions, on the basis, of sensors, thereby catering to the user-centered solutions. The leading vendors offer solutions which can accelerate the services of their company, along with the integration of digitalization in their business segments. The changing technological scenario are expected to adversely impact the vendors as customers expect continual innovations and upgrades in the services. Vendors such as **"Euronics"**, headquartered at Netherlands, is one of the leading providers for public washroom automation accessories segment. The washroom solutions offered by the company includes soap dispensers, sensor operated bidet, urinal sensors, hand dryer, washroom panels, paper dispensers, sensor taps, baby diaper changing station, fragrance dispensers and few more.

Touchless faucets

The touchless faucet or the motion sensing faucet/electronic faucet includes a proximity sensor with the mechanism which allows the valve to open to enable the water to flow. These touchless faucets operate with batteries and comprises of infrared sensors which helps in detecting the hand motion. Installation of the touchless faucets are largely gaining traction at the places such as hotels, airport lounges, restaurants, cruises and few more.

Few vendors such as "**OPHARDT**" have launched multifunctional touchless faucet, wherein, soap, water and disinfectant are all in one unit. These sensor-controlled products have wide range of applications in the kitchen and restroom areas. Increasing awareness for hand hygiene in different end-user segments and facilities is expected to significantly impact the overall demand for the touchless faucets in the market. By incorporating the latest technology faucet, the cross-contamination of germs is prevented, and the productivity of the particular area is also high.

Touchless trash bins

The touchless trash bins/automatic garbage bins are sensor-based and capable of working without any physical contact. These trash bins can easily be controlled via different mobile applications, thereby offering both ease of usage and higher degree of convenience to the guests. Rising demand for smart waste management and efficient waste disposal are expected to propel the demand for touchless trash bins. In addition to these, factors such as increasing need for convenience, as well as fast-changing lifestyle are increasingly boosting the market growth.

Various restaurants and hotels are engaged in the deployment of automatic garbage cans, particularly in the restrooms, thereby offering attractive services to the customers. In addition to these attractive features, incorporation of these touchless technologies in the businesses also ensure healthy and hygienic environment. With increasing technological innovations among players, the competition is expected to intensify in various product segments. For instance, market players such as **"EKO Group"** offers products which are designed to focus on the consumer's end user applications. The company is engaged in performing extensive research and development in order to cater to the changing market demand. EKO offers optimal sensor range for its product segments, thereby maximizing the style as well as usage to enhance the overall product experience to the customers.

FUTURE SCENARIOS OF HOSPITALITY INDUSTRY

For the hoteliers anticipating the future trends is expected to pose as a challenge on account of the upcoming wave of technologies. The path ahead for the businesses is unquestionably going to rely on the technologies which will further make the hotels and resorts much smarter, by offering personalized experiences to all its guests (Oracle Hospitality, 2016). In addition to adding "smartness" in its offerings, the technologies are also expected to create real human experience for the users.

Several technologies are expected to move into mainstream operation by 2025. For instance, the Robotic process automation (RPA), is expected to favourably impact the hotels' loyalty and booking processes. The Chatbot services are expected to form new notification channel for the guests by integrating with the purchasing and booking sections. Biometrics processes are anticipated to keep us safe and secure and the robots would rather help in delivering the breakfast to the rooms of the guests. With the re-invention of the technology, the on-site user experience has proven to be no exception. The staff-less hotels are likely to prove appealing to the travellers, thereby reducing the operating costs. With the shifting preferences towards the tech-enabled hotels, the Internet of Things technology is expected to prove as the solution for the businesses across the world.

The way we consume, and purchase hospitality services is thereby expected to witness significant changes as industry is undergoing the process of transformation in terms of its product offerings, services as well as the upgradation in infrastructure. The industry is expected to witness a shift in preferences from a system based on marketing and selling to creating actual value to the destinations. In 2025, the hospitality industry is expected to evolve with enhanced services and further, shift from sets of offerings to personalized suggestions for every guest on the basis, of different data points and variables.

CONCLUSION

Several new trends in the hospitality industry are expected to witness considerable rise in growth over the forecast period. Advent of technologies have led to enhanced energy efficiency, improvement in performance, minimization of production costs and miniaturization of the devices (Deloitte., 2022).

Nowadays, the manufacturers focus on identification of various opportunities in the market, thereby establishing new goals for the efficient usage of resources, and enhanced productivity. The current market scenario propels the vendors to alter and refine their value propositions to attain a strong market position. Introduction of differentiated products and solutions for varied application segments is one of the key strategies being adopted by the leading vendors. The key manufacturers are focusing on the expansion of their existing product line with innovativeness and smart technologies, thereby gaining an edge in the market. Sophisticated and advanced technologies enable the hospitality industry to evolve and change in an eloquent manner to the customers. With the evolving technologies, it is very essential for the managers to run and regulate their businesses in a leaner and smarter manner. The hotels which embrace smart technology is expected to offer more power to the guests, as the more the experience is controllable, the more the guest feels comfortable. Incorporating IoT solutions in the businesses is expected to offer much better environment and enhanced efficiency, thereby boosting their overall revenue as well.

This chapter discusses about the smart sensor-based technology trends in the hospitality industry, along with the future aspects of these Internet of Things Technology. It discusses about the smart technology insights which can significantly propel the performance, improvements in cost, customer experience, and gain profitability. The advancements in the industry enables these businesses to transform their offerings and services, thereby coping up with upcoming future trends. This chapter also discusses about the impact of IoT on businesses, guest experience, and the challenges associated with the same.

REFERENCES

Alcatel Lucent Enterprise. (2020, October). *The Internet of Things for the Hospitality Industry*. Solution Brief IoT in Hospitality. Retrieved from https://www.al-enterprise.com/-/media/assets/internet/documents/iot-for-hospitality-solutionbrief-en.pdf

Aluri, A. (2016). *The Future of IoT in Hospitality*. Tech Trends Connectivity. Retrieved from https://www.hftp.org/hitec/i/downloads/TechTalkIoT.pdf

Biz4Intellia. (2022). *Intellia IoT (An End-To-End IoT Business Solution)*. Intellia IoT. Retrieved from https://www.biz4intellia.com/

Booking global. (2018, April 17). *Where Sustainable Travel is Headed in 2018*. Retrieved from https://globalnews.booking.com/where-sustainable-travel-is-headed-in-2018/

Chauhan, D., & Saxena, P. (2021, December 19). *The Future of Hospitality - IOT will be the Next Big Thing*. SoftwareSuggest Blog. Retrieved from https://www.softwaresuggest.com/blog/iot-next-big-thing-future-hospitality/#

Coppol, S. (2019, November 18). *Redefining the Hospitality Industry with Internet of Things*. Compare the Cloud. Retrieved from https://www.comparethecloud.net/articles/redefining-the-hospitality-industry-with-internet-of-things/

Davis, E., & Bosma, T. (2020). *The Future of Travel & Hospitality*. Cognizant. Retrieved from https://www.cognizant.com/whitepapers/the-future-of-travel-and-hospitality-codex5236.pdf

Deloitte. (2022). *Smart Sensors and Supply Chain Innovation: Advanced Platforms Create Digital Insights*. Perspectives. Retrieved from https://www2.deloitte.com/us/en/pages/operations/articles/smart-sensors-and-supply-chain.html

Friedman, M., & Sen, S. (2017, June). *Hospitality in the Digital Era*. Retrieved from https://www.cognizant.com/whitepapers/hospitality-in-the-digital-era-codex2543.pdf

Igor. (2020, February 18). *IOT Smart Hotels: Enabling Innovation in the Hospitality Industry*. Igor. Retrieved from https://www.igor-tech.com/news-and-insights/articles/iot-smart-hotels-enabling-innovation-in-the-hospitality-industry

Imbardelli, A. P. (2019, April 29). *How Smart Rooms can Transform Hotel Brands*. LinkedIn. Retrieved from https://www.linkedin.com/pulse/how-smart-rooms-can-transform-hotel-brands-a-patrick-imbardelli

Kansakar, P. M., Munir, A., & Shabani, N. (2019). Technology in the Hospitality Industry: Prospects and Challenges. *IEEE Consumer Electronics Magazine*, *8*(3), 60–65. doi:10.1109/MCE.2019.2892245

Meisrow, R., & Blumenthal, J. (2019, April 18). *Why IOT is the "Smart" Solution for the Hospitality Industry*. PhocusWire. Retrieved from https://www.phocuswire.com/PwC-opinion-IoT-for-hospitality

van der Meulen, R. (2015, November 10). *Gartner says 6.4 Billion Connected "Things" Will be in Use in 2016, Up 30 Percent from 2015*. Newsroom Press Releases. Retrieved from https://www.gartner.com/en/newsroom/press-releases/2015-11-10-gartner-says-6-billion-connected-things-will-be-in-use-in-2016-up-30-percent-from-2015

Oracle Hospitality. (2016). *Hotel 2025*. Retrieved from https://www.oracle.com/webfolder/s/delivery_production/docs/FY16h1/doc31/Hotels-2025-v5a.pdf

Potter, K. (2019, March 3). *The Pros and Cons of IOT in the Hotel Industry*. Hotel Technology News. Retrieved from https://hoteltechnologynews.com/2019/03/pros-cons-iot-hospitality-industry/

Rangnekar, P. (2020, April 16). *How the Hospitality Industry is Enhancing with IOT Technology?* Biz4solutions. Retrieved from https://www.biz4solutions.com/blog/how-the-hospitality-industry-is-enhancing-with-iot-technology/

Short, T. (2015). *Guest Preferences for Technology Use in Hotels*. Software Advice. Retrieved from https://www.softwareadvice.com/hotel-management/industryview/technology-use-report-2015/

Simulation in Hospitality: Sparking Travelers' Imagination. (2021). *ediiie*. Retrieved from https://www.ediiie.com/hospitality.html

Team Trilyo. (2018, April 27). *What is the Role of Internet of Things (IOT) in Hotels*. Trilyo Blog. Retrieved from https://www.trilyo.com/blog/what-is-the-role-of-internet-of-things-iot-in-hospitality-industry/

Trivago. (2019, April 18). *Sustainable Hospitality: Eco-Friendly Industry Trends and Tips*. Trivago Business Blog. Retrieved from https://businessblog.trivago.com/sustainable-hospitality-trends-eco-friendly-hotel-tips/

KEY TERMS AND DEFINITIONS

Customers: An individual who purchases goods or services from a business or a shop.

Hospitality Industry: The hospitality industry includes a broader category of fields within the service industries which includes event planning, transportation, lodging, cruise line, airline, and other fields within the tourism industry.

Internet of Things: The interconnection via the Internet of computing devices embedded in everyday objects, which enables them to receive and send the data.

Sensors: Sensors are the modules, or devices, which helps to detect changes or events in its environment and accordingly responds and sends information to other electronics.

Smart Technology: Smart technology is a technology which uses big data analysis, machine learning, and artificial intelligence to provide cognitive awareness to the objects which were in the past considered as inanimate.

Touchless Technology: Touchless technology is anything which can operate without the need to physically touch the device.

Chapter 3
Smart Tourism Technologies in the Consumer Experience Journey

Yigit Yurder
Istanbul University, Turkey

Mehmet Altug Sahin
https://orcid.org/0000-0003-1048-1963
Istanbul University, Turkey

ABSTRACT

Smart technologies are becoming rapidly used in various industries successfully. The tourism industry stands as one of the evolving industries, benefiting from these smart technological developments such as virtual and augmented reality, robotics, and internet of things. Anticipatory, experiential, and reflective are the three main phases of the consumer behavior process in tourism, which are regarded as pre-travel, during travel, and post-travel phases in tourism. Smart tourism technologies are being implemented to enhance the tourist experience in these phases of their journey. This chapter aims to highlight the smart tourism technology applications in every phase of consumer experience by presenting examples from the tourism industry.

INTRODUCTION

In today's age, technology can be regarded as a vital force that people should pay attention to, as it has a significant impact on people's lives due to its continuous and rapid developments. These recent developments in Information and Communication

DOI: 10.4018/978-1-7998-6904-7.ch003

Technologies (ICTs) have changed the way how travelers plan, consume, create, perceive and share their experiences with other people (Wang et al., 2012). Experiences, in which ICTs are increasingly intertwined, also require new marketing and management approaches. Accordingly, tourism and hospitality businesses face both opportunities and challenges (Law et al., 2014).

Economy has evolved into an experience economy (Pine & Gilmore, 2011) where tourism stands as a pioneer example of it (Quan & Wang, 2004). Consumer experience has been a key research topic in tourism since the early 60's (Uriely, 2005). There have been mainly two approaches in experience researches in tourism. Marketing/management approach treats tourist experience as a consumer experience while social science approach regards tourist experience as a peak experience (Quan & Wang, 2004). We would like to note that, the marketing approach is being adopted in this chapter.

Tourist, consumer of tourism industry, creates a consumption set that will address the needs and it is expected to satisfy and match their expectations while generating experiences (Andersson, 2007). Consumer experience is a blend of many things. It consists both physical and emotional components; where physical component can be described as a company's performance while emotional component describes customer's emotions (Shaw & Ivens, 2002). Tourism companies aiming to succeed, require to enhance their customer's experiences in today's competitive environment (Mossberg, 2007). A wide variety of factors affecting the consumer experience should be taken into consideration with both a holistic approach and a narrow focus by tourism companies to create memorable positive experiences (Walls et al., 2011).

ICTs have started to play a very important role in providing competitive advantage to tourism organizations and destinations. It "*empowers consumers to identify, customize and purchase tourism products and supports the globalization of the industry by providing tools for developing, managing and distributing offerings worldwide*" (Buhalis & O'Connor, 2005). Websites of traditional destinations or agencies, personal blogs, review sites, social networks, are important technologies for the tourism industry. In addition, new technologies such as Virtual Reality (VR), Augmented Reality (AR), robotics and Internet of Things (IoT) have created Smart Tourism Technologies (STTs) together. Its effects were felt by all stakeholders in travel and tourism industry, from travel vendors to collection providers, from information brokers to the consumers themselves (C. D. Huang et al., 2017). For example, due to increasing usage frequency of smartphones and their applications, smart tourism developments increased rapidly. Afterwards, with the adoption of STTs its very easy to integrate network technologies, hardware and software to provide real time data which help to make more viable decisions for all stakeholders (Gretzel et al., 2015).

STTs started playing an important role in consumer experience with its different tools in tourism industry. Various implementation examples can be seen in anticipatory (pre-travel) phase (Xiang et al., 2015; Yoo et al., 2017), experiential (during travel) phase (Han et al., 2019; Nayyar et al., 2018) and reflective (post-travel) phase (Sun et al., 2016). The usage of these tools provides an opportunity to tourism and tourism related companies for enhancing consumer experience by providing additional services beyond their expectations.

In this chapter, aforementioned STTs being used in tourism industry will be introduced regarding the consumer experience stages. Examples of STTs will be shared in the tourism industry while also mentioning their possible contributions for the enhancement of consumers' experiences. Also, in this chapter these topics are being discussed:

- What are the effects of STTs on consumers' journey?
- What are the effects of consumers' involvement to the process?
- How can marketers maintain and increase quality of experiences with the help of STTs?

BACKGROUND

Consumer experience has a strong connection to consumer behavior and decision making. According to the most famous and widely accepted consumer behavior model (EKB model), there are five core stages of the decision-making process. The Engel Kollat Blackwell (EKB) model's five stages are problem recognition, search, alternative evaluation, choice, and outcomes (Darley et al., 2010). In the first stage, the need of a consumer also identified as a problem. The process starts with the feeling of difference between consumer's perception for an ideal status and given status. This difference forms a gap in the consumer's mind and a need or problem arises. After the problem recognition, the search for the information starts. This search aims to find the best potential problem solver. According to EKB model, consumers can search in two ways: internally or externally. The internal search consists of the memories and past experiences of the consumer; besides, the external search is gathering information from several sources like social media, family, and friends. When gathering all the information is completed, consumer begins to evaluate the options for the best problem solver. The consumers set priorities and compare the knowledge they have on specific services or products with what they consider most crucial. Not all consumers have the same evaluation criteria, some may evaluate price, some may evaluate quality or brand. The consumers then make their choice of what kind of product or brand to buy, where to buy (online or offline) and how they pay (credit card or cash). On the last step of the process, the consumers make post purchase evaluation of the product or service and decide how good or bad their experience was (Engel et al., 1986)

In today's age, the consumer experiences play a progressively important role in economic and social life. We are witnessing the rise of the experience economy. Like all the other service providers, tourist experience is surely one of the advanced examples of the experience economy (Quan & Wang, 2004). Experiences are the elements that enrich the tourism industry (Knutson et al., 2007). For this reason, it has become important to provide customers with an unforgettable experience rather than the benefit of the product. In the tourism industry, consumer experience comes to the fore because production and consumption occur simultaneously. Experience is created by presenting information, sensory and emotional elements to the client (Gupta & Vajic, 2000). Past studies have suggested that atmospheric stimuli, which are often at the center of consumer experiences, have two dimensions: human interaction and the physical environment (Ritchie & Hudson, 2009). In other words, tangible items experienced only in general consumption products may not be related to consumer experience generally; rather, the total combination of goods, services, and environment purchased and/or experienced is called the consumer experience (Lewis & Chambers, 2000).

ICTs have gained attention both from practitioners and scholars in recent years. One of the important areas in ICTs has been VR, which is being implemented in different areas of various industries including tourism (Guttentag, 2010). VR is computer-modeled 3D environments, in which the user can navigate and interact with (Bates-Brkljac, 2012; Gutiérrez A. et al., 2008), that results in real-times simulation of one or more of the user's five senses (Guttentag, 2010). While the navigation represents the exploration and moving around in 3D environments; interaction represents the ability of selecting and moving subjects in the same environment (Gutiérrez A. et al., 2008).

VR term was first introduced by Myron Kruger in the mid-1970s (Williams & Hobson, 1995) and a multisensory vehicle simulator called Sensorama that was introduced in 1962 by Morton Heilig, is considered as a milestone for VR technologies. This system offered 3 photographic slides, stereo sound, smell and wind generator. Yet, the user was just a passive observer in this system and had almost no interactivity with environment (Gutiérrez A. et al., 2008). VR systems try addressing the human senses using head-mounted displays (HMDs), personal computers (PCs), cave automation virtual environments (CAVEs) and workstations (Vince, 2004). However, the development of these systems requires complex technological infrastructures. While the usage of some senses are more established in VR systems, such as vision (computer graphics), hearing (3D sound synthesis) and touch (haptics); smell and taste lack focus due to technological complexities these systems require (Gutiérrez A. et al., 2008). Yet, thanks to the development of new technologies, new VR systems are expected to be able to address each one of these senses in the future (Guttentag, 2010).

AR is a variation of Virtual Environment which is also being used synonymously with VR (Azuma, 1997). AR combines reality and virtuality in a real environment while being interactive in real time and aligns real and virtual objects (Azuma et al., 2001). AR applications combines the real environment and 3D and/or 2D graphics by superimposing computer models on top of the real image (Kounavis et al., 2012). Even though VR has been one of the most outstanding technologies, immersion with the computer-generated virtual world does not offer its users an interaction with the real world. This caused AR's popularity rise in both academical and practical environments in recent years (Fritz et al., 2005).

The information provided in AR can be digitally manipulated (Kounavis et al., 2012) and it enhances the real-world experience, where VR replaces it (Fritz et al., 2005). Thanks to the recent advances observed in the technology in the last decades, AR applications can now be accessed from our smartphones too (Kounavis et al., 2012; Yovcheva et al., 2012).

Robotics started taking active role in working places where human workers has already been in action, thanks to the developments observed in Industry 4.0. A combination of robotics and human workforce provide hybrid and promising results in the applied areas (Sinha et al., 2020). While the first developed robots were more centered around industrial intentions, service and human oriented robots has started emerging in recent years with the development of mechanical engineering, electrical engineering and computer science (Tung & Law, 2017). Service robots are defined by the International Organization for Standardization as *"robots that performs useful tasks for humans or equipment excluding industrial automation applications"* in ISO 8373:2012; which are grouped into two categories, personal and professional service robots. Personal service robots are defined as service robots *"used for a non-commercial task, usually by lay persons"*, while professional service robots are defined as *"service robot used for a commercial task, usually operated by a properly trained operator"* (ISO, 2012). Service robots provide better mobility than industrial robots that are expected to be stationary due to their productional intentions. In addition, the social interaction level of service robots is higher (Murphy et al., 2017).

Robots can also be categorized based on their presence and embodiment levels; co-present robots (physical embodiment and physical presence), tele-present robots (physical embodiment and mediated presence) and virtual agents (virtual embodiment and mediated presence). A person can have a face-to-face physical interaction with co-present robots, a face-to-face mediated interaction with tele-present robots, digital interaction with virtual agents (Jamy Li, 2015).

From the beginning of 2000s, IoT has been the one of the most popular subjects of global interest. Simply put, IoT brings intelligence of the internet to physical products with the involvement of interconnected devices, systems that depend on the autonomous communication of physical devices in the network (Atzori et al., 2010), hence IoT makes all devices smarter and more connected (Ngyuen & De Cremer, 2016). According to Porter & Heppelmann (2014) these smart and connected products offer opportunities for more efficient product usage, new functions that cut across and transcend traditional product boundaries. Borgia (2014) commented that the expected great number of Internet enabled devices and the significant amount of available data open new ways to create services that will bring visible benefits to both companies and consumers.

The applications of the IoT extend over countless areas such as industrial automation, smart homes, smart cities and wearables therefore IoT provides numerous benefits to many industries and society (Bi et al., 2014). For example, wearable devices that helps to monitor heart and blood pressure, smart home devices and systems that connects to Wi-Fi for remote control and monitoring, automobiles with-in sensors and many more can be counted (Porter & Heppelmann, 2014). The broad area of IoT applications also effects the process of identifying needs and problems of consumers, and finding solutions that can satisfy their requirements by collecting strategic data with the help of devices that exchange data through internet. Therefore, IoT helps to improve communications, relations and become more proactive, responsive, efficient in customer services (Lo & Campos, 2018). After gathering the data from IoT devices, companies can transform a large amount of data into useful information with help of data mining and gain competitive advantage.

SMART TOURISM TECHNOLOGIES IN CONSUMERS' EXPERIENCE JOURNEY

Smart Tourism Technologies

Companies in the tourism industry need value-adding and innovative systems/ applications to create more engaging multimedia content in order to reach a wider audience (Fritz et al., 2005). Thus, STTs emerged with the help of technologies such as VR, AR, robotics and IoT.

VR and AR Technologies in Tourism Industry

VR has been used in many levels of tourism such as planning, managing, promoting, creating or transforming experiences (Loureiro et al., 2020). VR experience can be used as a complement of a real-world for enhancing the consumer's experience (Sussmann & Vanhegan, 2000), where some of its applications can even substitute for them (Guttentag, 2010). Tourism experiences can be manufactured according to the consumption choices of the consumers with the help of VR. These experiences can be enhanced on a level that was not possible to achieve before (Williams & Hobson, 1995).

From a marketing perspective, VR also provides opportunities for tour operators and travel agencies to revolutionize their promotion and sale procedures in the tourism industry (Williams & Hobson, 1995). These companies can reach out to their potential consumers of tourism destinations and offer them a pre-purchase VR experience, a sense of what can be purchased before reaching to that destination (Tussyadiah et al., 2017). 3D computer-modeled tourism sites and destinations in VR, are proven to be useful in the planning stage of a trip (Y. C. Huang et al., 2016).

Furthermore, recent researches in tourism industry reveal that, VR provides better results than 360° preview and images preview in mental imagery of the experience and presence prior to hotel stay which can enhance the tourism brand experience. VR can be used for promoting the consumers the possibilities of a lodging experience and pre-reviewing this experience. Marketers should develop the understanding that, the VR platforms can be strategically used in the development and marketing of new brands in the tourism industry (Bogicevic et al., 2019). Yet, from both the marketers' and the designers' perspective, modelling a VR experience – that convey the right message and propose new valuable experience to receiver – stands as an important challenge (Marchiori et al., 2017).

AR technologies provide the ability to deliver more personalized content and services to tourists while taking their personal needs into account (Kounavis et al., 2012) and provide opportunities for transforming travelers' experiences (Chung et al., 2018). Tourism related AR applications on smartphones offer services such as location-based information, access to variable updatable content, interactive annotations and additional information (Yovcheva et al., 2012). If the consumer has been satisfied with the application of AR, they are expected to have a positive attitude towards both that destination and intention of revisiting that site (Chung et al., 2018). AR technologies in archaeological sites and museums are also examples of positive outcomes (Fritz et al., 2005). A co-creation of AR applications in museums that involves all the stakeholders, can enhance the museum experience of visitors (Serravalle et al., 2019).

Even though these applications still lack some value-adding functionalities in tourism, they stand as a promising technology in the future (Yovcheva et al., 2012). Physical and sensory simulations, electroencephalography (EEG), longitudinal virtual experiences, deep brain simulations (DBS), well-being employment, wearables, Artificial Intelligence (AI), brain-computer interfaces (BCI) are the topics that are expected to increase their importance in tourism and related areas as a consequence of technological developments in both VR and AR (Loureiro et al., 2020).

Robotics in Tourism Industry

Robotics research in travel, tourism and hospitality has been an important topic that significantly rose its importance and expected to grow in line with the adoption of robotics in the future (Ivanov et al., 2019). Robots in tourism industry bring new experience opportunities to consumers, while also reducing the labor costs associated with the human work-force (Bhimasta & Kuo, 2019). Robots and service automation devices are being currently used in hotel, restaurants, meeting and events, theme and amusement parks, car rental companies, airports, travel agencies and tourist information centers, museums and art galleries with various forms of implementation (Ivanov et al., 2017).

Robotics in hospitality industry has started playing crucial rolls in various levels. Productivity (industrial robots), accessibility (tele presence devices) and service augmentation (service chatbots and robotic assistance devices, AI bots) are the main drivers of these adoptions (Alexis, 2017). Even though the service robots are quite expensive to own by tourism companies in the current state (Ivanov & Webster, 2019); with the technological developments in the future, improved robot capabilities and decreased purchasing and maintaining costs will make them an option for the human work-force (Ivanov et al., 2017).

Travel chatbots offer an easier and more compact alternative for travel apps. The biggest deficiency of the travel apps is that each one them help the tourists on a different purpose. For example, one app provides information for the restaurants in a particular area or another app offers weather forecast yet the other manages hotel bookings and flights. Chatbots don't create a mess in consumers' smartphones, instead they use only existing platforms like messengers or browsers that almost everyone has on their smartphones. Chatbots make possible to search for assistance or information with a simple question in your own words (Farkash, 2018). Another common name used for chatbots is virtual agents (Melián-González et al., 2019). Chatbots, user-interactive computer programs using natural languages, was developed in 1960s with an intention of mimicking human conversation and entertainment purposes (Atwell & Shawar, 2007). Companies are currently using these systems on their websites and mobile applications for providing different services to their users (Melián-González et al., 2019).

IoT in Tourism Industry

The developments observed in ICTs, gave cities the opportunity to coordinate their activities and services. Thus, information was linked and made the cities more accessible and enjoyable from the perspectives of both visitors and residents (Buhalis & Amaranggana, 2013). ICTs enabled the transformation from traditional cities to smart cities. Smart cities became more efficient and responsive with the involvement of IoT and Big Data which stand as key technologies for the implementation of ICTs in smart cities (Mohanty et al., 2016).

Smart cities are expected to have at least one of the smart components such as smart transportation, smart health care, smart energy and etc. (Mohanty et al., 2016). Smart cities can encourage the transformation of Smart Tourism Destinations (Buhalis & Amaranggana, 2013), with the involvement of IoT. Tourism components such as tourism infrastructure, tourism information data and tourism service platforms can speed up the transformation process of smart tourism (Guo et al., 2014). Thereby, smart tourism destinations can enrich tourist experience, while also creating a competitive advantage for the city (Buhalis & Amaranggana, 2013).

The Tourist Experience and Smart Tourism Technologies

According to researches, travel experience is a linear process and identified the tourist experience in three phases (Craig-Smith & French, 1994; Park & Santos, 2017; Wang et al., 2012);

- The anticipatory phase (pre-travel stage) where the tourists researches information, evaluate the alternatives, plans the travel / visit.
- The experiential phase (during travel stage) where the physical experience happens.
- The reflective phase (post-travel stage) where the tourists produce a judgement for their past travel experience and reflects their experiences to others.

From the perspective of the travel stages, travel consist pre-travel, on-site and post travel consumer experiences in tourism industry; which are also being used as pre, during and after travel stages synonymously. Anticipations and expectations play role in pre-travel stage where information collection and negotiation take place. Core experience –experiences based on collected information– and extended experience –personal experiences beyond expectations– takes place in the during travel stage. The post-travel stage consist recollection, which emphasizes on unique personal experiences and sharing it with other possible future travelers (Park & Santos, 2017).

According to Wang et al. (2012) the usage of smart technologies like IoT, AR, VR, smartphones and apps are considered as important sources because they provide useful information about the event / activity not only in the anticipatory phase but also in the experiential phase and the reflective phase. Therefore, the smartphones and the other smart technologies the user of the smartphone has access to such as apps, IoT, AR and VR creates an enjoyable experience and satisfies several functional and emotional needs such as pleasure, novelty, creativity and. information search.

The focus in this book chapter is technological innovations in the tourism industry to improve consumers' experiences, both the physical environment and the interaction. In today's advancing technology, companies get a lot of help from these technologies in their marketing efforts to understand consumer needs better and provide them satisfying experiences. Technologies such as VR, AR, robotics and IoT have become widespread in the tourism sector as well as in other sectors. The places and contributions of these technologies in the tourist experiences' three phases (anticipatory, experiential, reflective) will be discussed with examples.

Anticipatory phase

The first phase, the anticipatory phase, is the stage where expectations are formed and it is the first step in the consumers' journey. During this stage, the image and perception of the destination formed in the mind of the consumer has a very important place when evaluating alternatives. However, at this phase, the destinations and plans that consumers choose are rarely completely what they want. Because there are travel agencies and companies that always direct and influence these travel plans (Potgieter et al., 2016). In this phase, perceptions and diffusion of information have a substantial influence on the selection of destination and holiday type.

There are multiple examples of 360° VR systems applied in the hospitality industry and Shangri-La Hotels and Resorts is one of them. They are providing 12 different VR experiences in various destinations and hotels such as Hong Kong, Beijing and Shanghai (Shangri-La Hotels and Resorts, n.d.).

Navitaire, an Amadeus company that provides innovative solutions in both airline and rail industries (Navitaire, n.d.); shared their new VR system which enables travel search and booking experience, a first in this manner. This system encourages its users to search for travel destinations on a 3D spinnable projection of the world, look for flights to or from the chosen destination by setting a departure and arrival date, make class and seat selections while traveling through a 3D modeled plane, ask for options for rental cars from different rental companies with a 3D modeled projection of a car and make a payment for this entire planned trip just by using a VR headset. In the future, the hotel booking experience is expected to be added to this system, also with the social media integration which will provide users to view destinations their Facebook friends traveled before (Vallantin, 2017).

VR systems are also being used in mobile applications in tourism industry. Travel World VR is a downloadable mobile phone application where its users can experience tours (London bike and bus, Pompeii, Tahiti snorkel safari), destinations (Venice, Rome, Florence, etc.), hotels (Club Med, Sunset, etc.), attractions (One World Trade Center) and cruise lines (Quark Antarctica, MSC Lirica, etc.). This app provides a technological infrastructure to its users to experience these services either by using a VR headset or just by using their mobile phones (Travel World VR, n.d.).

Pepper, which was first introduced in Innorobo Paris in 2016, was developed by Softbank Robotics as an outcome of a humanoid robot project. It was designed and programmed mainly for helping travel consultants and also welcoming guests and interacting with them. Pepper asks questions to its users and uses deep learning AI algorithms using the answers they provide. Pepper creates a profile depending on these algorithms and offers a shortlist of possible travel destinations (Holi, 2016). Amadeus and TUI Germany teamed up and launched a project in early 2017 by positioning Pepper in a real time environment. This project aimed bringing solutions to challenges and issues retail travel agencies face in the real world. Every result of these interactions and reactions of users were recorded and analyzed for later development of this prototype, which has been identified as a continuing innovative journey by the company (Monhof, 2018).

Companies like Expedia.com and Skyscanner.com are providing chatbot services in social media platforms like Facebook. Another online travel agency, Booking.com has also been using chatbot services for helping their customers to find and book an accommodation. Before providing chatbot service, prototypes went through some tests to deliver a better chatbot service. First, they tried suggested responses. However, the answers became biased due to these suggestions that chatbot delivered and took away the freedom of choosing from customers. Then they tried pretending to be a bot for collecting answers and building a pattern of the flow. Although this method provided positive results, it also had its flaws such as making the conversation too human. To overcome this issue, they built a structure for possible conversations around that flow and wrote scripts for the chatbot (Khanchandani, 2019).

Experiential phase

The second phase, the experiential phase, essentially covers all tourism experiences that occur at the destination. As part of the overall travel experience, the outward journey to the destination and the impressions of the return journey constitute the very important parts of this phase (Potgieter et al., 2016). In this phase, consumers expect to be freer and more flexible.

Henn na Hotel, world's first hotel staffed by robots, offers its services with three different concepts. They aim to increase efficiency by using robots as the main staff in operational services and facial recognition system for a keyless stay, reduce electricity usage with motion sensors and radial panel air conditioning systems, reduce waste by removing unnecessary amenities (Henn na Hotel, n.d.). However, due to some encountered difficulties such as front desk robots' failures in check-in processes, in-room robot Churi's inability to meet guest requests, etc. (Bhimasta &

Kuo, 2019); Henn na Hotel, which first launched with 80 robots, had to abandon half of their 243 operating service robots that low-performed. Yet, the company plans to continue their services in their newer branches with the successful robots while also providing necessary updates (Gale & Mochizuki, 2019).

YOTEL, a brand that embraces technology for creating unique experience, offers their services in hospitality industry (Yotel, n.d.). They can be viewed as pioneers with their efficient use of technology in the hospitality industry. Self-check-in mobile apps, adjustable smart beds and techno wall are some of the successful technological achievements of the brand (Hochman, 2018). In addition to these systems, they use robots to enhance their guests staying experience. A luggage storage robot named YOBOT is stationed in brand's New York City hotel. Customers that arrive early or want to check-out late and discover New York City can leave their luggage to YOBOT, which offers its services as a complimentary (Perry, 2015). Another robot the brand uses is YO2D2, a customized guest service robot –a butler– created by Savioke. There are camera and other sensors embedded into YO2D2. Thus, it can greet guests, safely navigate through crowds, use the elevator, identify the room door, make the delivery, interact with guests by using its built-in touchscreen and can even dance. Guests are asked for a review instead of a tip. If the review is positive, YO2D2 shimmies and shows its enjoyment (Hochman, 2018).

Connie is another robot used in the hotel industry. Connie, Watson enabled concierge robot, is presented as an outcome of a collaboration between Hilton Worldwide and IBM in 2016. Connie is designed to interact with the guest and inform them about local touristic attractions, dining recommendations, hotels feature and amenities in a friendly and informative manner. As the interaction level increased, Connie could learn, adapt and improve its recommendations. Hilton collected questions and Connie's answer to later analyze and make improvements in their guests experiences before, during and after their stays (Hilton, 2016).

KLM partnered up with SME Bluebotics, various universities, Centre National de la Recherche Scientifique (CNRS) and Amsterdam Airport Schiphol in a robotics project named Spencer which was co-funded by European Commission and coordinated by University of Freiburg. This project focused on assisting the transfer passengers and providing them a guidance assistance. It is aimed to reduce the missing of connection flights which might be caused by delays, restricted transfer times, getting lost in airport and language barriers. Spencer aids the transfer passengers to their gates, avoid the obstacles in its route and provide them the necessary information while making sure the passengers are following him. Once Spencer reaches the gate, it informs the passengers with a message displayed on its screen. Spencer is expected to reduce the workload of KLM employees and canalize their works towards other issues. It is also expected to present pro-active solutions in unexpected situations, recognize groups and respond correctly to their behaviors, recognize emotions.

Spencer was tested in early months of 2016 to guide some KLM passengers to their departure gates and succeeded in these various tests (KLM, 2015, 2016).

SITA, a company that specialized in air transport ICT, aims to provide solutions for the air transport industry. The company presented Leo as one of their newest innovations. Leo is a fully autonomous and self-propelling baggage robot that can offer services such as checking-in, printing bag tags and carrying up to 2 baggage that weighs a maximum total of 32 kg. Leo can travel through crowded places in while avoiding obstacles that might intervene. It welcomes passengers upon their arrivals at the airport and offers its services. While introducing Leo the company noted, using robotics could be the new way on how the baggage can be handled by airports and airlines (SITA, n.d.).

In recent years, travelers' faces started replacing IDs and boarding passes thanks to the facial recognition systems implemented in airports. These systems are being applied in international flights to reduce the workload of ground services. They can shave up to %10 off time by providing security and boarding services. Travelers are not obliged to use this service and can opt out. Airlines like JetBlue, Delta and Air France are active users of these systems in different destinations. Even though these systems might fail on a rarely basis and strict rules need to applied for the security reasons, they are expected to be implemented by airlines in new destinations (McCartney, 2019).

Robotics are actively used in food industry to achieve success in different operational tasks. A robotic dishwasher developed by Dishcraft Robotics can identify, isolate and clean each dish that was loaded into dish washing robot. Dishcraft reduces waste and increases efficiency in food services. It uses machine learning to identify, if the dish has been cleared properly. Dishcraft places clean dishes in racks and passes them into the sanitizing machine. It helps the companies to reduce their costs by minimizing the usage of water, broken dishes and requirement of labor (Dishcraft, n.d.). Servi is a food service robot that was developed by Bear Robotics. Servi can navigate through narrow and crowded places, stop immediately and change its course once an obstacle is detected by its cameras and laser sensors, shows versatility by overcoming multiple tasks (serving, bar running, drink delivery, bussing) and even sing a happy birthday song to customers. Company has developed its first model in 2017 and Servi is the fourth generation enhanced model (Bear Robotics, n.d.). Flippy, Miso Robotics robot, operates as an autonomous kitchen assistant that can work on grill or fryer, provide cooking consistency, be collaborative with the kitchen staff, washable, recognize items and change its tools upon the requests. Flippy has 3D and thermal scanners with AI which collaboratively provides a brain. Also, it has safety sensors and can be integrated with POS systems of restaurants (Miso Robotics, n.d.).

Royal Caribbean International has been running their cruise operations with an innovative approach since 1969. In the recent years, they started offering new features and activities to their guest in their ships. Bionic Bar is one of them, where two robotic arm bartenders prepare the drink guests order from an app. Guest can grab their drinks they ordered from the bar or prefer to be served to their table by a human waiter. Guest can choose from either from a standard menu or choose from 30 spirits, 21 mixers (sodas, juices, syrups, etc.) and create their own combinations. Robotic arm bartenders can muddle, stir, shake and strain; and they can also produce a drink in two minutes and serve up to 1.000 drinks a day. Royal Caribbean International cooperated with Makr Shakr for this project. Makr Shakr collaborated with Marco Pelle, a principal dancer at the New York Ballet, to analyze his dance moves and simulate them into these robotic arms to acquire more human-like movements (Royal Caribbean International, 2016). Two70 is another robotic feature company uses for enhancing stage performances. This stage has an ocean view of 270-degree angle that transform into a multimedia theater with the help of dancing robotic screens once the show starts. This dynamic movement of robotics arms adds a unique experience to stage performances of the company (Royal Caribbean International, n.d.-c).

LIFEWTR, a premium water company, and Marriott Hotels launched an in-room guest experience project in late 2018, Art Through Technology, which utilizes AR technology. This project was launched as LIFEWTR's seventh design series, where they collaborated with various artists. As a result of this project, 18 unique virtual artworks were embedded on the bottles of LIFEWTR. The cooperation between LIFEWTR and Marriott provided guests the opportunity of transforming the walls of their rooms using AR technology. Guests need to follow some steps to enjoy this experience. First, they are required to scan the custom tags of a designed bottle, which can be found in their rooms in cooperated Marriott Hotels. Then the users will be directed to Facebook in-app, that requires a camera for the AR experience. After successfully scanning the tag, guests can begin customizing their room from any of these 18 artworks of various designers. They can also use these artworks and color them to their likings. Guests can also share their unique experience and even their customized designs through social media platforms (LIFEWTR, 2018, 2019).

Rooms of Hub Hotel from Premier Inn, located in London Covet Garden's, comes with an AR ready map on the walls. Guests can experience this interactive map which provides information about the area the hotel located around (Killham, 2015). In addition to this stylish décor painted on the wall, the room provides a smart TV which can be connected to guest's mobile phone and mirror it on the screen (Premier Inn, n.d.).

Royal Caribbean International also uses VR in their various cruise activities. Jump Rally is a bungee adventure game where a VR set is being used while the user is being attached to a real bungee trampoline. Players try to stop power orbs in the virtual reality, while also jumping on the trampoline in real time. The game gets more difficult if the players succeed. The play time of the game is expected to be two to four minutes and high scores of players are being displayed on the daily leaderboard (Royal Caribbean International, n.d.-b). Another virtual reality game company provides is Bass Bouncer, a product of Sky Pad. Bass Bouncer is built around music in a futuristic world and expected to be two to four minutes per game. It provides a musical journey to the players and at the end of the game users get to hear the music they created (Royal Caribbean International, n.d.-a).

Nexos by Igor provides a complete smart building solution by making hotels smarter and more customizable using IoT. The company aims to improve guest experience, increase safety and security, provide proactive maintenance, sustainability solutions. This enables automated check-in, smart locks and security solutions, wayfinding solutions by navigation, smart room customizations (e.g., lighting, room and water temperature, curtains and shades, smart TVs, room services). The company installed their systems in several GR8 Hotels located in Netherlands to provide more personalized guest experience (IGOR, n.d.).

Marriott International Teams, with the co-operation of Samsung and Legrand, presented IoT hotel room named IoT Guestroom Lab in late 2017 that aims to deepen the personalized guest experience. It is designed to enhance hotel room guest experience and create more efficient room design to meet the personal needs of the guests to the best, while also maintaining and contributing to Marriott's sustainability efforts and goals. This Lab has been defined as a smart room where multiple responsive IoT systems, devices and applications can communicate with each other. Guests can set an alarm, start a yoga session, ask for an additional housekeeping service, start the shower using their desired temperature levels with just using voice activated virtual assistant or using pre-set customer profiles on mobile app. Thanks to this project, hotel operations are expected to be more personalized and be more centered around the personalized preferences of hotel guests. These rooms are expected to be used in real hotel environments in the near future (Marriott International, 2017).

Reflective phase

Finally, third phase, the reflective phase, takes place after the trip. The satisfaction of the consumer and the trip's perceived value can be explained as, consumer satisfaction is the extent of overall pleasure and joy felt by the consumer, resulting from the ability of trip experience to meet the consumer's desires, expectations and needs in associate to the trip, and the perceived value of the trip is the consumer's assessment of the net worth of the trip according to consumer's evaluation of benefits and costs (Chen & Tsai, 2007). Finally, in this phase, consumers relive their unique experiences in their minds and describe them in great detail, share them with the other people and the companies.

As a result of the common adoption of Web 2.0 and applications by companies and consumers, user-generated content (UGC) and social media have become a mass phenomenon. Therefore, the wide usage of UGC and social media carve out a strong relationship between information technologies and tourism industry. According to this strong relationship, it is not a surprise that travel-related topics are listed as most popular issues in social media (Öz, 2015). From a perspective of tourism marketing, these applications and social media are mostly used as evaluation platforms for consumer experiences (Morosan et al., 2014). The role of social media is enabling customers who are in two different phases, to talk to each other (pre-travel and post-travel).

Tripadvisor, one of the leading UGC sharing platforms of the tourism industry, had 859 million reviews in 2019 in their website. By the end of 2014, Tripadvisor had 200 million entries and by the end of 2019 entry total more than quadrupled (Statista Research Department, 2020). In addition to Tripadvisor, there are a lot of social media platforms (Facebook, Instagram, Twitter, etc.) and OTA websites (Expedia, Booking, etc.) UGCs can be shared. The sum of these UGCs - both the reviews and photos- can lead up to a tremendous level of data, which plays an active role in Big Data research in the tourism area. (Jingjing Li et al., 2018). As a result of Big Data analytics, new products, strategies and related topics can be offered in the tourism industry and both consumers and companies can benefit from it mutually.

Hilton teamed up with Oculus for Business to train their team members and build empathy by providing an innovative learning experience with the use of VR technology. Hilton Learning Team gained help from SweetRush for developing these training modules. Corporate Team Members, whom have probably never been in a service operation, are presented with training modules for building empathy and appreciation for the front-line service operations of the hotel. These modules included setting up room service trays under two minutes, completing nine tasks of the guest room cleaning checklist with 60+ assignments, checking-in guests and delivering the benefits of Hilton's membership program, Hilton Honors. They also built a role-playing model for Hotel Team Members with multiple scenarios that present challenges from virtual guests. Hotel Team Members are expected to offer proactive solutions and successfully handle the presented challenges in front desk, meeting room, room service, breakfast service and departure operations. The interaction results of the virtual guest experience vary depending on whether these scenarios are handled poorly, resolved correctly, or handled beyond expectations; which eventually helps to build empathy with their guests too. Thanks to this project, Hilton was able to reduce their training times from 4 hours (in-class training) to 20 minutes (VR training). Hilton's 87% of their Team Members who received VR training, changed their behaviors as an outcome of this process (Oculus for Business, 2020).

COVID-19 and Smart Tourism Technologies

COVID-19 pandemic is the biggest global health crisis of the 21st century and the greatest challenge the humanity has faced since World War Two (UNDP, n.d.). The large number of infections and deaths has forced governments to implement various restrictions, such as travel controls, closures of schools and many institutions, to contain the spread of the epidemic in the world (Karabulut et al., 2020). The tourism industry stands as one of the most affected industries, due to both external and internal travel limitations and border closures. According to the previous studies (Blake et al., 2003; Kuo et al., 2008; Rosselló et al., 2017), diseases have significant effects on tourism like decline on demand and decrease on expenditure. Therefore, the pandemic prompted the tourism industry to take many measures and make crucial changes.

One of the first actions taken against the pandemic is the robots that use ultraviolet rays to kill bacteria and viruses in airports (San Antonio International Airport and Heatrow). The robot emits intense bursts of UV light at 67 Hz to destroy viruses in high touch areas and hard to clean surfaces. The robots can be used on elevators, restrooms, baggage area, ticket counters and gate areas (Xenex, n.d.). Another example is, U.S hotel companies improve their automation technologies such as digital room keys that can be used with the help of customers' smart phones and the voice- activated AI assistants to minimize contact between hotel staff and guests (Castellanos, 2020).

The appearance of COVID-19 pandemic caused lockdowns in multiple industries and one of them has been tourism. The ending of the pandemic can't be foreseen, but it emerged a topic that will possibly play a significant role, technology. If the appropriate technology solutions will be adapted, more sustainable forms of tourism are likely to emerge (Zeng et al., 2020).

CONCLUSION

Today, Information and Communication Technologies (ICTs) offer many opportunities for the tourism and travel industry to develop itself and enhance the experience of the consumers. Within the frame of technological development that makes a huge impact on the tourism and travel industry, this chapter presents the most recent smart tourism applications and explains how these technologies affect the travel experience of the consumers. Smart Tourism Technologies (STTs) in this chapter refer to any forms of ICTs that consumers interact with such as Virtual Reality (VR), Augmented Reality (VR), robotics and Internet of Things (IoT). To enhance the consumer experience, STTs collect and harness data from social environment, physical infrastructure and human bodies with the help of advanced technologies (Yoo et al., 2017). STTs such as mobile applications, smart tools that can connect to the internet, and learning machines, also aim to make the consumer experience more satisfactory.

The focus in this book chapter is STTs which improve consumers' experiences in both the physical environment and the interaction process. Technologies such as VR, AR, robotics and IoT have become widespread in the tourism sector as well as in other sectors. The places and contributions of these technologies in the tourist experiences' three phases (anticipatory, experiential, reflective) become very efficient in so many ways. If we sum up the experience phases briefly, in the anticipatory phase (pre-travel stage) where the tourists research information, evaluate the alternatives, and plan the travel/visit. In the experiential phase (during travel stage) where the physical experience happens. Finally, in the reflective phase (post-travel stage) where the tourists produce a judgement for their past travel experience and reflect their experiences to others.

In the anticipatory phase, technological developments in the tourism industry are mainly focused on VR and robotics. VR systems are being offered to potential consumers to provide a pre-experience opportunity. They are being used to share the experience of a hotel guest, act as a travel search and booking engine and are being used by mobile applications to promote tours, hotels and destinations. Robots are mainly being adapted by travel agencies as virtual agents, which are also named as chatbots.

While robotics plays a key role in the experiential phase; AR, VR and IoT are also being adapted in the tourism industry. Recently, companies operating in the hotel, travel and food industries started using robotics actively in their operational services. For example, robots help guests to check-in, carry and store luggage, provide guest services, guide through airports, serve/cook foods and beverages, wash the dishes and even provide entertainment. AR is being used to enhance the guest room experience, while VR in cruises is being used to improve the enjoyment level and experience of their guests with the help of gamification. Finally, IoT enables automated check-in, smart locks and security solutions, wayfinding and smart room customizations.

In the reflective phase, consumer reviews play a crucial role in today's environment. User generated content became a trustful resource for both providing a self-assessment of a trip and sharing it with others via social networking sites, content communities, collaborative web sites. In addition to these platforms, online travel agencies act as an intermediary between the future and the past travelers. Post travel reviews of the consumers from UGC platforms deliver information for both consumers and companies. Big Data, which will create a foundation for marketing research, can be generated by combining these data. As an outcome of these marketing researches, companies can offer better services for the future customer satisfaction.

Consumers' travel behaviors and preferences are expected to change with the emergence of the COVID-19 pandemic. Health-related travel issues became a primal subject in the tourism industry. Robotics are mainly used to provide some solutions for these issues and technology is expected to play a primary role in the future.

REFERENCES

Alexis, P. (2017). R-Tourism: Introducing the Potential Impact of Robotics and Service Automation in Tourism. *Ovidius University Annals: Economic Sciences Series.*

Andersson, T. D. (2007). The Tourist in the Experience Economy. *Scandinavian Journal of Hospitality and Tourism*, *7*(1), 46–58. Advance online publication. doi:10.1080/15022250701224035

Atwell, B., & Shawar, E. A. (2007). Chatbots: are they really useful? LDV-Forum: Zeitschrift Für Computerlinguistik Und Sprachtechnologie. 10.1.1.106.1099

Atzori, L., Iera, A., & Morabito, G. (2010). The Internet of Things: A survey. *Computer Networks*, *54*(15), 2787–2805. Advance online publication. doi:10.1016/j.comnet.2010.05.010

Azuma, R. (1997). A survey of augmented reality. In *Presence, Teleoperators and Virtual Environments*. doi:10.1162/pres.1997.6.4.355

Azuma, R., Baillot, Y., Behringer, R., Feiner, S., Julier, S., & MacIntyre, B. (2001). Recent advances in augmented reality. *IEEE Computer Graphics and Applications*, *21*(6), 34–47. Advance online publication. doi:10.1109/38.963459

Bates-Brkljac, N. (2012). Virtual Reality. Academic Press.

Robotics, B. (n.d.). *Products*. Retrieved November 11, 2020, from https://bearrobotics.ai/products/index.html

Bhimasta, R. A., & Kuo, P. Y. (2019). What causes the adoption failure of service robots? A case of henn-na hotel in Japan. *UbiComp/ISWC 2019- - Adjunct Proceedings of the 2019 ACM International Joint Conference on Pervasive and Ubiquitous Computing and Proceedings of the 2019 ACM International Symposium on Wearable Computers*. 10.1145/3341162.3350843

Bi, Z., Da Xu, L., & Wang, C. (2014). Internet of things for enterprise systems of modern manufacturing. *IEEE Transactions on Industrial Informatics, 10*(2), 1537–1546. Advance online publication. doi:10.1109/TII.2014.2300338

Blake, A., Sinclair, M. T., & Sugiyarto, G. (2003). Quantifying the impact of foot and mouth disease on tourism and the UK economy. *Tourism Economics, 9*(4), 449–465. Advance online publication. doi:10.5367/000000003322663221

Bogicevic, V., Seo, S., Kandampully, J. A., Liu, S. Q., & Rudd, N. A. (2019). Virtual reality presence as a preamble of tourism experience: The role of mental imagery. *Tourism Management, 74*, 55–64. Advance online publication. doi:10.1016/j.tourman.2019.02.009

Borgia, E. (2014). The internet of things vision: Key features, applications and open issues. Computer Communications. doi:10.1016/j.comcom.2014.09.008

Buhalis, D., & Amaranggana, A. (2013). Smart Tourism Destinations. Information and Communication Technologies in Tourism 2014. doi:10.1007/978-3-319-03973-2_40

Buhalis, D., & O'Connor, P. (2005). Information communication technology revolutionizing tourism. *Tourism Recreation Research, 30*(3), 7–16. Advance online publication. doi:10.1080/02508281.2005.11081482

Castellanos, S. (2020, August 5). Hospitality Industry Turns to Tech to Lure Guests Back. *CIO Journal*. https://www.wsj.com/articles/hospitality-industry-turns-to-tech-to-lure-guests-back-11596636001

Chen, C. F., & Tsai, D. C. (2007). How destination image and evaluative factors affect behavioral intentions? *Tourism Management, 28*(4), 1115–1122. Advance online publication. doi:10.1016/j.tourman.2006.07.007

Chung, N., Lee, H., Kim, J. Y., & Koo, C. (2018). The Role of Augmented Reality for Experience-Influenced Environments: The Case of Cultural Heritage Tourism in Korea. *Journal of Travel Research, 57*(5), 627–643. Advance online publication. doi:10.1177/0047287517708255

Craig-Smith, S., & French, C. (1994). *Learning to ive with tourism*. Pitman Publishing Pty Limited.

Darley, W. K., Blankson, C., & Luethge, D. J. (2010). Toward an integrated framework for online consumer behavior and decision making process: A review. *Psychology and Marketing, 27*(2), 94–116. Advance online publication. doi:10.1002/mar.20322

Dishcraft. (n.d.). *Solutions*. https://dishcraft.com/solutions

Engel, J. F., Blackwell, R. D., & Miniard, P. W. (1986). Consumer behavior (5th ed.). Prentice-Hall, Inc.

Farkash, Z. (2018). *Travel Chatbot — How Chatbots Can Help City Tourism.* Cahtbots Magazine. https://chatbotsmagazine.com/travel-chatbot-how-chatbots-can-help-city-tourism-a2f122c0896d

Fritz, F., Susperregui, A., & Linaza, M. T. (2005). Enhancing Cultural Tourism Experiences With Augmented Reality Technologies. *The 6th International Symposium on Virtual Reality, Archaeology and Cultural Heritage VAST.*

Gale, A., & Mochizuki, T. (2019, January 14). Robot Hotel Loses Love for Robots. *The Wall Street Journal.* https://www.wsj.com/articles/robot-hotel-loses-love-for-robots-11547484628

Gretzel, U., Sigala, M., Xiang, Z., & Koo, C. (2015). Smart tourism: Foundations and developments. *Electronic Markets, 25*(3), 179–188. Advance online publication. doi:10.100712525-015-0196-8

Guo, Y., Liu, H., & Chai, Y. (2014). The embedding convergence of smart cities and tourism internet of things in China: An advance perspective. *Advances in Hospitality and Tourism Research, 2*(1), 54–69.

Gupta, S., & Vajic, M. (2000). The Contextual and Dialectical Nature of Experiences. In J. A. Fitzsimmons & M. J. Fitzsimmons (Eds.), *New Service Development: Creating Memorable Experiences* (pp. 33–51). SAGE Publications, Inc., doi:10.4135/9781452205564.n2

Gutiérrez, A. M. A., Vexo, F., & Thalmann, D. (2008). Stepping into virtual reality. Stepping into Virtual Reality. doi:10.1007/978-1-84800-117-6

Guttentag, D. A. (2010). Virtual reality: Applications and implications for tourism. *Tourism Management, 31*(5), 637–651. Advance online publication. doi:10.1016/j.tourman.2009.07.003

Han, D.-I. D., Weber, J., Bastiaansen, M., Mitas, O., & Lub, X. (2019). *Virtual and Augmented Reality Technologies to Enhance the Visitor Experience in Cultural Tourism.* doi:10.1007/978-3-030-06246-0_9

Henn na Hotel. (n.d.). *Henn na Hotel General Concept.* Retrieved September 26, 2020, from https://www.h-n-h.jp/en/concept

Hilton. (2016). *Hilton And IBM Pilot "Connie," The World's First Watson-Enabled Hotel Concierge.* EMEA Hilton Newsroom. https://newsroom.hilton.com/corporate/news/hilton-and-ibm-pilot-connie-the-worlds-first-watsonenabled-hotel-concierge

Hochman, D. (2018, February 23). Robot Butlers Are Taking Over Hotels (Good News: No Tipping Required). *Forbes Life*. https://www.forbes.com/sites/davidhochman/2018/02/23/robot-butlers-are-taking-over-hotels-good-news-no-tipping-required/#645ed6e7636c

Holi, L. (2016). *Could robots become the next travel retailers?* Amadeus. https://amadeus.com/en/insights/blog/robots-become-next-travel-retailers

Huang, C. D., Goo, J., Nam, K., & Yoo, C. W. (2017). Smart tourism technologies in travel planning: The role of exploration and exploitation. *Information & Management, 54*(6), 757–770. Advance online publication. doi:10.1016/j.im.2016.11.010

Huang, Y. C., Backman, K. F., Backman, S. J., & Chang, L. L. (2016). Exploring the Implications of Virtual Reality Technology in Tourism Marketing: An Integrated Research Framework. *International Journal of Tourism Research, 18*(2), 116–128. Advance online publication. doi:10.1002/jtr.2038

IGOR. (n.d.). *Nexos for Hospitality*. Markets and Case Studies. Retrieved December 15, 2020, from https://www.igor-tech.com/filesimages/Case Studies/Sell Sheet_Hospitality.pdf

ISO. (2012). *ISO 8373:2012(en), Robots and robotic devices*. ISO Online Browsing Platform. https://www.iso.org/obp/ui/#iso:std:iso:8373:ed-2:v1:en

Ivanov, S., Gretzel, U., Berezina, K., Sigala, M., & Webster, C. (2019). Progress on robotics in hospitality and tourism: a review of the literature. Journal of Hospitality and Tourism Technology. doi:10.1108/JHTT-08-2018-0087

Ivanov, S., & Webster, C. (2019). Robots in tourism: A research agenda for tourism economics. Tourism Economics. doi:10.1177/1354816619879583

Ivanov, S., Webster, C., & Berezina, K. (2017). Adoption of robots and service automation by tourism and hospitality companies. *Revista Turismo & Desenvolvimento, 27*(28), 1501–1517.

Karabulut, G., Bilgin, M. H., Demir, E., & Doker, A. C. (2020). How Pandemics Affect Tourism: International Evidence. SSRN *Electronic Journal*. doi:10.2139/ssrn.3579530

Khanchandani, P. (2019). *Prototyping Booking.com's Chatbot for Better Conversations*. Booking.Design. https://booking.design/prototyping-booking-coms-chatbot-for-better-conversations-4f080ad27d8d

Killham, E. (2015). *Hub hotel has smart rooms you run from your wrist*. Cult of Mac. https://www.cultofmac.com/328325/hub-hotel-has-smart-rooms-you-run-from-your-wrist/

KLM. (2015). *Robot "Spencer" to guide KLM passengers at Amsterdam Airport Schiphol*. Newsroom.

KLM. (2016). *Spencer robot completed tests guiding KLM passengers at Schiphol*. Newsroom.

Knutson, B. J., Beck, J. A., Kim, S. H., & Cha, J. (2007). Identifying the dimensions of the experience construct. *Journal of Hospitality & Leisure Marketing*, *15*(3), 31–47. Advance online publication. doi:10.1300/J150v15n03_03

Kounavis, C. D., Kasimati, A. E., & Zamani, E. D. (2012). Enhancing the tourism experience through mobile augmented reality: Challenges and prospects. *International Journal of Engineering Business Management*, *4*, 10. Advance online publication. doi:10.5772/51644

Kuo, H. I., Chen, C. C., Tseng, W. C., Ju, L. F., & Huang, B. W. (2008). Assessing impacts of SARS and Avian Flu on international tourism demand to Asia. *Tourism Management*, *29*(5), 917–928. Advance online publication. doi:10.1016/j.tourman.2007.10.006 PMID:32287724

Law, R., Buhalis, D., & Cobanoglu, C. (2014). Progress on information and communication technologies in hospitality and tourism. *International Journal of Contemporary Hospitality Management*, *26*(5), 727–750. Advance online publication. doi:10.1108/IJCHM-08-2013-0367

Lewis, R. C., & Chambers, R. E. (2000). *Marketing leadership in hospitality. Foundations and practices*. John Wiley.

Li, J. (2015). The benefit of being physically present: A survey of experimental works comparing copresent robots, telepresent robots and virtual agents. *International Journal of Human-Computer Studies*, *77*, 23–37. Advance online publication. doi:10.1016/j.ijhcs.2015.01.001

Li, J., Xu, L., Tang, L., Wang, S., & Li, L. (2018, October). Big data in tourism research: A literature review. *Tourism Management*, *68*, 301–323. Advance online publication. doi:10.1016/j.tourman.2018.03.009

LIFEWTR. (2018). *LIFEWTR® and Marriott Hotels Join Forces to Show Travelers Inspiration Behind Every Door*. https://www.multivu.com/players/English/8456651-pepsico-lifewtr-marriott-hotels-art-inspiration-augmented-reality-ar/

LIFEWTR. (2019). *Art Through Technology Inspires The LIFEWTR Brand's Newest Design Series*. https://www.multivu.com/players/English/8513451-lifewtr-art-through-technology/

Lo, F. Y., & Campos, N. (2018). Blending Internet-of-Things (IoT) solutions into relationship marketing strategies. *Technological Forecasting and Social Change*, *137*, 10–18. Advance online publication. doi:10.1016/j.techfore.2018.09.029

Loureiro, S. M. C., Guerreiro, J., & Ali, F. (2020). 20 years of research on virtual reality and augmented reality in tourism context: A text-mining approach. Tourism Management. doi:10.1016/j.tourman.2019.104028

Marchiori, E., Niforatos, E., & Preto, L. (2017). Measuring the Media Effects of a Tourism-Related Virtual Reality Experience Using Biophysical Data. Information and Communication Technologies in Tourism 2017. doi:10.1007/978-3-319-51168-9_15

Marriott International. (2017). *Marriott International Teams with Samsung and Legrand to Unveil Hospitality Industry's IoT Hotel Room of the Future, Enabling the Company to Deepen Personalized Guest Experience*. Marriott International News Center. https://news.marriott.com/news/2017/11/14/marriott-international-teams-with-samsung-and-legrand-to-unveil-hospitality-industrys-iot-hotel-room-of-the-future-enabling-the-company-to-deepen-personalized-guest-experience

McCartney, S. (2019, August 14). Are You Ready for Facial Recognition at the Airport? *The Wall Street Journal*. https://www.wsj.com/articles/are-you-ready-for-facial-recognition-at-the-airport-11565775008

Melián-González, S., Gutiérrez-Taño, D., & Bulchand-Gidumal, J. (2019). Predicting the intentions to use chatbots for travel and tourism. *Current Issues in Tourism*. Advance online publication. doi:10.1080/13683500.2019.1706457

Robotics, M. (n.d.). *Flippy*. Retrieved November 11, 2020, from https://misorobotics.com/flippy/

Mohanty, S. P., Choppali, U., & Kougianos, E. (2016). *Everything you wanted to know about smart cities. IEEE Consumer Electronics Magazine*. doi:10.1109/MCE.2016.2556879

Monhof, D. (2018). *Humanoid robot teams up with TUI Germany*. Amadeus. https://amadeus.com/en/insights/blog/humanoid-robot-teams-tui-germany

Morosan, C., Bowen, J. T., & Atwood, M. (2014). The evolution of marketing research. *International Journal of Contemporary Hospitality Management*. Advance online publication. doi:10.1108/IJCHM-11-2013-0515

Mossberg, L. (2007). A Marketing Approach to the Tourist Experience. *Scandinavian Journal of Hospitality and Tourism*, *7*(1), 59–74. Advance online publication. doi:10.1080/15022250701231915

Murphy, J., Hofacker, C. F., & Gretzel, U. (2017). Robots in Hospitality and Tourism: A Research Agenda. *EReview of Tourism Research ENTER 2017*. http://3ws1wk1wkqsk36zmd6ocne81.wpengine.netdna-cdn.com/files/2016/12/RN107.pdf

Navitaire. (n.d.). *Navitaire & Amadeus*. Retrieved October 9, 2020, from https://www.navitaire.com/amadeus

Nayyar, A., Mahapatra, B., Le, D. N., & Suseendran, G. (2018). Virtual Reality (VR) & Augmented Reality (AR) technologies for tourism and hospitality industry. *IACSIT International Journal of Engineering and Technology*, *7*(2.21), 156. Advance online publication. doi:10.14419/ijet.v7i2.21.11858

Ngyuen, B., & De Cremer, D. (2016). *The Fairness Challenge of the Internet of Things*. The European Business Review. https://www.europeanbusinessreview.com/the-fairness-challenge-of-the-internet-of-things/

Oculus for Business. (2020). *Building empathy to enhance hospitality*. https://business.oculus.com/case-studies/hilton/

Öz, M. (2015). Social media utilization of tourists for travel-related purposes. *International Journal of Contemporary Hospitality Management*, *27*(5), 1003–1023. Advance online publication. doi:10.1108/IJCHM-01-2014-0034

Park, S., & Santos, C. A. (2017). Exploring the Tourist Experience: A Sequential Approach. *Journal of Travel Research*, *56*(1), 16–27. Advance online publication. doi:10.1177/0047287515624017

Perry, R. (2015). *Everyone has baggage...* https://www.yotel.com/en/blog/everyone-has-baggage

Pine, B. J. I., & Gilmore, J. H. (2011). *The Experience Economy, Updated Edition*. Harvard Business Review Press.

Porter, M. E., & Heppelmann, J. E. (2014). How smart, connected products are transforming competition. Harvard Business Review.

Potgieter, M., Jacobs, M., & Geldenhuys, S. (2016). Incentive travellers' proclivity to revisit an incentive destination. *African Journal of Hospitality, Tourism and Leisure*, *5*(3), 1–21.

Inn, P. (n.d.). *FAQs | Technology at hub by Premier Inn*. Retrieved November 11, 2020, from https://www.premierinn.com/gb/en/faq/hub-technology.html

Quan, S., & Wang, N. (2004). Towards a structural model of the tourist experience: An illustration from food experiences in tourism. *Tourism Management*, *25*(3), 297–305. Advance online publication. doi:10.1016/S0261-5177(03)00130-4

Ritchie, J. R. B., & Hudson, S. (2009). Understanding and meeting the challenges of consumer/tourist experience research. *International Journal of Tourism Research*, *11*(2), 111–126. Advance online publication. doi:10.1002/jtr.721

Rosselló, J., Santana-Gallego, M., & Awan, W. (2017). Infectious disease risk and international tourism demand. *Health Policy and Planning*, czw177. Advance online publication. doi:10.1093/heapol/czw177 PMID:28104695

Royal Caribbean International. (n.d.-a). *Bass Bouncer*. Retrieved November 5, 2020, from https://www.royalcaribbean.com/cruise-activities/bass-bouncer

Royal Caribbean International. (n.d.-b). *Jump Rally*. Retrieved November 5, 2020, from https://www.royalcaribbean.com/cruise-activities/jump-rally

Royal Caribbean International. (n.d.-c). *TWO70®*. Retrieved November 5, 2020, from https://www.royalcaribbean.com/cruise-activities/two-seventy

Royal Caribbean International. (2016). *Robot Bartenders Shake Things Up At Sea*. https://www.royalcaribbean.com/blog/robot-bartenders-shake-things-up-at-sea/

Shangri-La Hotels and Resorts. (n.d.). *Experience the virtual world of Shangri-La Hotels and Resorts*. Retrieved November 10, 2020, from http://www.shangri-la.com/vr/en/index.htm

Shaw, C., & Ivens, J. (2002). Building Great Customer Experiences. Building Great Customer Experiences. doi:10.1057/9780230554719

Sinha, N., Singh, P., Gupta, M., & Singh, P. (2020). Robotics at workplace: An integrated Twitter analytics – SEM based approach for behavioral intention to accept. *International Journal of Information Management*, *55*, 102210. Advance online publication. doi:10.1016/j.ijinfomgt.2020.102210

SITA. (n.d.). *Leo, SITA's baggage robot*. Retrieved October 31, 2020, from https://www.sita.aero/innovation/sita-lab/intelligent-machines/leo-sitas-baggage-robot

Statista Research Department. (2020). Number of user reviews and opinions on TripAdvisor worldwide from 2014 to 2019. statista.com.

Sun, Y., Song, H., Jara, A. J., & Bie, R. (2016). *Internet of Things and Big Data Analytics for Smart and Connected Communities*. IEEE. doi:10.1109/ACCESS.2016.2529723

Sussmann, S., & Vanhegan, H. (2000). *Virtual reality and the tourism product: substitution or complement*. Ecis.

Travel World, V. R. (n.d.). *About*. Retrieved November 10, 2020, from https://www.travelworldvr.com/about/

Tung, V. W. S., & Law, R. (2017). The potential for tourism and hospitality experience research in human-robot interactions. International Journal of Contemporary Hospitality Management. doi:10.1108/IJCHM-09-2016-0520

Tussyadiah, I. P., Wang, D., & Jia, C. (2017). Virtual Reality and Attitudes Toward Tourism Destinations. Information and Communication Technologies in Tourism 2017. doi:10.1007/978-3-319-51168-9_17

UNDP. (n.d.). *COVID-19 pandemic*. Retrieved December 3, 2020, from https://www.undp.org/content/undp/en/home/coronavirus.html

Uriely, N. (2005). The tourist experience. Conceptual developments. *Annals of Tourism Research*, *32*(1), 199–216. Advance online publication. doi:10.1016/j.annals.2004.07.008

Vallantin, C. (2017). *An Amadeus company, Navitaire, unveils the world's first Virtual Reality travel search and booking experience*. Amadeus. https://amadeus.com/en/insights/press-release/an-amadeus-company-navitaire-unveils-the-worlds-first-virtual-reality-travel-search-and-booking-experience

Vince, J. (2004). Introduction to Virtual Reality. Introduction to Virtual Reality. doi:10.1007/978-0-85729-386-2

Walls, A., Okumus, F., Wang, Y., & Kwun, D. J. W. (2011). Understanding the consumer experience: An exploratory study of luxury hotels. *Journal of Hospitality Marketing & Management*, *20*(2), 166–197. Advance online publication. doi:10.1080/19368623.2011.536074

Wang, D., Park, S., & Fesenmaier, D. R. (2012). The Role of Smartphones in Mediating the Touristic Experience. *Journal of Travel Research*, *51*(4), 371–387. Advance online publication. doi:10.1177/0047287511426341

Williams, P., & Hobson, J. P. (1995). Virtual reality and tourism: Fact or fantasy? *Tourism Management*, *16*(6), 423–427. Advance online publication. doi:10.1016/0261-5177(95)00050-X

Xenex. (n.d.). *Our Clients*. Retrieved December 4, 2020, from https://xenex.com/our-clients/

Xiang, Z., Magnini, V. P., & Fesenmaier, D. R. (2015). Information technology and consumer behavior in travel and tourism: Insights from travel planning using the internet. *Journal of Retailing and Consumer Services*, *22*, 244–249. Advance online publication. doi:10.1016/j.jretconser.2014.08.005

Yoo, C. W., Goo, J., Huang, C. D., Nam, K., & Woo, M. (2017). Improving travel decision support satisfaction with smart tourism technologies: A framework of tourist elaboration likelihood and self-efficacy. *Technological Forecasting and Social Change*, *123*, 330–341. Advance online publication. doi:10.1016/j.techfore.2016.10.071

Yotel. (n.d.). *Our DNA: Mission Statement*. Retrieved September 28, 2020, from https://www.yotel.com/en/about-yotel/dna

Yovcheva, Z., Buhalis, D., & Gatzidis, C. (2012). Overview of smartphone augmented reality applications for tourism. e-Review of Tourism Research.

Zeng, Z., Chen, P. J., & Lew, A. A. (2020). From high-touch to high-tech: COVID-19 drives robotics adoption. *Tourism Geographies*, *22*(3), 724–734. Advance online publication. doi:10.1080/14616688.2020.1762118

Chapter 4
Technological Innovations in the Hospitality and Tourism Industry

Anusha Thakur
https://orcid.org/0000-0001-8761-2250
University of Petroleum and Energy Studies, India

ABSTRACT

The hospitality and tourism industry is one of the most diversified and biggest industries in the world. Increasing trends of technology and innovation are expected to favorably influence the future scenario of the hospitality and tourism industry. Growing adoption of new technologies in these industries over the past few years has significantly reformed the services provided by the same. One of the major goals of the upgradation of services by the hotels or leisure segments is digitalization of services and personalization of customer experiences. In this chapter, state-of-the-art technologies currently being used in the hospitality and tourism sector have been discussed. This chapter further emphasizes how information technology is proving significant in keeping pace with the competitiveness in terms of services in the hospitality and leisure segments. In addition to all of the above, the chapter also focuses on the trends, impacts, and opportunities of information technology applications in the industry.

INTRODUCTION

In today's scenario, the hospitality and tourism industries play a significant role, particularly in the economies of developed and industrialized regions. With the

DOI: 10.4018/978-1-7998-6904-7.ch004

increasing technological advancements, hospitality and tourism businesses are emphasizing on changing the overall content of the environment. The hospitality sector accounted for approximately 10 percent of the GDP across the globe in 2019 (In 2019, there were nearly 700,000 resorts and hotels across different regions in the world). Factors such as globalization, consolidation, and technological advancements are expected to impact these businesses across the world. In addition to these, large-scale mergers and acquisitions, cross-border investment, and data analytics are also anticipated to impact the overall landscape of the same.

Several industries are considerably being impacted by the technological revolution, hospitality industry being one of them. In a dynamic commercial environment, profitable excellence requires offering individualized, and robust value propositions, and addressing their needs as they evolve. Over the past few years, value proposition within the hospitality industry, is expected to have evolved from the pure product play such as management services, restrooms, restaurants, and hotel rooms, with customized level of services. Hospitality industry is closely linked to the tourism industry, wherein, the major goal is to offer desired products and services to the travelers across different regions of the world. The hospitality businesses emphasize on the efforts for maintaining a great quality guest service. This includes the improvisation of efficient and effective ways for maintaining the customer needs and requirements. The leading vendors are engaged in identifying different market trends and opportunities in their businesses, thereby enhancing their overall profitability. Identification of latest trends, enables the manufacturers to emphasize on introducing differentiated products and solutions in their businesses, thereby also gaining an edge in the market. Innovation is one of the key aspects for hospitality businesses which caters to the changing requirements and necessities, and at the same time remain relevant according to the current scenario. Innovations are introduced in new processes, new services, and new products, which arises in these new blends. The industry is undergoing rapid transformation, which is profoundly being shaped by new technologies as well as the needs of diversified range of travelers. Innovativeness in technologies, enables the hospitality businesses to offer their guests with greater convenience, customization, and seamless experiences. Few of these innovations includes hotel leverage technologies such as application development and artificial intelligence which offers the guests with an array of differentiated services. As smart products are making their way to consumers' mainstream lifestyle, the vendors are investing in this arena to get better returns. The opportunities offered by smart products are seemingly game-changing for companies seeking a new level of efficiencies. From various operations to marketing services to guest experience, latest trends of incorporating smart technology in different sectors provides opportunities for the businesses to reach new levels and gain profitability. Implementation of smart technology thereby, is expected to change the way, travelers are opting to travel across the world, and the way, businesses operate. Further research and development, in various application segments allows the manufacturers to continually roll-out enhanced features and solutions in their product line-up to cater to the changing market scenario.

LITERATURE REVIEW

Over the past few years, services have had a vital role in the economies of technologically advanced nations (Song, 2009). Introduction of innovative business models and technologies is significantly engaged in transforming the environment. Service innovation is best defined as the "introduction of something innovative into the life, timing, organization, as well as placement of what can be defined as the collective as well as individual processes which relates to the customers" (Cerović, 2019, September).

In today's scenario, every industry is enduring a technology-driven transformation, including the hospitality industry as well. Innovative technologies have exponentially boosted the growth in market for the tourism and hospitality industries. Innovation is one of the key aspects for hospitality businesses which caters to the changing requirements and necessities, at the same time being relevant according to the time as well.

One of the other key factors catering to the success of these businesses is the quality of services offered by them. In today's competitive world, it is very essential to keep in pace with latest technological trends across different sectors in the hospitality industry, to compete in the market. The hospitality businesses thereby emphasize on substantial efforts for maintaining a great quality guest service. Whether it is customer relationship management, social media, online booking systems, all of these systems are offering a new experience to the customers across the industry. Maintaining the quality of customer services are one of the crucial factors in the industry. Digital technology is making its way into every aspect of the industry: hotel operations, guest services, communications, revenue management, distribution, CRM and marketing. These dynamic and enhanced processes occurring in the industry are the outcome of agile enrichment, which has further revolutionized globetrotting.

Few of these innovations include mobile self-check-in, phone-as-key-cards, mobile booking, electronic luggage tags, lobby media panels, self-service check-in kiosks, guest device connectivity tools, hotel service optimization systems, tablet menus, smartphone boarding passes, table reservation systems, wearable devices to identify guests, voice over internet protocol phones which are interconnected with the hotel's ecosystem. These innovations might take place directly as a front-line service or as the backend service (which are particularly invisible to the customers).

In few of the hotels or hospitality services, the guests are engaged in using their mobile phones to make dinner plans or interact with the latest gadgets in the hotel lobby. Such advancements enable the consumer to witness new experiences by offering them a new way of discovering local tourist attractions, thereby enhancing their overall hotel experience.

The hospitality industry largely comprises of different businesses and services which are linked to the customer satisfaction and leisure activities. Hospitality industry involves addressing consumer satisfaction, thereby adhering to the needs and requirements of the customers. These businesses thereby focus on the ideas of pleasure, luxury, experiences, and enjoyment for its customers. Hence, excellent and efficient quality of service is significantly one of the prime facets in the hospitality business. Various consumer related challenges can be addressed by dealing with different services being offered by the businesses in different sectors such as amenities, travel, and accommodations.

Sectors in the Hospitality Industry

The different sectors within the hospitality industry includes:

- *Accommodation:* Accommodation sector in the hospitality sector mainly deals with the customers with any place to stay temporally. This is mostly associated with the tourism industry, wherein the customers are engaged in booking their lodgings and trips. This segment includes hotels, breakfasts & bed, hostels, resorts, and serviced apartments.
- *Travel and Tourism:* The hospitality offerings are mainly linked to the customer experience, satisfaction, and pleasure. The cross-over between hospitality and tourism industry is largely dependent on the services, in comparison to end-products (Revfine.com, 2021). This segment includes Online Travel Agencies (OTA), tour operators, travel agents, car rentals, cruises, and casinos.
- *Food and Drinks:* Apart from being the necessities, the food and drinks segment also offer the people a way of spending in the leisure time, and socialize. The food and drinks segment provides a wider range of customers, to the passers-by, ex-pats, locals, and tourists. This segment includes bars & cafes, catering, restaurants, tea & coffee shops, and nightclubs.

In order to keep pace with the changing consumer preferences, it has hence, become very essential for the hospitality businesses to focus on the digitally advanced technologies. Upgradation in any of the above sectors of the hospitality industry requires a platform by enabling partnerships between the business segment and the guests. Adoption of latest technologies in the hospitality businesses offers excellent travel experience to the guests with an improvement in managerial and operational efficiency of the business as well.

Evolution of "Data Analytics" in the Hospitality Industry

Evolution and implementation of data analytics is expected to transform the way hospitality companies perform their businesses. The businesses are increasingly using larger amounts of data, with more advanced techniques, which helps them to make better decisions across the different areas, which thereby improves their performance. Implementation of such techniques focuses on the areas of revenue growth, such as better customer relationship management and improved guest experience. Analytics is expected to change the way hospitality companies approach transactions, thereby favorably impacting the overall market outlook.

Analytics in Hotel Industry

Usage of analytics is engaged in directly impacting the revenue for hospitality and tourism sector companies. In the revenue segment of these industries, the companies are expected to leverage the data from a multitude of sources such as competitor sales, air traffic, future reservations, as well as online search behavior in order to optimize the pricing as well as forecast the demand. With the usage of analytics in their business models, it is becoming easier for the hoteliers to understand the guest expectations and their behaviors, and accordingly develop their strategies to gain profitability. Analytics is thereby expected to significantly enhance the overall guest experience. For instance, inclusion of analytics in the businesses enables the app developers to improve the algorithms and user interface behind the booking procedure by understanding consumer expectations. Proper understanding of the consumer preferences thereby empowers the businesses to offer custom-made recommendations with a personalized experience.

Analytics in transactions

Usage of advanced and big data analytics offers deeper insight at a bigger level. In today's scenario, with the advent of technology, the businesses emphasize on offering insights into customer activities, and proper distribution networks. This level of analysis is performed via Machine Learning (ML) and data mining processes, hence, enhancing the customer understanding as well as operational trends. In addition to this, the businesses working in line with these, often establishes/positions themselves in the competitive market with financial strength.

Social media is considerably one of the areas in transaction analytics. The analytic tools in the social media segment, offers the investors with an overall view of the customer sentiments such as launching of new products, understanding of brand affinity, and increasing marketing effectiveness. Further, visualization tools, such as Tableau, also enables the dealmakers to understand the data sets, to explore different dimensions and gain success in the market.

In order to manage the businesses effectively, the hotel professionals are increasingly focusing upon the adoption of advanced tools and frameworks, thereby catering to the changing market scenario. One of such tools increasingly being adapted by the businesses includes the usage of Robotic Process Automation (RPA). The inclusion of these configured advanced software/tools, enables the hoteliers to automatically perform their business processes. Few of such benefits includes efficient and quicker operations with minimized risks of errors in processing the data. This technology is prominently gaining traction among the varied hotel functions. For instance, in cases of customer service segments such as room service, check-in, and reservations, this technology plays a significant role. In addition to this, this can also be used in the other segments such as procurement, payroll, accounting, human resource, vendor payments, along with the administration of learning developments. RPA technology helps in detecting and creating alerts in case of errors, which disrupts the information flow, and hence, leads to delays. Hence, implementation of RPA provides the businesses with opportunities to make the processes more cost-effective, reliable, and efficient.

Since, the hospitality industry captures vast volumes of data sets, effective usage of analytics is intensely expected to impact the way these businesses are working. Embracing the data analytics concept is not only anticipated to enhance the guest experience, but also boost the bottom as well as top line results for both hospitality and tourism businesses. Hence, implementation of data analytics concept is expected to help the hotel industry understand the customers and manage them effectively.

Technological Trends in The Hospitality and Tourism Industry

Emerging trends in innovation, luxury, sustainability, as well as technology are all expected to favorably impact the future of hospitality industry. In today's scenario, it is becoming very much essential for the hoteliers to proactively enquire about, monitor and familiarize themselves with the "Next Generation Technologies", which have already paved their way in the hospitality industry. These technologies include Internet of Things (IoT), Artificial Intelligence (AI), Chat Bots, voice assistants, blockchain, and robotics. Wider adoption and implementation of these technologies can be attributed to the increasing trends of digitally obsessed consumers across different regions in the world. Few of these technologies are illustrated below:

Next Generation Technologies

Voice Search/Voice Assistants

One of the latest technological trends witnessing an increase in the market is voice control technology, which in particular, poses to be significantly beneficial to the hotel owners. This application includes integration of hotel brand Central Reservation Systems (CRS), with the voice assistants such as Alexa, Apple Home Pod, Amazon, and Google Assistant to enhance the customer services. Voice search/ Voice Assistants are considerably impacting and revolutionizing the booking processes. Nowadays, the online travel user demands for an intuitive and seamless, travel planning experiences, which comprises of the ability to search as well as book hotels through voice search with their favorite digital voice assistant. Usage of the voice search technology owes the potential to enhance the user booking experience much faster and easier.

In addition to these, these assistants enable the travel customers to explore the accommodations, or in some situations also help in booking hotels. Amazon Alexa, together with Kayak.com, provides voice search-powered hotel bookings to the travel customer with a Kayak.com account, although there being a few notable restrictions to the technology.

Few iterations of the voice-assisted technology in the hospitality and travel industry includes:

- **Mobile App-based Digital Voice Assistants:** These dedicated assistants help in navigating a particular product, application, and services, which are considerably used by online retail firms for offering customer support.
- **Voice Search Assistants:** Microsoft Cortana, Google Assistant, Apple's Siri, Amazon's Alexa, and other speech-recognition agents embedded with the concept of artificial intelligence (AI).
- **Home Digital Voice Assistants:** These include Google Home and Amazon's Alexa which have large scale usage in millions of households.
- **Smart TV-based Voice Assistants:** These enable the consumers to search for various video and movie content via their televisions. Direct TV and Amazon FireTV voice-activated remote are few examples in this segment.

Virtual assistants are increasingly becoming a vital component in the overall smart device industry, thereby absolutely altering the way customers interact with their devices. With the increasing expansion of industries and the growing advancements in technology, the companies are emphasizing towards the enhanced usage of "smart technology". Introduction of smart technology in the hospitality

Figure 1. Number of digital voice assistants in usage across the world (2019-2024) (in billion units)
Source: (Vailshery, 2021, January 22) https://www.statista.com/statistics/973815/worldwide-digital-voice-assistant-in-use/

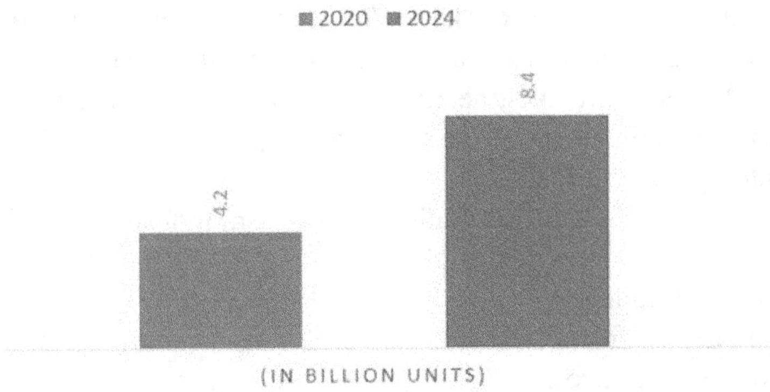

and leisure industries enables the guest users to communicate with their connected smart devices in a similar manner as in case of smartphones.

As per the above-mentioned statistics, the growth is being boosted by faster implementation across different industry verticals, ranging from banking to healthcare sectors, wherein, the companies are in pace to release their own voice technology integrations as per the consumer demand.

Artificial Intelligence (AI)

Growing number of travelers with significant technological changes in the hospitality industry forces the businesses to offer differentiated products and solutions in order to keep pace with the changing competitive scenario. As a result, the business processes and practices need to be reanalyzed at all levels of businesses to cater to the user experiences and accordingly achieve customer loyalty and memorable guest experiences. In order to personalize the services, the industries are focusing on embedding the Artificial Intelligence (AI) technologies into its customer services and operations.

The hospitality and tourism industry are engaged in the adoption of modern and advanced technology along with a higher degree of AI and Machine Learning. The concept of AI enables the brands to enhance the customer experience, customer services, and marketing. Few of the solutions offered by the technology includes customer service (chat bots), personalization (one-to-one marketing, one-to-one pricing), database management (single-view customer data) and loyalty programs.

AI owes the ability to predict the guest preferences and identify the data patterns by increasing accuracy. The technology owes the capability to offer invaluable insights which has not yet been experienced by the hoteliers and travel companies, and are significantly gathered and analyzed with the help of the relevant customer information. In addition to all of these, AI employs cognitive learning to accept the traditional human functions at an optimum cost with enhanced efficiency and profitability. As per the reports, nearly 70 percent of the Online Travel Agencies and approximately 60 percent of the hotels are expected to adopt the concepts of Artificial Intelligence.

Factors Driving AI in the Hospitality and Tourism Industry includes:

- Operational Efficiency
- Revenue Improvement
- Customer Experience

With the growing number of travelers, it is becoming an essential factor for the service providers to distinguish themselves with highly tailored services. The process of understanding the consumer preferences and monetizing the guest data has significantly propelled the introduction of data analytics Machine Learning poses to be the best source to learn and act on the basis of past experiences, and accordingly enhance their operational efficiency in order to cater to the changing consumer needs and requirements.

Incorporation of Machine Language and Artificial Intelligence in the Hospitality and Tourism Industry has widespread consequences for inspiring the end-user experience of the guests. The applications driven by the ML and AI technologies allows the businesses to analyze the guest reviews, personalize the services, and provide virtual assistance.

Context Sensitive Personalized Services

Customized services enable the users to have a pleasant and seamless experience. In order to attain the personalization level, the data relevant to customer preferences, satisfaction levels, and purchase patterns from various professional network or other sources are decoded. One of the significant challenges here poses to be, to collect data from various systems, and accordingly, draw inferences from the same.

- Machine Language technology enables the companies to identify patterns from larger sets of data, gather predictive data and accordingly understand the consumer behavioral patterns to simplify Frequent Flyer Points (FFP) redemption.

- Artificial Intelligence technology allows to create hyper-personalized campaigns, in order to generate customer retention and revenue opportunities.

Figure 2. Applications of AI in travel and tourism industry
Source: (Infosys, Role of AI in Travel and Hospitality Industry, 2018) https://www.infosys.com/industries/travel-hospitality/documents/ai-travel-hospitality.pdf

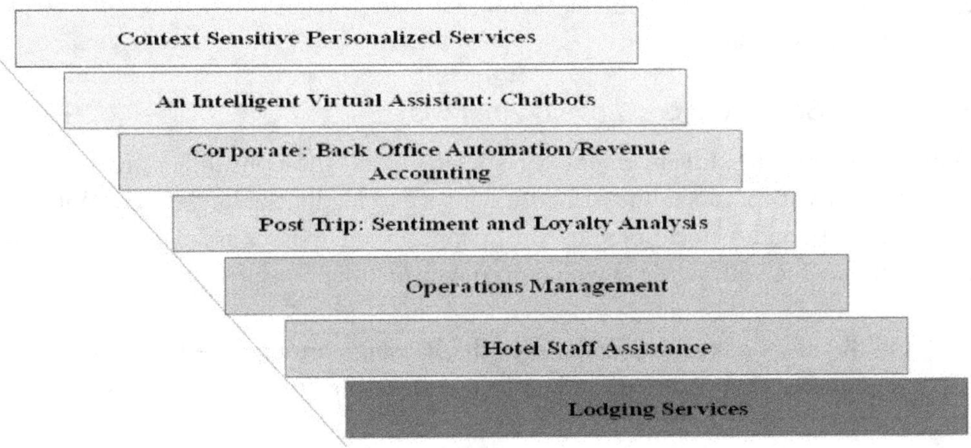

Chatbots: An Intelligent Virtual Assistant

Chatbot is a piece of software which is engaged in conducting discussions via textual or auditory methods. Two types of Chatbots includes Voice-based Chatbots and Text-message based Chatbots. Text-message based chatbots offers messages in the form of text messages in reply to the query of consumers. Voice-based chatbots offers messages in the form of voice-based messages in reply to the query of consumers (Kumar R. L., 2018); (Kumar V. K., 2016, March).

One of the distinguished features of Chatbots comprises of giving multiple answers to a single question. Chatbots comprises of inbuilt programs which can easily recognize the keywords inside the questions which activates the number of responses in reply to a single question (Makar, 2014).

In today's scenario, most of the airports and airlines are expected to implement AI-based intelligent chatbot services. These chatbots are expected to scan the keywords and accordingly reply with identical keywords from the learning database. The software will also process, record as well as monitor the events around the guests.

Corporate – Backoffice Automation and Revenue Accounting

Robotic Process Automation is considerably one of the vital tools which owes the capability to execute the repetitive and mundane tasks performed by the humans.

- ML helps in automating the classification of invoices, which otherwise is performed by an accountant.
- AI enables to identify the demand, plans the human resource requirements, and forecasts the prices.

Post Trip – Sentiment and Loyalty Analysis

With the increasing advancements of social media, the consumers are likely to express their viewpoints in the form of reviews and feedbacks on the web. With the help of the ML and AI technologies, the hospitality businesses are expected to scale the features which attracts and which do not attract the consumers.

- On the basis of recommendations, the hoteliers are expected to accordingly personalize their services and also create personalized smart videos for their guests.
- Sentimental analysis enables the businesses to get insights related to the negative, neutral, and positive views by analyzing the emotional behavior.

Operations management

According to the International Air Transport Association (IATA), it is estimated that nearly 7.2 billion passengers will be in transit by 2035, thereby posing to be a challenge in order to maintain smoother operations with larger capacity of passengers. The turnaround activities of the flights are expected to require an intelligent and smarter systems in order to assess and monitor the delays in departure.

- Machine learning technology can create a virtual assistant which can analyze the typical baggage breaking and mishandling points, and environments, and hence trace the baggage of the user.
- On the basis of previous patterns and data along with the current operational situations, the risks of delayed departures can be assessed by ML.
- In addition to all of these, ML and predictive analytics also owes the capability to predict the passenger flow, thereby avoiding overcrowding of the airports.

Hotel staff assistance

Hotel staffs such as housekeeping, front-desks, and concierge generally need to personal contacts with the guest users. With the help of Property Management Systems (PMS) and Point of Scale (PoS), AI can reveal the required guest services to cater to the changing customer demands. Incorporation of PoS and PMS, provides real-time tracking, confirmation, and booking. The technology is also engaged in offering real-time recommendations by analyzing the preferences of the users.

Lodging services

AI and ML helps in integrating real time feedbacks into the workflows, which helps in determining the satisfaction level of the guests.

- AI-powered smart devices such as mirrors allows the users to read the news, and check the maps and weather forecasts.
- Artificial Intelligence technology also helps in keeping a tab on the amenity maintenance schedules.

Blockchain

The usage of blockchain technology in the various aspects such as marketing, distribution, and operations segment of the hospitality industry is significantly gaining traction. Blockchain has three main profound uses in the hospitality sector which includes procurement, universal digital IDs, and secure contracting.

Internet of Things (IoT)

In today's scenario, implementation of IoT concept plays a significant role in the hotel industry. Inclusion of these strategies into their properties, enables the businesses to minimize unnecessary labor and costs, and at the same time increase the guest satisfaction, as well as productivity. These technologies are used in the hotel security, customer service segment such as hotel lobby, concierge, and room service, and operations segment such as A/C management and power supply. Few trends of IoT in the hospitality industry includes:

- *Guest-room Automation:* Guest-room automation offers a way for the hotels to distinguish themselves by keeping in pace with smart home technology in order to make the customer experience a comfortable one.
- *Hyper-Personalization:* Personal data offers a personal touch to the hotel guests, thereby making their stay special and comfortable at the same. For

instance, an IoT platform offers the advantage of memorizing a guest's detailed comfort preferences, thereby automatically setting up the room according to their requirements (such as choice of TV channels, lighting, temperature).

- *Predictive Maintenance:* The concept of predictive maintenance enables the hoteliers to recognize the hazardous trends, and hence, alert the staffs to maintain the same.
- *Mobile Engagement:* Mobile plays a significant role in the hospitality industries. In case of the hotels, guests can use their phone for submitting the requests to the front desk or as the key to their rooms. Mobile engagement is also the house tool via the usage of an Enterprise Asset Management (EAM) and Computerized Maintenance Management Systems (CMMS) application. In addition to all of these, engineers also have the access to the work requests or orders which are required to be performed on the property, thereby minimizing the need for a paper, and hence, improving the guest satisfaction.

With such growing digital advancements, the consumers are propelling the hoteliers to adopt latest technology, in order to acquire, engage, retain, and offer services to the digitally-obsessed customers across different digital touch points. The hospitality sector is thereby being transformed into a 100 percent digital technology-enabled industry, which is driven by mobile, blockchain, IoT, online, cloud, AI tools and applications.

Hence, implementation of the smart technology in the hospitality and tourism businesses offers potential services to enhance the user experience and expand the customer base. Smart technology offers great savings and revenue opportunities, thereby increasing the overall profitability of the businesses. Inclusion of such advanced technologies in the businesses, offers seamless experience to the customers with enhanced level of convenience, and flexibility. Various IoT sensor technology and services are illustrated below:

Smart Energy Management

With the rapidly changing regulatory scenario, factors such as market consolidation and increasing demand for connectivity is increasingly gaining prominence in the different regions of the world. Several businesses are expected to witness growth in the energy-efficient systems in different applications. Businesses, such as the hotel industry is significantly increasing their requirement for the energy management technology, in order to cater to the technological needs of the individual hotel guest rooms. Such needs and requirements include software algorithms, sensors, and distributed control systems. For instance, implementation of smart energy management systems is expected to reduce the energy costs in hotel businesses

Figure 3. Overview of the HVAC industry in different regions of Europe
Source: (Zoric) https://www.business-sweden.com/globalassets/insights/reports/trade/hvac.pdf

by nearly 20 percent thereby producing some of the fastest payback periods as well as increasing the overall resale value of any hotel (Attala, 2019). The energy management systems are witnessing a dynamic period of transition from room-centric HVAC control, autonomous, to the integrated energy control, monitoring as well as management (Stephens, 2014). Implementation of smart HVAC systems uses machine learning algorithms and occupancy sensors to optimize the HVAC energy consumption and analyze the demand load patterns. Various regions are expected to witness an increase in the market for HVAC systems, owing to the increase in its demand across the different businesses. For instance, the analysis of HVAC systems in Europe is illustrated below:

Analysis of HVAC Systems in Europe:

Europe is the hub to most of the major HVAC manufacturers in the world. Germany accounts for a significant market share in the HVAC industry. In addition to this, countries such as Italy, the Netherlands, the UK, Spain, and Scandinavia, are also anticipated to witness considerable rise in demand for the HVAC products.

Smart lighting technology enables the hoteliers to automate the consumption, understand their energy needs, as well as get easily adapted to the real-time changes. In addition to this, the smart lightning systems enables the hotels to track the occupancy patterns, set preferred lighting times, as well as improve the overall energy consumption. Energy efficiency factor is also significantly being backed up by various regulatory measures being introduced in the different regions. For instance, The Energy Performance Buildings Directive is engaged in promoting energy efficient solutions with various incentive schemes. Increasing demand for unitary air conditioners is also anticipated to witness significant rise in demand, particularly in the non-residential construction activities. Implementation of the innovative energy management systems are expected to cut the energy costs for various commercial buildings by nearly 65 per cent (Kapiki, 2010, January). Likewise, they can also be personalized to manage energy consumption effectively and hence, cater to the individual needs in one product.

In addition to all of these, installation of smart meters across the businesses, helps in unlocking the energy data, thereby upgrading any country's energy grid and helping them to pave the way for technologies such as electric vehicles, battery storage and solar panels. Hence, in order to support the rollout of smart meters for the hospitality industry, the governments of different countries are emphasizing towards the development of energy management innovations such as web platforms and apps. These apps and platforms are engaged in turning the energy data into energy saving actions for businesses such as hotels, restaurants, and pubs.

Robotic and Building Automation Technology

The labor-intensive positions in the hotels which involves structured or repetitive work such as customer service, housekeeping, along with the restaurant & bar staffs is expected to witness widespread adoption of automation, robotics, along with other AI-powered devices over the years.

In today's scenario, robots, artificial intelligence, and automation technologies (RAIA), are expected to act as a substitute for the human labor (Ivanov S. W., 2017, May 17-19). With the introduction of RAIA technologies, role of labor in the economy is expected to be marginalized, thereby leading to a massive transformation (Dong, 2014, November), (LaGrandeur, 2017). Implementation of the RAIA technologies is expected to significantly impact the tourism industry. The businesses nowadays, are emphasizing towards offering elite, and highly-touched human delivered tourism services. In today's scenario, few hotels are engaged in offering enhanced experiences, wherein, they have the robots to deliver the items via room services. These systems offer ease in delivering the products and services, thereby enhancing their quality of services and digital experiences of the customers. Furthermore, increasing usage of robots, requires the hospitality and tourism businesses to invest in robot-friendly facilities. Personal digital assistants with the capabilities of predictive analysis can book trips even before done by the humans.

In addition to this, hospitality services brought by the developments in IoT-enabled systems such as digital concierge, automated check-in and check-out services, and keyless entry services, are increasingly gaining traction in the market. Hence, with the implementation of automated services, the service providers as well as the guests are all expected to be benefitted. Further, building automation technology is also expected to enhance the managerial and operational efficiency. For instance, the infrared scanners help in minimizing/reducing disruptions in the housekeeping services. Usage of infrared scanners directly detects the presence of an individual in room, and tells the staffs whether it is occupied or unoccupied, thereby improving the overall customer experience.

Hence, automated and robotic services are anticipated to increase the productivity and efficiency of the business activities. However, there are a few challenges which are expected to restrain the demand for these services. These includes skill deficits, higher costs, along with significant changes related to the cultural differences. The integration of robotic technology requires the consideration of balance between the roles of the human employees and the robots to offer better-quality guest experience and environment.

Body Area Sensors

One of the key trends in the IoT revolutions, includes the smart and wearable devices. Wireless Body Sensor Network (BSN), is a collection of wearable sensor nodes which owes the capability to communicate (with themselves, other sensors, as well as smart devices). These sensor nodes further have wireless transmission, computation, sensing capabilities, and storage capacity. These wearable devices gather the data such as fitness activities, location, heart rate, and temperature from the users. In addition to these, the body sensor networks, are the key enablers for several other application segments such as: e-Fitness, e-Sport, e-Social, and e-Wellness. Furthermore, these sensor networks are also used to monitor the vital signs non-invasively and continuously, wherein, the tiny wireless sensors are integrated with the garments or placed on the skin (Fortino).

With the proper analysis of the gathered data gathered via the body area sensor networks, it becomes easier to offer services such as automatic adjustment of in-room lighting (on the basis of sleeping cycle of the users), automatic adjustment of the in-room temperature (on the basis of body temperature), as well as the meal suggestions (on the basis of the user's fitness goals). Further inclusion of these sensors also enables the hospital service providers to provide special facilities to the guests as per the medical devices they use. For instance, these service providers owe the capability to filter out the sugary meal and high carbohydrate options for high cholesterol and diabetic meal options for the patients suffering from the same.

Big Data and Big Data Protection

The capability to collect, interpret, and simultaneously respond to the information gathered poses to be very essential for the hospitality and tourism businesses. With the implementation of big data concept, the hotel and other companies in the hospitality industry are focusing on ways to generate more revenue with enhanced customer experience as well.

The big data concept is expected to be beneficial to the hospitality industry in the following ways:

- *Revenue Management:* The big data concept enables the businesses with the revenue management strategy. Implementation of this strategy enables the hoteliers or other companies to carry out the predictive analysis, thereby allowing them to accurately antedate the levels of demand for their services among the consumers. Few of these services include the current booking rates, past occupancy rates, as well as information related to the holiday season and various events. Big data upon collection from these services, are widely used for the prediction of demand. This predictive analysis enables the businesses to accordingly optimize the rate of their services, and hence, increase their profitability.
- *Customer Experience:* Hotels have varied types of interactions with the potential as well as current customers, thereby offering opportunities to gather large amounts of data. Big data concept is expected to prove significant to the hospitality industry. This can be attributed to the fact that the big data concept enables the hoteliers to easily identify the customer experiences, in terms of their reviews/opinions. Customers experience information is usually gathered via their reviews and rating on different websites, feedbacks on social media, and other relevant information. When the data is analyzed, information such as services used by the customers, and also being requested by them, are often revealed. Hence, the detailed analysis of gathered data leads to a proper understanding among businesses in terms of customer requirement and their experience, and hence, accordingly strategize to improve if required. In addition to this, the technology enables the hotel owners to make decisions related to new services, along with the services that are no longer required. Furthermore, big data enables the businesses to easily make decisions associated to partnerships with other businesses, such as restaurants or bars, local taxi and online travel agency (OTAs).
- *Competition Scouting:* The concept of big data offers an idea about the existing players or the entry of new players and the competition posed by them in the market. The implementation of this technology enables the

businesses to know about the outperformance and also key shortcomings of their competitors and accordingly capitalize upon the same. Hence, with the availability of vast information, the businesses owe the capability to gain profits in their revenue and simultaneously market share as well.

The innovative technologies intertwining themselves into different sections of the contemporary human existence, enables the hospitality industry to move towards elevated long-term productivity and higher efficiency. These innovative trends are engaged in offering immense savings and improvements for the hospitality industry businesses; thereby enabling the hotel developers to accordingly plan their staffing requirement, management services and infrastructural development.

From boosting the brand-to-guest communication, to the growing guest satisfaction across the board, to accelerating the crucial as well as basic processes within the company, to optimizing the payroll expenses, these advancements in technology are expected to reshape the hospitality industry holistically (Miller, 2019). Hence, implementation of such advanced methods empowers the businesses to uphold a competitive edge in today's scenario.

RESEARCH METHODOLOGY

This study comprises of various information which are gathered via different secondary sources such as research papers, journals, company annual reports, trade reports, and articles. Varied innovations across different businesses were particularly considered. In addition to these, the study was based on different strategies, initiatives and measures taken towards the technological development processes by the different regions.

Research Questions

What are the different innovative changes impacting the overall market for hospitality and tourism industry?

- What are the different challenges or hurdles which the companies are facing to implement these technologies?
- Evolution of data analytics in the hospitality industry.
- To identify the impact of COVID-19 on the overall hospitality businesses over the forecast period.

Significance of the Study

This study aims to provide an in-depth analysis and insights of different types of technologies, and solutions adopted by the businesses to keep in pace with the changing competitive scenario. Further, this study focuses on different topics such as: -

- To understand the different IoT sensor technologies and services being incorporated in the industries.
- This study offers a brief overview of the next generation technologies.
- To get an insight on the different technological trends in the hospitality industry.

Scope of the Study

- This study provides an emphasis on the overview of travel and hospitality industry in the regions of North America (US), and Europe (France).

ANALYSIS

"North America": US Region

Unparalleled growth driven by strong economy, rising digital innovations, increasing consumer purchasing power, are few factors expected to contribute to the overall growth of the hospitality and tourism sector in the North American region. With the increase in number of travelers in the region, the attractive destinations in the regions are also expected to expand owing to the increasing competitiveness in the market.

The graph mentioned below illustrates the travel industry in US in terms of the revenue generated from the travel bookings in the region.

The airline and hotel industry are expected to witness significant growth, owing to the increasing advancements in the market, to offer excellent experience to their customers. With the growing modernization in the region, industries are focusing more on the seamless experience of consumers in the region. This focuses on creating a well-balanced, cleaner, and safer environment to the consumers. Hotels, airports, cruises, restaurants are all focusing on the smooth-running, and well-paved business with the incorporation of smart technology in the same. In order to cater to the factors such as water conservation, hassle-free experience, and convenience to the workers at these end-user segments, these businesses are focusing on advancing towards the touchless technologies in different product segments in the lobby areas, rooms, and washrooms (in terms of faucets, soap dispensers, and towel dispensers).

Figure 4. US travel industry gross bookings and revenue, by segment (2009-2017) (in $ billion)

Source: (Langford, 2019) https://www2.deloitte.com/content/dam/Deloitte/us/Documents/consumer-business/us-consumer-2019-us-travel-and-hospitality-outlook.pdf

In addition to this, one of the key factors impacting the demand for technological advancements in the hospitality and tourism sector in US is the increasing demand for smart products and solutions which offers enhanced efficiency and convenience to the consumers and businesses. Hence, the businesses in the North American region are engaged in redefining the guest experiences and relationships to gain a competitive edge in the market. The hotel and leisure segments emphasize on incorporating such technologies which can easily match the lighting, complimentary drinks, temperature of a room, and entertainment settings according to their guest's preferences even before they enter the room. Incorporation of latest techniques are expected to enable the vendors to gain profitability from their products in the market.

Moreover, the COVID-19 pandemic has led to an increase in demand for touchless solutions across different end-user segments, in order, to offer protection to the individuals across the same, thereby further accelerating the need for enhanced technologies.

Table 1. Tourism expenditure in France - 2019

Domestic/International Spending	Value
Domestic Tourism Expenditure	$ 150.77 billion
International Tourism Expenditure	$ 57.37 billion

Source: (NationMaster, 2019) https://www.nationmaster.com/nmx/timeseries/france-international-tourism-expenditures

"Europe": France Region

France is one of the significant tourist destinations in the world. Both the international as well as the domestic tourism are expected to significantly impact the economy in France. With the increasing tourism industry in the country, the occupancy rate of the hotels in France is also anticipated to witness surging rise in demand. The table mentioned below illustrates the tourism expenditure in France:

The hotel industry in France witnessed nearly 89.4 million arrivals in 2018 (Ministry for Europe and Foreign Affairs, 2019, May) https://www.diplomatie.gouv.fr/en/french-foreign-policy/tourism/the-action-of-maedi-in-promoting/ . Hence, the arrival of private accommodation space poses to be an important factor to be considered. The hotels in the region are increasingly emphasizing on the innovative concepts and strategies to cater to the expectations of the travelers. Technological innovations are constantly shaping the hospitality industry. Innovation poses to be a key factor for the hospitality businesses to remain relevant as well as to cater to the different needs and expectations of diversified range of travelers. Advancements in technology offers excellent experience and convenience to the customers. One such advancement includes the implementation of automated solutions in the rooms, receptions, housekeeping, kitchens and restrooms of the hotels and restaurants. Usage of sensor-based products offers convenience and seamless experience to the guests.

Impact of COVID-19 on the Hospitality and Tourism Businesses

The ongoing pandemic has impacted not only the life and health of the people but has also brought the global economy in turmoil. The major markets including the US, Italy, Spain, France, the UK, Germany, Canada, and China are the most affected ones. Various cities in these countries are under extreme lock-down state with mobility near to zero.

The pandemic has led to severe global economic disruptions, postponement or cancellation of sporting, religious, and cultural events, and widespread fears of supply shortages. Considering macro-economic factors, the global hospitality and tourism market, is in doldrum. The declining record of hotel occupancy rate witnessed an exponential decline. This can mainly be attributed to the stay-at-home orders, quarantine rules, and travel restrictions being imposed by different regions in order to protect the people across the world from contracting the virus. The crisis has significantly hindered the travel plans of individuals across the world. With regard to the impact on revenue of hotel and tourism businesses, it is thereby becoming essential for the resort management and owners to inquire about their policies, and check their insurance providers in order to cater to the current scenario.

Changes to Remain Post-COVID

Technology poses to be a primary component in boosting the overall value proposition of the hotels and restaurants. In a post COVID scenario, technology and the concepts of "digitalization" is expected to play an essential role in transforming the overall hospitality segments.

Increasing demand for sanitization and elevated health practices are expected to pose as an essential factor for the consumers. Adoption of touchless/contactless check-in as well as check-out processes along with the digital concierge/support are anticipated to gain traction. Hence, proactive analysis of sanitation procedures, consumer behavior and e-commerce adoption are significantly one of the key aspects among the businesses (Landman, 2021). Apart from these, several other precautionary measures need to be taken by various users as well as businesses in order, to protect themselves. Implementation of e-commerce and self-service processes is also anticipated to be one of the core factors for the hospitality businesses. Digital solutions which can classify the touchpoints and confirm minimum physical contact among the employees and guests at every point of contact is expected to be of prime focus. For instance, implementation of sensor-activated products in the lobby areas, washrooms, restaurants, rooms might be few of these. Since, the customer behavior is unpredictable, it is hence, very much essential for the hoteliers and marketers to focus on analyzing the current consumer trends and their needs, in order to gain their trust back and increase their customer base.

CHALLENGES

In today's scenario, there are a lot of technological advancements with the growing need to work with a host of vendors in their service efforts and guest acquisition. Every different service segment ranging from revenue generation to guest services (which includes front desk, engineering operations, and housekeeping services), vary from each other significantly. Every segment has its own databases, technology, vendors and tools, which are completely different from other segments. Hence, few restraints expected to impact the industry from becoming a digitally driven market holistically is explained below:

Data Management

Analysis of the guest data and aggregation of the same, is an essential part of the hospitality service chain. With the introduction of latest technologies and service platforms in the hospitality businesses, the data volume is expected to witness an exponential rise. Personalization of guest experience is also anticipated to favorably impact the increase in data volume. With the increase in traction of personalized services in the hospitality industry, it is thereby becoming an essential factor for the Hospital Service Providers (HSP) to offer great quality services and at the same time, maintain up-to-date and accurate behavioral and preference records of the guests. These service providers can collect the data via personal guest devices and the guest-facing systems, connected to the hotel networks. The BoH management systems in hotels must owe the capability to properly manage the influx of wide variety of guest data from different sources. In order to offer such tailor-made services to the guests, the BoH systems need to analyze the guest requirement or preference profile with data related to the state of environment detected from IoT sensors/devices. This concept is expected to place a burden on the BoH management systems which can only be tackled via the usage of specialized technological infrastructure. In addition to this, sharing of relevant data from these guest profiles across the different inter-organization and intra-organization systems, poses to be a challenge which requires both decentralized and centralized data management approaches.

Privacy and Security

In order to offer the guests with highly personalized services, it is very much essential for the service providers to track guest locations, their behavior, and preferences. In addition to this, these providers must also guarantee safe storage of the guest data in order to protect them from societal, economic, and physical threats. The point-of-sale terminals and guest-facing systems are considerably the most vulnerable systems in hotel businesses for security attacks. Hence, it is very much essential for these systems to ensure proper security, while interactions with guests, thereby avoiding data leaks and theft. Security in the hotel businesses prevents the hackers from accessing the guest data by attacking individual guest devices linked to the network. In addition to this, enhanced security level prevents the hackers from reprogramming the IoT systems of hotels from any type of malicious activities. Hence, addition of strong security protocols for improving the guest interaction essentially requires significant computing resources. Furthermore, a decentralized computing platform plays a crucial role to adhere to such requirement, which might pose to be challenging at times.

The data fragmentation and technology in the hospitality sector is also expected to pose as a challenge and restraint for the overall industry growth. This can be attributed to the fact that the hospitality industry is a highly fragmented industry with various technological needs and deficiencies which requires smart solutions. These are a few major obstacles for the hospitality and tourism businesses which are expected to prevent these from becoming a complete technologically-driven industry.

CONCLUSION

The "advanced mode of life" adopted by today's technology-enabled consumers is propelling the hospitality industry to increase the adoption of digital technologies. In today's scenario, the hospitality businesses are focused upon investing in the Guest Services Technology, Guest engagement, retention as well as acquisition technology. The hotels and leisure segments nowadays, are proactively partnering with other providers, thereby leading to enhanced and loyal networks among the airlines, hotels, retailers and restaurants. Investing in the advanced technological application segments enables businesses to involve the tourists at the possible touchpoints, thereby opening avenues for the hoteliers in the market. It is expected that most of the technologies by 2025, will be moved to the mainstream operations. For instance, RPA is expected to significantly impact the booking processes of the hotels and restaurants, thereby benefiting the segments in a profitable way. Implementation of chatbots in travel and hospitality, as well as restaurant segments are expected to offer customized and seamless guest experience.

This chapter discusses about the technological trends in the hospitality and tourism industry, along with the future aspects of these Next Generation Technologies. The advancements in hospitality and tourism industry enables these businesses to restructure their service platform in order to cater to the modern technological landscape and gain a competitive edge in the market. The chapter gives a brief insight about the travel and hospitality industry in the regions of North America and Europe. In addition to these trends, few of the challenges are also discussed in this chapter. These challenges need to be overcome in order to offer future-proof solutions with enhanced technological aspects to the hospitality industry.

REFERENCES

Attala, J. (2019, March 26). *10 Ways Smart Technology is Reshaping the Hotel Industry*. Hotel Management. Retrieved from: https://www.hotelmanagement.net/tech/10-ways-smart-technology-reshaping-hotel-industry

Cerović, M. P. (2019, September). The Importance of Innovation in Designing a Luxury Tourism Offering: The Perspective of Tourism Destination Stakeholders. *Tourism in Southern and Eastern Europe: Creating Innovative Tourism Experiences: The Way to Extend the Tourist Season.*

Dong, X., & McIntyre, S. H. (2014, November). The Second Machine Age: Work, Progress, and Prosperity in a Time of Brilliant Technologies. *Quantitative Finance, 14*(11), 1895–1896. doi:10.1080/14697688.2014.946440

Fortino, G. G. (n.d.). *Advances in Body Sensor Networks: Sensors, Systems, and Applications*. MDPI.

Infosys. (2018). *Role of AI in Travel and Hospitality Industry.* Infosys.

Ivanov, S. W. (2017, May 17-19). Adoption of Robots and Service Automation by Tourism and Hospitality Companies. *INVTUR 2017*, 1-19.

Kansakar, P. M. (2018, February). Technology in Hospitality Industry: Prospects and Challenges. *IEEE Consumer Electronics Magazine.*

Kapiki, S. (2010, January). Energy Management in Hospitality: A Study of the Thessaloniki Hotels. *Economics and Organization of Future Enterprise*, 78-97.

Kumar, R. L., Li, A., & Wang, W. (2018). Learning and Optimizing through Dynamic Pricing. *Journal of Revenue and Pricing Management, Springer, 17*(2), 63–67. doi:10.105741272-017-0120-2

Kumar, V. K. (2016, March). Sanative Chatbot for Health Seekers. *International Journal of Engineering and Computer Science, 5*(3), 16022 - 16025.

LaGrandeur, K. a. (2017). *Surviving the Machine Age - Intelligent Technology and the Transformation of Human Work.* Springer.

Landman, P. (2021). *Top Hospitality & Hotel Trends of 2021 and Beyond.* Retrieved from Xotels.com: https://www.xotels.com/en/hotel-management/hotel-trends

Langford, G. W. (2019). *US Travel and Hospitality Outlook.* Deloitte.

Makar, M. a. (2014). *Automatic Message Selection with a Chatbot.* Academic Press.

Miller, E. (2019). *5 Innovations that are Reshaping the Hospitality Industry.* Retrieved from: https://innovationmanagement.se/2019/01/24/5-innovations-that-are-reshaping-the-hospitality-industry/

Ministry for Europe and Foreign Affairs. (2019, May). *The Ministry's Actions to Promote Tourism.* Ministry for Europe and Foreign Affairs.

NationMaster. (2019). *France - International Tourism Expenditure.* Retrieved from: https://www.nationmaster.com/nmx/timeseries/france-international-tourism-expenditures

Revfine.com. (2021). *Hospitality Industry: The No. 1 Hospitality Information Guide.* Retrieved from: https://www.revfine.com/hospitality-industry/

Song, L. S. (2009). A Staged Service Innovation Model. *Innovation Technology Entrepreneurship & Marketing*, *40*(3), 571–599.

Stephens, M. (2014). *Energy management for hospitality industry to reach $970M by 2020.* Retrieved from: https://insights.ehotelier.com/news/2014/03/13/energy-management-for-hospitality-industry-to-reach-970m-by-2020/

Vailshery, L. S. (2021, January 22). *Number of digital voice assistants in use worldwide from 2019 to 2024 (in billions).* Statista.

Zoric, M. a. (n.d.). *Achieving Growth in Europe's Evolving HVAC and Filters Industries.* Business Sweden.

KEY TERMS AND DEFINITIONS

Customers: A person who purchases goods or services from another.

Hospitality: Relationship between a guest and a host, wherein the host receives the guest with some amount of goodwill, including the reception and entertainment of guests, visitors, or strangers.

Information Technology: Information technology is the use of computers to create, store, process, exchange, and retrieve all kinds of electronic data and information.

Innovation: Practical implementation of ideas which result in the introduction of new goods or services or improvement in offering goods or services.

Leisure: The time when you are not working or doing other duties.

Tourism: The commercial organization and operation of holidays and visits to places of interest.

Chapter 5
Impact of Technology in Sustainable Tourism Development:
Virtual Reality

Ali Yuce
https://orcid.org/0000-0003-4700-2006
Cappadocia University, Turkey

ABSTRACT

Tourism has been one of the engines powering the thriving economies, leveraging the quality of lives and transforming the people's sociocultural status worldwide. There is no doubt that innovative technologies are critical and significantly efficient for the economies during a hard time and in blossoming times. Moreover, it is also vital to understand and present some essential issues that harm tourism development. Thus, this chapter strives to offer innovative technology-oriented alternative strategies and practical implications to benefit all the stakeholders at the destination and the consumers by focusing on the long-term benefits. This chapter also aims to understand how technology can take a role to stimulate visitors' motivation and behaviour towards a touristic product or a destination.

INTRODUCTION

VR technology has set an irreversible mark between the classical and modern-day tourism product marketing approach. Travel agencies and customers use to book their destination thru two-dimensional images, which mostly resulted in disappointment

DOI: 10.4018/978-1-7998-6904-7.ch005

due to the unrealistic based information that manipulated the customers' intention towards a tourism service and product. Unlike the classical method, VR technology provides three-dimensional vivid and more realistic digital media that capture and engage the users by creating a sense of presence' feeling. Providing a lifelike impression with computer-generated 3D images are the core dimension of the VR environment that evolved the marketing technology (Choi, Hickerson, & Lee, 2018). Hence, VR transforms the way how we travel, the way how we promote our products, the way how we operate our businesses, and how we visit attractions (Guttentag, 2010; I. P. Tussyadiah, Wang, Jung, & tom Dieck, 2018; Yuce, Arasli, Ozturen, & Daskin, 2020).

VR is a digitization environment that depicts a destination or a site where users are taken away without leaving their seat. VR can be defined as a gadget that allows the users to feel and react to the stimulated or depicted (virtual) environment with all the senses as if they are in the actual place (Hudson, Matson-Barkat, Pallamin, & Jegou, 2019; Laurell, Sandström, Berthold, & Larsson, 2019). VR has been used in many areas, including education (Bower, DeWitt, & Lai, 2020), medical (L. Li et al., 2017), marketing (Pestek & Sarvan, 2020), and tourism (Choi et al., 2018). With the advent of using VR in the tourism industry, VR offers unlimited opportunities for consumers and suppliers due to changes in the way the consumer's shop and the way suppliers market their products (Rather, 2020). For example, from the consumers' perspective, VR is a decision-making technology that allows potential consumers to try the product or service before buying it. From the suppliers' aspect, VR poses an excellent opportunity to confront many challenges during economic downturns, visiting susceptible sites or coping with the health-related crises like the CORONA-19 epidemic (Rather, 2020). This epidemic caused fundamental drops in visitation during the first year of the pandemic in 2020 (S. Li, Tian, Lundberg, Gkritzali, & Sundström, 2020). Using such a useful technology also presents alternative strategies for those who cannot travel due to age, financial, and health issues (Otoo, Kim, & Choi, 2020).

Scholars have concentrated on investigating the impact of VR technology on businesses, organizations, and customers to understand whether VR is beneficial for all involved in the tourism industry (Errichiello, Micera, Atzeni, & Del Chiappa, 2019; I. P. Tussyadiah et al., 2018). Even though Virtual Reality (VR) is not a new buzzword like 'Humanoid Robots' (Murphy, Gretzel, & Pesonen, 2019; I. Tussyadiah, 2020), yet it presents several benefits for the travel industry. As VR technology gets familiar within our world, most major industries have placed their interest in cutting the marketing costs and maximizing their benefits by offering alternative opportunities to the consumers. Tourism organizations, travel companies, and destinations have implemented VR technology into their system to increase service quality, enhance efficiency, and improve customer satisfaction while reducing the overall business operational costs such as advertisement expenses (Yuce et al., 2020). VR also allows the companies to increase their revenue significantly because of VR's unique functionalities like allowing virtual tours of destinations worldwide. Moreover, VR has also been interested in multidiscipline areas such as higher education, art, psychology, entertainment, and marketing (Kaimal et al., 2020; Ludlow, 2015; Radianti, Majchrzak, Fromm, & Wohlgenannt, 2020).

It is also important to note that as advanced technology gets more sophisticated, one can predict that tourism can generate more significant and unprecedented outcomes in the coming decades. Sufficient to say that implementing advanced information and communication technologies into tourism industry also improves customers' satisfaction due to their touristic experience using innovative mediums such as computers, tablets, and smartphones (Law, Buhalis, & Cobanoglu, 2014). As mentioned above, tourism is the accelerator and vital source of thriving economic development, employment, and socializing between different cultures and societies until the most recent shocking pandemic called COVID-19. Tourism generates a considerable amount of the global GDP (10.3%), which means $8.9 trillion; 1 in every ten jobs (330 million jobs) worldwide and 1 in every four positions are created during the past five years in the tourism sector in 2019 ("Economic Impact Reports," 2019). But, there is a question; can those digital figures create sustainable tourism in particularly at the vulnerable and sensitive touristic sites?

Besides all those influencing marketing statistics above, however, the tourism industry is one of the most volatile sectors with a weak and vulnerable immune system to different social, political, climate change, and economic downturns besides COVID-19 pandemics. Tourism organizations and governments need to develop more proactive and novel approaches, applicable and functional strategies to avoid and confront the possible direct or indirect of these issues. Information and communication technologies present alternative and suitable plans by implementing innovative technologies to strengthen the tourism industry's immune system from worsen and ease the industry's pain.

Immersive technologies stand at the forefront of emerging technologies, whether we live in an epidemic term or live in a chaotic situation. First of all, immersive technologies are double eco-friendly innovative technologies. It creates ecological and economic benefits that can play a crucial role in reducing operational losses and providing a more sustainable ecosystem. Second, immersive VR settings allow both service providers and consumers to satisfy their visitation needs by being transported to the destination during the hard-times, including COVID-19. Finally, accessibility and affordability also present a unique opportunity for those with a disability and older adults (Appel et al., 2020). Nevertheless, keep in mind that tech-driven solutions may not be a permanent cure for the industry's challenges currently due to uncontrollable human greed.

THE ROLE OF VR WITH CURRENT TOURISM ISSUES

On the other hand, the tourism sector requires investors to have efficient and sufficient vision, a high degree of responsibility and sensibility, and, most importantly, awareness of the importance of the tourism resources besides the determination to multiply their high volume of capital. Sustainable tourism development is beyond predictable touristic affairs. In other words, tourism is not only tourism. It depends on many internal and external factors such as climate change, overpopulation, political, economic, and health issues called a disaster for the tourism industry. These factors make tourism one of the most fragile, unpredictable, and sensitive sectors among other industries. For instance, today, the tourism industry has been suffering dramatically due to undergoing the COVID-19 pandemic like many other service industries. Although organizations, governments, and the tourism industry are trying to make profound changes in response to the COVID-19 pandemic, confronting the challenges such as a dramatic drop in visiting numbers has not been easy because of the high risk and high rate of contagiousness of COVID-19. While utilizing the most functional digitalized tool, tourism will inevitably help both at a global and local scale to recover the economies during the post-COVID-19 era. Depends on the familiarity of VR technology during post- COVID-19, its role has never been understood in great detail. Since the epidemic influences every facet of our lives, we became obliged to use the technologies whether we like using technologies or not.

The tourism industry differs from other industries in many ways, such as having unique sources that are not imitable or replaceable. Moreover, short-term oriented tourism plans result in devastating ecological issues in the long term. The core sources of tourism, such as historical, environmental, and natural, are also vulnerable due to the uncontrolled, unplanned, or greedy-oriented tourism development projects. Like any other economic object, these sources have a limitation that does not last forever. Most other industries' products are not dependent on only certain raw materials to produce their goods. Integration of technology enables manufacturers to deliver them with alternative raw materials even if the existing sources disappear. However, it is hard to say the same thing for tourism products. Again, the value of some destinations with a rich history, unique culture, and untouched natural resources is immeasurable. A castle that remained from ancient Greeks, an item in a museum in Mesopotamia, a historical palace in Rome arrived from the civilizations is an irreplaceable cultural and natural heritage for all people worldwide. Therefore, implementing the most applicable technology in the tourism service is vital to ensure these treasure's sustainability for the destinations.

One of the most critical priorities of achieving a sustainable tourism goal is preserving touristic destinations' value for the long term. Authorities, investors, and managers consider that tourism resources are crucial, vulnerable, and irreplaceable.

These resources are essential, and they should be carried out for future generations instead of devouring them in predatory-oriented tourism. On the other hand, current literature emphasizes that digitalization and sustainable tourism are inseparable elements for the countries' economic growth. They also argue that adaptable technologies increase visitors' number to sites or attractions. Although more visiting numbers and increased GDP are a good indicator for the current economic growth, the overall impact of tourism in each industry layer, particularly in the local business chain, is not distributed fairly. Eco-friendly VR technology can help avoid these problems by converting attraction into 360 degrees as a responsible and future-oriented tourism development process.

Moreover, those numerical figures do not reveal how total GDP is distributed among industry and society. In light of this brief criticism, this study emphasizes how VR can be implemented into the tourism industry's long term benefits rather than seasonal benefits of the touristic activities. Newer perspectives and strategies can help leverage economic growth with future-friendly tourism development plans that promote fair distribution from Small and Medium Enterprises (SMEs) to giant global companies. A practical solution can be transforming the tourism products into digital format to ensure the tourism sector survives during the hard-times and meets the demands of in the battle of highly competitive marketing place while preventing and preserving the tourism resources.

A well-organized planning effort is vital to keeping the sources as attractive as the object's originality. Any form of tourism development should ensure that tourism sources are essential for the public and generations to come. Being shortsighted or neglecting the value of the resources for the long term poses irreversible negative images for tourism development. Hence, tourism plans cannot focus on a handful of investors' daily base economic benefits. With or without the organizations and government authorities, the local public should know how valuable their resources remained as a treasure for them. Educating people about the importance of the environment and natural resources raises awareness to protect these destinations adequately. Public awareness is an essential motive for technology innovation (Hjalager, 1996, 2002). Public awareness can preserve historical sites and many other crucial touristic destinations, whether recognized as so-called 'World Heritage' or not in the long term, by utilizing efficient technology. The more extraordinary awareness people have, the healthier and more sustainable tourism development will exist for a long time with the advanced technology.

ICT AND TOURISM

The question is how the tourism industry can preserve its dynamic functionality and role in creating economic opportunities. Moreover, what innovative technology can help reduce the cost of unprecedented situations or challenges during the health crises that immensely shock each industry layer? Another problem is how to control and manage unwanted environmental issues due to overpopulation and the higher cost of living at destinations. Answering these questions might be subjective, but these are the questions that shape the tourism future. From the objective point of view, tourism sources are vulnerable if they are not managed and planned if suitable and emerging technologies are not implemented as driving forces in the tourism industry.

Furthermore, environmental, biological, and sociocultural degradation gradually speeds up the destinations' vulnerability and depletes the core sources that play an essential role in influencing the visitors' decisions. Technology itself may not help or prevent the social, economic, political related negative impact or prevent the COVID-19. However, technology can help some tourism industry organs, such as museums and sites, to implement emerging technologies such as Virtual Reality. While this chapter argues how past and current trends of the ICT's have impacted business and society, we also have to mention that tourism resources are remarkably as vital as the catalyzer of the economic growth that influences our quality of life. Therefore, it is crucial to understand how VR is essential to achieve sustainable tourism to maximize all stakeholders' benefits at the local, national, and international levels.

DIGITAL TOURISM

Most studies in the literature argue that advanced technology transforms individuals' every aspect of their lives, from transportation, information, accommodation, and entertainment to communication. The impact of information and communication technologies (ICT) on tourism development has been growing substantially and gradually for the last several decades, particularly after the invention of the internet during the 1990s. ICT also evolves the way organizations and companies perform their businesses since the world turned into virtually a flat place and virtually a small social village. ICT allows companies and organizations to deliver their messages, marketing plans, brochures, and reservation information at their fingertips to the other side of the world in a second. According to a study conducted by yougov (Beck, 2016), the role of using VR technology in tourism has been significantly increasing. Tourism stores with VR settings receive more potential customers (64%) than the the ones without using the VR technology. On the other hand, according to the figure 1, 31% of the participants stated that VR technology can help them to

Figure 1. Study conducted by YouGov: Travel booking trends MENASA
Source: http://www.virtual-reality-in-tourism.com/

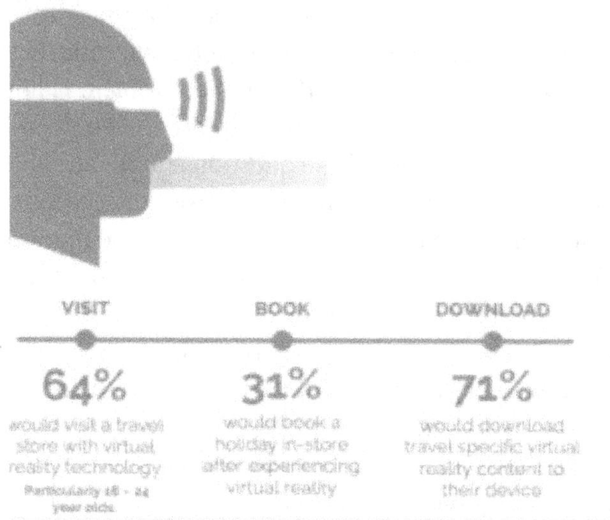

book their tourism product after experiencing the destination via VR (Beck, 2016). Finally, 71% of the participants declared that they are interested in downloading tourism related VR contents (Beck, 2016).

As was mentioned above, tourism is an essential source of economies for the world. But, technology is the core motor that drives tourism development. As a critical element in the tourism sector, ICT influences tourism organizations from the human resources department to marketing their brands worldwide (Gonzalez, Gasco, & Llopis, 2020). *Moreover, a tourism development without implementing and preparing the latest or future innovative technologies such as big data, artificial intelligence, automation, and virtual technologies cannot be considered as a smart tourism development that will last for the generations to come and rip the best benefit for the destinations' shareholders.* These are the technologies that promise to open a newer era for our daily and working lives. The question is, who will be able to afford these technologies in the future while small and medium-sized tourism enterprises (SMEs) suffer to afford the current technologies. The advantage of VR is respectively and relatively affordable technology comparing many current and prospective technologies.

What is the role of the emerging technologies to help the tourism industry to survive while it struggles with adverse situations? An extensive literature review argues how beneficial technology is to tourism and how IT improves our daily lives. Adapting the right innovation is crucial for the companies to achieve their organizational goal, improve their efficiency, and face the challenges in a highly volatile, risky, and competitive industry (Nicolau & Santa-María, 2013). Some of the most driving breakthrough innovations, such as; Internet, Web 1.0 and Web 2.0, and social media, have transformed our social lives, opened up a new era, and helped companies and organizations remarkably for their business operations within the tourism industry (Buhalis, 2003). Designing, utilizing and integrating the most suitable technology in businesses profoundly affected consumers and suppliers. Advanced technology can play a significant role in reducing loss. The current digitalization trend has a considerable impact on demand and supply. The effect of technology on tourism might be remarkable and charming for today.

VIRTUAL REALITY AND TOURISM

Virtual reality is not a brand-name technology. VR has been in the industrial market for decades, but its marketing role has not been discovered in the tourism industry. VR technology is a handy tool in destination marketing. However, awareness of the possibilities that VR offers has not been understood with great detail for both supply and demand sides. Digital awareness is as crucial as public and supplier awareness for creating efficient and sustainable tourism development to rip the most economic benefits for the current and future generations. As the public gets aware of VR's functionality, digital marketing of the destinations' can offer better opportunities for all the stakeholders.

On the other hand, VR technology without a VR application is a useless element. Developing, designing sophisticated applications that focus on the industry and customers' needs is also crucial to improving this technology's overall benefits. Therefore, to have a more significant impact on users and the new business models, the development process of VR applications require designers and developers to keep up with the changing tourists' profile, including demographic structure along with their consumer behaviours.

The post-COVID 19 era has pushed us to live in a virtual world by being socially isolated from the even most beloved ones and changed the way we operate our businesses. Social and cultural activities are cancelled, indoor and outdoor visitor numbers dropped like a skyrocket, and concerts and festivals were halted suddenly due to the pandemic. Nevertheless, thanks to VR technology, many organizations kept their businesses and activities running around the clock.

Each enterprise focuses on selling its products or services via a distribution chain. However, despite how valuable and quality a product or a service we have, if we do not use effective, constructive, creative, and convincing marketing strategies, it will be challenging to reach out to a wide range of consumers that will directly maximize our benefits. Consumers buy their food, clothing and replace anything they own by shopping online and in-person in a classical shopping style. Unlike ordinary business models, tourism products differ from ordinary items in the stores; in the tourism sector, consumers need to buy the touristic product to visit them. In other words, while products are shipped to customers in a typical business style, tourism requires consumers to be sent to the touristic products.

With the unexpected crisis, most industries experience extreme suffering, and they try everything to survive during the stagnation. Unlike the products in the market, in-store or online, tourism products are stationary. One used to travel for access to a tourism product or service in-person. But, not any longer. Availability of accessing touristic products using VR reality through websites transformed how we purchase or access them.

Furthermore, organizations also have changed their century-old marketing mediums with the most current ones. This device is called VR, even though unfamiliarity and unawareness of VR. In light of this, VR technology using the internet brought newer perspectives to the tourism market that made tourism products available around the clock and all year long without any time limitation. Society gets used to using the VR technology that makes the attractions' contents accessible easily and available for us to visit touristic sites without even leaving our homes and time constraints.

VIRTUAL TOURISM

Virtual tourism is a new phenomenon that impresses the users' perception significantly with the relevant visual information it provides. Virtual reality creates an emotional intensity that results in immersion in the brain cells thru the virtual images and contents. So, how can we define VR? There is not a standard definition of VR in the literature. However, it can be defined as VR is a computer-generated high-quality digital image in which participants are manipulated with the sense of immersion as if they are transferred from the physical world to the illustrated virtual environment during the real-time interactivity (Ng, Ma, Ho, Ip, & Fu, 2019; Slater, 2009; Suh & Prophet, 2018).

Due to the constant development of innovative technologies and their transformational impact in the tourism industry, tourism organizations began using VR to improve visitor satisfaction and business capacity (Jung, tom Dieck, Lee, & Chung, 2016). Navío-Marco, Ruiz-Gómez, and Sevilla-Sevilla (2018) argued that understanding the tourists' behaviour, such as their wishes and motivation, is a crucial element that will help design and implement an innovation that will orbit around the tourist-centricity. In other words, VR can be seen as a device that makes a dream come true by moving us to the virtual world without actually leaving a physical place by fooling the brain (LaValle, Yershova, Katsev, & Antonov, 2014). VR's implementation in real-life situations began influencing the organizational progress, user satisfaction, and engagement, particularly after integrating the portable VR headsets with smartphones in which VR contents and applications are found quickly. The adaptation of emerging technologies, including VR, helps organizations keep up with the current or advanced technologies for marketing, communication, psychology, and business operations and helps them for entertainment and training purposes (Avidar, 2018; Guttentag, 2010). The constant development of information, communication, and virtual reality technologies promise common implementation areas throughout the tourism sector (Cheong, 1995; Wei, Qi, & Zhang, 2019). However, it demands careful consideration and planning to discover what factors influence consumers' desire to purchase tourism destinations.

Using VR, customers will have the opportunity to experience what they will encounter in the destination they like to visit. They will have pre-knowledge about the culture they will meet. They will have ideas about what type of food and drink are available. They will figure out whether public transportation is available for them or not. Visitors will decide where they want to visit before they arrive at the destinations. Based on all these preferences, creating images will be more useful and practical for tourism organizations. The technological achievement will help the tourism sector to identify what type of images are being transmitted to individuals' minds that attract their mental process about their decision-making process. However, when the theoretical grounding of image representation's impacts on tourists' perception is investigated, unfortunately, it is found that the existing literature could neither generate a sufficient theoretical framework nor the precise and clear conceptualization that helps to understand how customers' emotional thoughts, attitudes, and intentions are changed with the attribution of the destination image (Kock, Josiassen, & Assaf, 2016; Stylidis, Belhassen, & Shani, 2015). Although immersive VR settings consisting of cutting-edge technologies (Guttentag, 2010) suggested, VR based innovative technologies may require reformulation to enable the adoption and adaptation of the newer generations towards the VR environment. Unfortunately, most of the innovations fail in the marketplace due to lack of understanding of the relationship between the sector and the technology (Heidenreich & Kraemer, 2015; Heidenreich & Spieth, 2013; Rodríguez Sánchez, Williams, & García Andreu, 2020; Yim & Park, 2019).

TYPES OF VR

There are two major types of VR: Untethered one is commonly known as Head-Mounted and Tethered displays. Head-mounted or untethered VR is a mobilized version, and it does not require any digitalized station. On the other hand, the head-mounted one is categorized as immersive, while the tethered one is classified as non-immersive technologies. Tethered VR is connected to computerized devices such as a personal computer, server, or automated platform (Suh & Prophet, 2018). In any form, virtual reality is composed of software and hardware that provide the immersed virtual environment in which users are engaged and attracted to VR content.

Nevertheless, some critical questions are raised, such as how a technology -even if built with state-of-the-art equipment and program, can replace real experiences, create a positive attitude, and shift consumers' behaviour. In the tourism context, VR denotes an interactive virtual medium that enables individuals to create simulated experiences of unreal environments using a head-mounted display (HMD) of a VR device (Perry Hobson & Williams, 1995). VR mounted headsets use 360-degree video to take viewers to a virtually constructed world, allowing users to view and have full information in six areas of tourism, including "planning and management, marketing, entertainment, education, accessibility, and heritage preservation" (Guttentag, 2010). While immersed technology gets more sophisticated with mobilized hardware such as smartphones (Hassan & Jung, 2018), content and the program creates a more dynamic, authentic virtual environment with realistic and indistinguishable images and information from reality. Therefore, the mobile version of VR tends to significantly alter the customer's decision-making process towards a tourism destination (Buhalis & Sinarta, 2019; Dorcic, Komsic, & Markovic, 2019).

HISTORICAL ASPECT OF VR

Each transformation of the production process and production tools had a remarkable impact on the communities' lives, from the clustered population in the cities to communication mediums. Although industry 4.0, the fourth industrial revolution, has begun with the rapid development and occupation of the Internet of Things (IoT), its transformation has been continuing to evolve, and it changes every aspect of our lives more significantly. Digital transformation enabled scientists to invent more efficient methods and develop technological tools to capture and *manipulate* the users' psychological and state of mind (Moseley, 2016) in multidisciplinary areas and industries. Besides trying to ease people's lives and improve the organizations' efficiency and productivity, scientists have been trying to develop a system to stimulate or induce people's basic instincts with the destinations and result in the constant digitalization development. The digital transformation era consists of innovative digital tools that aim to improve the efficiency of both human and nonhuman sources of an organization, empower the managers and owners to boost an organization's productivity to compete with the highly demanding business world.

In the historical context of VR, VR has been evolved significantly after the prevalence of IoT. The momentum of IoT catalyzed using technologies in a broader area without any limitation at an instant moment. But, breathtaking progress or advancement has not been achieved over a night. It took several decades and after the ups and downs of many inventions. On the other hand, despite the authors and sources are not agreed on the birthdate of the VR and by whom, Steinicke (2016) and Mihelj, Novak, and Beguš (2014) assert that VR historical root goes back to the 1960s. For these authors, Morton Heilig, the cinematographer, and filmmaker is the virtual reality father. Morton Heilig foresaw that the viewers would be immersed in future movies through the virtual environments.

Having such a vision-enabled Morton Heilig earned a patent of the first machine that allowed him to ride a bicycle virtually (Mihelj et al., 2014). The device they invented is called the '*Sensorama*' (the 1950s), which allowed the user to display "three-dimensional city display, heard the sounds of the city, felt the wind and vibration of the seat, and could even smell the scents of the city" (Mihelj et al., 2014). Sensorama is a first recognized VR environment composed of a "cabinet with a 3D display, vibrating seat, and scent producer" (Steinicke, 2016). It was invented by Stanley Weinbaum, a science fiction writer; the dream that he mentioned in his book called *Pygmalion's Spectacles* (1935) come true (Mihelj et al., 2014). In their book, Stanley Weinbaum wrote that the readers might taste, smell, touch as if they are in the story.

Following Sensorama, the Philco HMD (1965) was the first head-mounted device that allowed the users to experience "a video from a real, distant location" (Mihelj et al., 2014). Two engineers invented the Philco HMD (Head Mounted Display) helmet, and they called it *Headlight* (Steinicke, 2016). The helmet consisted of "a single CRT element and magnetic tracking system" with "a closed camera system for telepresence setups" (Steinicke, 2016).

In 1965, Ivan Sutherland's vision resulted in inventing the 'Sketchpad,' the VR development process. The sketch is termed "The Ultimate Display" by Sutherland, and it was connected to the virtual environment (Mihelj et al., 2014; Steinicke, 2016). The sketch is the first computer graphical user interface that allowed the user to draw graphic images interactively. The program they used allowed the user to control graphics with his body's movement (Steinicke, 2016; Sutherland, 1968), Sutherland's headset is eventually resulted in a more advanced development called "Sword of Damocles" (1968). Sword of Damocles was equipped with two small screens to create three-dimensional images. Still, it did not allow active interaction with the virtual environment (Mihelj et al., 2014). A virtual environment capable of responding to its users' movements required a more sophisticated system than previous VR versions, even though they were also a big step in the VR development process. Next, Myron Krueger is the first inventor who created a virtual environment that interactively involved with the users' movements and actions (the 1970s).

Nevertheless, the transformation of VR has not yet been completed. Moreover, scientists have been working to take the current system next level, where VR technology can influence the users with more immersions in more effective virtual environments. Today, VR is used through mobile phones that allow VR applications to be used in every field of our lives. Its usage area varies from marketing to the training of flight and driving simulators, from the medical field to engineering.

VR AND SENSE OF PRESENCE

A sense of presence is one of the most vital and critical concepts of VR technology. As (IJsselsteijn, De Ridder, Freeman, & Avons, 2000) asserted, a sense of presence depends on two core elements that are objective and subjective components. Sense of presence or 'being there' "is a subjective experience that is created, bringing of these two parameters in combining in a mediated environment" (IJsselsteijn et al., 2000). Virtual reality and sense of presence have been among the researchers' most trending phenomena for several decades (IJsselsteijn et al., 2000; Steinicke, 2016). Despite VR's definition differs as its transformation has not been completed due to fast-tracking technological development.

Virtual reality is based on the factor that creates a cognitive connection with realistic images and videos. Cognitive relationships with the presence phenomenon occur due to interacting with the VR environment (Carassa, Morganti, & Tirassa, 2004). VR illustrates images, videos, games, and movies in a 3D format that impacts users' perceptions, making them feel like they are depicted (Yuce et al., 2020). VR allows users to control the system by physical movement, speech, or text depending on the user's characteristics, preferences, and platform's goal. The sense of presence created by VR is the primary element that serves as a motivator for a particular attitude or behaviour (Carassa et al., 2004; Ludlow, 2015). However, as stated in Marchiori, Niforatos, and Preto (2018), identifying the VR technology factors or characters will improve the users' perceptions of the destinations and services.

Steinicke (2016) argues that a sense of presence is a susceptible and fragile concept that depends on different factors, such as hardware and software challenges. In the most recent form, a sense of presence is the core factor that allows the user to engage or detach from the virtual reality environment due to the quality or lack of efficient hardware and software development presence. With an improved sense of presence, VR can create attractiveness and engagement with its features mentioned in the definition (Yuce et al., 2020). A virtual reality system's innovative hardware and unique software, along with high resolution, fast-tracking, update rates with low latency, has the ultimate impact on the aspect of the sense of presence (Yuce et al., 2020). This book chapter defines the sense of presence as 'being transported from a real-world environment to the illustrated virtual setting for a particular time with the awareness of being lost.'

On the other hand, given that a sense of presence is a critical function of the virtual reality system, it directly helps individuals develop cognitive engagement with the environment depicted (Nash, Edwards, Thompson, & Barfield, 2000). Sense of presence is defined as "sense of being there" or "the feeling of being there" (McMahan, 2003) or "being there" in a subjective manner that is illustrated by the system (Barfield, Zeltzer, Sheridan, & Slater, 1995; IJsselsteijn et al., 2000; Lessiter, Freeman, Keogh, & Davidoff, 2001; Mihelj et al., 2014). Murray and Murray (2017) and McMahan (2003) argue that immersion is a metaphorical term often used interchangeably with 'presence' that indicates transporting from the real-world to the psychologically immersed environment, called, in essence, virtual reality.

DISCUSSION AND CONCLUSION

This chapter argued the role of virtual reality in the context of tourism. Virtual reality is more than a marketing tool that has great potential to transform how we travel. Tourism is one of the most volatile, sensitive, and vulnerable industry foreseen is due to unpredictable and uncontrollable political, social, economic, and public health-related challenges or issues. Hence, this study intended to extend VR technology's applicability beyond the current scope of the literature. A majority of the studies suggest using VR as a marketing tool. A majority of the studies suggest using VR as a marketing tool to achieve sustainable tourism objectives. This study argued that VR is an alternative tool to take virtual tours worldwide as millions of people have grounded on their sofas during the unprecedented COVID-19 pandemic era and unpredictable, chaotic situations. Furthermore, this study also emphasized that VR can be part of eco-friendly tourism affairs for susceptible sites and overpopulated destinations. As VR technology gets more sophisticated with creating higher sensory feelings, the applicability of VR will be spread to even broader tourism subsectors that will help SME's to survive in this highly competitive tourism world.

Moreover, virtual reality plays a vital role in maximizing the financial originated outcomes and minimize tourism's negative impacts on sensitive destinations since it provides eco-friendly virtual tourism opportunities. The impacts of technology are exceedingly and remarkably transforms our social lives and reshape our century-old habits. Digitalization is a boon that adds unimaginable and unthinkable values to our businesses regarding the accessibility of information with a great deal of convenience. Our lives heavily became dependent on technology since it virtually enters our bodies' cell with the rapid development of digitalization in our modern era. Virtual reality creates significant advantages for us, mainly using smartphones, computers and tablets. Keeping in mind that they do not necessarily and automatically create positive impacts on the destinations due to focusing on short-term tourism development benefits. Nevertheless, despite the increasing popularity and broad range of applicability of VR in the tourism sector, the evolution of the development process has not been completed yet. Developers of this technology need to enhance the sensory features that will enrich the consumers' senses for touch, taste, and smell with more comfortable and user-friendly headsets.

THEORETICAL IMPLICATIONS

The paper's theoretical framework aimed to provide some critical contributions for all the stakeholders in this conceptualized study. First, this chapter drew attention to tourism development paradigms that focus only on creating the rising statistical data trends and patterns while ignoring tourism's future, as suggested in many previous studies. Hence, this study pointed out that the current understanding of using innovative technologies -as a traditional marketing medium- that focuses on increasing the total number of visitors cannot solve the current issues that the tourism industry faces. Furthermore, unlike most of the studies, this chapter proposed integrating VR technology and the long-term tourism benefits will be more beneficial not only for the current generations and current businesses but also for future generations and future businesses. Tourism developers and beneficiaries usually focus daily and only economic benefits of the touristic affairs rather than on future goals and benefits. Therefore, this chapter emphasized the importance of virtual reality on sustainable tourism development in preserving the touristic destinations' natural, historical, and environmental values. In other words, instead of designing and developing short-term tourism development by implementing the latest innovative and digitalized tools will minimize the potential economic risks and losses during the shocking crisis days.

Second, this chapter also extends the current understanding of the applicability of VR from an ordinary marketing medium to an extent in which VR is used as virtual tourism that allows the users to access the touristic sites without the barriers such as time, money, and distance. The COVID-19 outbreak has been causing dramatic disruptions at a tragic level, including but not limited to drops in tourist arrivals that worsens the unemployment rates and more. In light of this brief information, pandemic and other unpredictable, risky situations offer watershed moments for the immersive technologies in the context of tourism. *VR can help organizations cope with this unusual situation by presenting an alternative travelling approach to allow the potential consumers experience the real-world environment in the case of visiting museums, castles and other touristic sites as they are forced to stay at their houses.*

Finally, this conceptualized chapter suggests that determining the right technology is crucial to achieving an organization's goals and objectives. Understanding the relationship between the sector's needs and the consumers' expectations can help design, develop, and implement more suitable and effective technological solutions to maximize the businesses' overall potential and improve visitors' satisfaction. VR technology frequently in the progress of evolution that influences every layer of the tourism branches. This study agreed with most of the existing literature that VR is a cost-effective technology that allows the organization to promote its products with less marketing expenses than the classical approaches. VR allows the consumers to have virtual experiences with 'try before you buy' option using real images and videos from the consumers' perspective. Allowing potential visitors to experience authentic images and videos rather than using the photo manipulation approach influence their perception and creates more satisfied and pleased travellers.

PRACTICAL IMPLICATIONS

How can practitioners get the most out of VR technology? Since VR is a substitute marketing tool that creates unique advantages for organizations to reach a wider area and more people with much less marketing expenses, hotel managers, destination authorities can implement VR to promote their products for those with disability and seniors. Next, VR technology provides an alternative proactive opportunity for online virtual tours of the museums, cultural sites, cities, and other attractions of the destinations during hard-times such as the COVID-19 pandemic. Finally, tourism planners or decision-makers can consider using VR technology to sustain their susceptible tourism resources for the next generations.

LIMITATIONS AND FUTURE STUDIES

Besides the contributions and recommendations of this chapter, some limitations need special attention for future studies. First of all, VR offers many opportunities for the industry. However, many organizations are not recognized or aware of the functionality and importance of VR technology yet. Therefore, this study recommends future studies to discover what functionality of VR influences the visitors' decision-making progress. Second, this study suggested implementing VR to create more sustainable tourism development for fragile destinations. However, future studies can measure the consumers' perception of their virtual tourism experiences to investigate whether VR creates a sufficient impact on users or stimulates their desire for a tourism service.

REFERENCES

Appel, L., Appel, E., Bogler, O., Wiseman, M., Cohen, L., Ein, N., Abrams, H. B., & Campos, J. L. (2020). Older adults with cognitive and/or physical impairments can benefit from immersive virtual reality experiences: A feasibility study. *Frontiers in medicine*, *6*, 329. doi:10.3389/fmed.2019.00329 PMID:32010701

Avidar, R. (2018). Engagement, interactivity, and diffusion of innovations. *Handb. Commun. Engagem, 505*.

Barfield, W., Zeltzer, D., Sheridan, T., & Slater, M. (1995). Presence and performance within virtual environments. *Virtual environments and advanced interface design*, 473-513.

Beck, J. (2016). *Statistics about VR in travel and tourism*. Retrieved 2/20/2021, from http://www.virtual-reality-in-tourism.com/statistics-about-vr-in-travel/

Bower, M., DeWitt, D., & Lai, J. W. (2020). Reasons associated with preservice teachers' intention to use immersive virtual reality in education. *British Journal of Educational Technology*, *51*(6), 2214–2232. doi:10.1111/bjet.13009

Buhalis, D. (2003). eTourism: Information technology for strategic tourism management. Pearson Education.

Buhalis, D., & Sinarta, Y. (2019). Real-time co-creation and nowness service: Lessons from tourism and hospitality. *Journal of Travel & Tourism Marketing*, *36*(5), 563–582. doi:10.1080/10548408.2019.1592059

Carassa, A., Morganti, F., & Tirassa, M. (2004). *Movement, action, and situation: Presence in virtual environments*. Academic Press.

Cheong, R. (1995). The virtual threat to travel and tourism. *Tourism Management*, *16*(6), 417–422. doi:10.1016/0261-5177(95)00049-T

Choi, Y., Hickerson, B., & Lee, J. (2018). Investigation of the technology effects of online travel media on virtual travel experience and behavioral intention. *Journal of Travel & Tourism Marketing*, *35*(3), 320–335. doi:10.1080/10548408.2017.1340224

Dorcic, J., Komsic, J., & Markovic, S. (2019). Mobile technologies and applications towards smart tourism–state of the art. *Tourism Review*, *74*(1), 82–103. doi:10.1108/TR-07-2017-0121

Economic Impact Reports. (2019). https://wttc.org/Research/Economic-Impact#:~:text=Over%20the%20past%20five%20years,10.3%25%20of%20global%20GDP

Errichiello, L., Micera, R., Atzeni, M., & Del Chiappa, G. (2019). Exploring the implications of wearable virtual reality technology for museum visitors' experience: A cluster analysis. *International Journal of Tourism Research, 21*(5), 590–605. doi:10.1002/jtr.2283

Gonzalez, R., Gasco, J., & Llopis, J. (2020). Information and communication technologies and human resources in hospitality and tourism. *International Journal of Contemporary Hospitality Management, 32*(11), 3545–3579. doi:10.1108/IJCHM-04-2020-0272

Guttentag, D. A. (2010). Virtual reality: Applications and implications for tourism. *Tourism Management, 31*(5), 637–651. doi:10.1016/j.tourman.2009.07.003

Hassan, A., & Jung, T. (2018). *Augmented reality as an emerging application in tourism marketing education. In Virtual and Augmented Reality: Concepts, Methodologies, Tools, and Applications*. IGI Global.

Heidenreich, S., & Kraemer, T. (2015). Passive innovation resistance: The curse of innovation? Investigating consequences for innovative consumer behavior. *Journal of Economic Psychology, 51*, 134–151. doi:10.1016/j.joep.2015.09.003

Heidenreich, S., & Spieth, P. (2013). Why innovations fail—The case of passive and active innovation resistance. *International Journal of Innovation Management, 17*(05), 1350021. doi:10.1142/S1363919613500217

Hjalager, A.-M. (1996). Tourism and the environment: The innovation connection. *Journal of Sustainable Tourism, 4*(4), 201–218. doi:10.1080/09669589608667268

Hjalager, A.-M. (2002). Repairing innovation defectiveness in tourism. *Tourism Management, 23*(5), 465–474. doi:10.1016/S0261-5177(02)00013-4

Hudson, S., Matson-Barkat, S., Pallamin, N., & Jegou, G. (2019). With or without you? Interaction and immersion in a virtual reality experience. *Journal of Business Research, 100*, 459–468. doi:10.1016/j.jbusres.2018.10.062

IJsselsteijn, W. A., De Ridder, H., Freeman, J., & Avons, S. E. (2000). *Presence: concept, determinants, and measurement.* Paper presented at the Human vision and electronic imaging V.

Jung, T., tom Dieck, M. C., Lee, H., & Chung, N. (2016). *Effects of virtual reality and augmented reality on visitor experiences in museum. In Information and communication technologies in tourism 2016*. Springer.

Kaimal, G., Carroll-Haskins, K., Berberian, M., Dougherty, A., Carlton, N., & Ramakrishnan, A. (2020). Virtual Reality in Art Therapy: A Pilot Qualitative Study of the Novel Medium and Implications for Practice. *Art Therapy: Journal of the American Art Therapy Association*, *37*(1), 16–24. doi:10.1080/07421656.2019.1659662

Kock, F., Josiassen, A., & Assaf, A. G. (2016). Advancing destination image: The destination content model. *Annals of Tourism Research*, *61*, 28–44. doi:10.1016/j.annals.2016.07.003

Laurell, C., Sandström, C., Berthold, A., & Larsson, D. (2019). Exploring barriers to adoption of Virtual Reality through Social Media Analytics and Machine Learning–An assessment of technology, network, price and trialability. *Journal of Business Research*, *100*, 469–474. doi:10.1016/j.jbusres.2019.01.017

LaValle, S. M., Yershova, A., Katsev, M., & Antonov, M. (2014). *Head tracking for the Oculus Rift.* Paper presented at the 2014 IEEE International Conference on Robotics and Automation (ICRA). 10.1109/ICRA.2014.6906608

Law, R., Buhalis, D., & Cobanoglu, C. (2014). Progress on information and communication technologies in hospitality and tourism. *International Journal of Contemporary Hospitality Management*, *26*(5), 727–750. doi:10.1108/IJCHM-08-2013-0367

Lessiter, J., Freeman, J., Keogh, E., & Davidoff, J. (2001). A cross-media presence questionnaire: The ITC-Sense of Presence Inventory. *Presence (Cambridge, Mass.)*, *10*(3), 282–297. doi:10.1162/105474601300343612

Li, L., Yu, F., Shi, D., Shi, J., Tian, Z., Yang, J., ... Jiang, Q. (2017). Application of virtual reality technology in clinical medicine. *American Journal of Translational Research*, *9*(9), 3867. PMID:28979666

Li, S., Tian, W., Lundberg, C., Gkritzali, A., & Sundström, M. (2020). Two Tales of One City: Fantasy Proneness, Authenticity, and Loyalty of On-Screen Tourism Destinations. *Journal of Travel Research*.

Ludlow, B. L. (2015). Virtual reality: Emerging applications and future directions. *Rural Special Education Quarterly*, *34*(3), 3–10. doi:10.1177/875687051503400302

Marchiori, E., Niforatos, E., & Preto, L. (2018). Analysis of users' heart rate data and self-reported perceptions to understand effective virtual reality characteristics. *Information Technology & Tourism, 18*(1), 133–155. doi:10.100740558-018-0104-0

McMahan, A. (2003). Immersion, engagement and presence. *The Video Game Theory Reader, 67*, 86.

Mihelj, M., Novak, D., & Beguš, S. (2014). *Virtual reality technology and applications*. Academic Press.

Moseley, R. (2016). *Immersive brain entrainment in virtual worlds: actualizing meditative states. In Emerging Trends and Advanced Technologies for Computational Intelligence*. Springer.

Murphy, J., Gretzel, U., & Pesonen, J. (2019). Marketing robot services in hospitality and tourism: The role of anthropomorphism. *Journal of Travel & Tourism Marketing, 36*(7), 784–795. doi:10.1080/10548408.2019.1571983

Murray, J. H., & Murray, J. H. (2017). *Hamlet on the holodeck: The future of narrative in cyberspace*. MIT Press.

Nash, E. B., Edwards, G. W., Thompson, J. A., & Barfield, W. (2000). A review of presence and performance in virtual environments. *International Journal of Human-Computer Interaction, 12*(1), 1–41. doi:10.1207/S15327590IJHC1201_1

Navío-Marco, J., Ruiz-Gómez, L. M., & Sevilla-Sevilla, C. (2018). Progress in information technology and tourism management: 30 years on and 20 years after the internet-Revisiting Buhalis & Law's landmark study about eTourism. *Tourism Management, 69*, 460–470. doi:10.1016/j.tourman.2018.06.002

Ng, Y.-L., Ma, F., Ho, F. K., Ip, P., & Fu, K. (2019). Effectiveness of virtual and augmented reality-enhanced exercise on physical activity, psychological outcomes, and physical performance: A systematic review and meta-analysis of randomized controlled trials. *Computers in Human Behavior, 99*, 278–291. doi:10.1016/j.chb.2019.05.026

Nicolau, J. L., & Santa-María, M. J. (2013). The effect of innovation on hotel market value. *International Journal of Hospitality Management, 32*, 71–79. doi:10.1016/j.ijhm.2012.04.005

Otoo, F. E., Kim, S., & Choi, Y. (2020). Understanding senior tourists' preferences and characteristics based on their overseas travel motivation clusters. *Journal of Travel & Tourism Marketing, 37*(2), 246–257. doi:10.1080/10548408.2020.1740136

Perry Hobson, J., & Williams, A. P. (1995). Virtual reality: A new horizon for the tourism industry. *Journal of Vacation Marketing*, *1*(2), 124–135. doi:10.1177/135676679500100202

Pestek, A., & Sarvan, M. (2020). Virtual reality and modern tourism. *Journal of Tourism Futures*.

Radianti, J., Majchrzak, T. A., Fromm, J., & Wohlgenannt, I. (2020). A systematic review of immersive virtual reality applications for higher education: Design elements, lessons learned, and research agenda. *Computers & Education*, *147*, 103778. doi:10.1016/j.compedu.2019.103778

Rather, R. A. (2020). Customer experience and engagement in tourism destinations: The experiential marketing perspective. *Journal of Travel & Tourism Marketing*, *37*(1), 15–32. doi:10.1080/10548408.2019.1686101

Rodríguez Sánchez, I., Williams, A. M., & García Andreu, H. (2020). Customer resistance to tourism innovations: Entrepreneurs' understanding and management strategies. *Journal of Travel Research*, *59*(3), 450–464. doi:10.1177/0047287519843188

Slater, M. (2009). Place illusion and plausibility can lead to realistic behaviour in immersive virtual environments. *Philosophical Transactions of the Royal Society of London. Series B, Biological Sciences*, *364*(1535), 3549–3557. doi:10.1098/rstb.2009.0138 PMID:19884149

Steinicke, F. (2016). *Being really virtual*. Springer. doi:10.1007/978-3-319-43078-2

Stylidis, D., Belhassen, Y., & Shani, A. (2015). Three tales of a city: Stakeholders' images of Eilat as a tourist destination. *Journal of Travel Research*, *54*(6), 702–716. doi:10.1177/0047287514532373

Suh, A., & Prophet, J. (2018). The state of immersive technology research: A literature analysis. *Computers in Human Behavior*, *86*, 77–90. doi:10.1016/j.chb.2018.04.019

Sutherland, I. E. (1968). A head-mounted three dimensional display. *Proceedings of the December 9-11, 1968, fall joint computer conference, part I*.

Tussyadiah, I. (2020). A review of research into automation in tourism: Launching the Annals of Tourism Research Curated Collection on Artificial Intelligence and Robotics in Tourism. *Annals of Tourism Research*, *81*, 102883. doi:10.1016/j.annals.2020.102883

Tussyadiah, I. P., Wang, D., Jung, T. H., & tom Dieck, M. C. (2018). Virtual reality, presence, and attitude change: Empirical evidence from tourism. *Tourism Management*, *66*, 140–154. doi:10.1016/j.tourman.2017.12.003

Wei, W., Qi, R., & Zhang, L. (2019). Effects of virtual reality on theme park visitors' experience and behaviors: A presence perspective. *Tourism Management*, *71*, 282–293. doi:10.1016/j.tourman.2018.10.024

Yim, M. Y.-C., & Park, S.-Y. (2019). "I am not satisfied with my body, so I like augmented reality (AR)": Consumer responses to AR-based product presentations. *Journal of Business Research*, *100*, 581–589. doi:10.1016/j.jbusres.2018.10.041

Yuce, A., Arasli, H., Ozturen, A., & Daskin, M. (2020). Feeling the Service Product Closer: Triggering Visit Intention via Virtual Reality. *Sustainability*, *12*(16), 6632. doi:10.3390u12166632

Chapter 6
Museum Experience and Digital Consumption:
The Role of Mobile Augmented Reality in Tunisia's Cultural Heritage

Khouloud Mouihbi
IHEC of Carthage, Tunisia

Nedra Bahri-Ammari
IHEC of Carthage, Tunisia

Daouda Coulibaly
ISTEC Paris, France

ABSTRACT

This study explores the role of mobile augmented reality (MAR) technology in enhancing user experience (UX) in the context of cultural heritage. Data were collected via individual semi-structured interviews, according to a theoretical model of UX. The findings show that MAR can enhance UX in cultural heritage museums when the product's features and characteristics match users' needs. A contemporary challenge in the cultural heritage industry is to find creative ways to attract new visitors and enhance their museum experience. Therefore, this study was conducted to improve museum curators' understanding about the UX of MAR apps to satisfy their visitors' expectations.

DOI: 10.4018/978-1-7998-6904-7.ch006

INTRODUCTION

In the sub-sector of cultural heritage tourism, innovative methods and new technological tools are continuously being sought to engage and enhance visitors' experience (Tscheu and Buhalis, 2016). Cultural heritage museums are among the most visited destinations in the world owing to their cultural and historical richness. These technologies, as well as the museums themselves, also play a very important role in the development of smart cities. Nevertheless, many museums provide limited access to information in certain exhibits owing to a lack of space, preservation and restoration of artefacts, which could cause the quality of visitor experience to decline even more. However, interactive technology (Pauget and Dammak, 2019) can be used to overcome these limits and provide users with the opportunity to fully appreciate cultural heritage museums, thereby breaking the boundaries of space and language.

Augmented reality (AR) is one of the most revolutionary technologies of recent years. AR has the ability to reshape the layout of museum displays and to impact users' behaviour (Arnaboldi et al., 2021) by enhancing real-world objects and spaces with digital information that is overlaid in the real world (Milgram and Kishino, 1994).

Contemporary museum visitors are no longer passive; they have certain expectations and want to actively participate in their museum experience. Therefore, they are now demanding information and entertainment through innovation and creativity (Hoyer et al., 2020), in addition to multisensory stimulation (Li et al., 2019; Pine and Gilmore, 1999). AR can be used as a tool to enhance museum experience and change how museums are seen, especially for young visitors, by displaying virtual content without disturbing the real environment. Its navigation functions have made AR technology very attractive to the tourism industry, especially in the sub-sector of cultural tourism (Van Krevelen and Poelman, 2010). Some authors found that AR has great potential and it can be considered a powerful means for museums to engage their visitors by showcasing information in a captivating and creative way (Neuburger and Roman, 2017). Therefore, AR has been selected by tourism practitioners as a mechanism through which enjoyable and interactive experiences can be created, as in the case of the Manchester Art Gallery (Leue et al., 2015), Dublin AR (Han et al., 2013), and Greevor Tin Mine museum in Cornwall UK (Jung et al., 2016). The implementation of AR in museums has also enabled the improvement of historical knowledge.

In this way, technology like mobile augmented reality (MAR) can be used as a tool to enhance user experience (UX) in the context of cultural heritage museums. However, understanding its role in transforming cultural tourism experiences is important for it to be successfully deployed. The aim of this chapter is to explore and to get a better understanding of the different aspects of UX in MAR, as this could have an impact on enhancing the experience of a cultural heritage museum visitor.

Definition of Augmented Reality

Milgram and Kishino (1994) defined AR as a kind of a display that allows the real world to be enriched by virtual entities, while AR is any situation in which the virtual world is augmented by real-world objects. According to Chylinski et al., (2020) and Azuma (1997), AR does not replace reality, it just improves it, and he defined AR by comparing it to virtual reality (VR) to make it clearer and to show exactly what the difference between the two of them is. He stated that AR is a form of VR, commonly known as virtual environment technologies, which completely engulfs the user within a virtual environment. This is contrary to AR, which enables the user to see the real environment in a way that is enhanced by virtual objects. Flavián et al., (2019) identified the following characteristics of AR: combination of the real and virtual worlds (Physical and virtual coexist); and an interaction that takes place in real time and is displayed in 3D. Azuma et al. (2001) believed that AR is neither limited to a certain type of display technologies (head-mounted displays or HMDs) nor to the sense of sight. In fact, AR can be used to enhance the smell, hearing and touch of its users. It can even be applied to replace or increase users' lost senses through sensory restoration, for example by increasing the sight of a user with poor vision or augmenting hearing for deaf users through visual clues. Sherman and Craig, (2002) defined AR as an exception of VR: 'A type of virtual reality with additional information, otherwise imperceptible to the human system (Qin et al, 2021), is made perceptible and registered with the display of the physical world'. In contrast, Craig (2013) did not consider AR a technology, but simply a medium. The justification for this is that AR links ideas between humans and computers, and to implement AR, technology must be used. Thus, he defined it as a medium in which virtual information is interactive in real time and is superimposed on the real world in both temporal and spatial registration. According to Craig (2013), there are four fundamental aspects of AR:

- The physical world is enhanced by virtual information displayed in view of the real world.
- The information is superimposed in registration with the real world.
- The information showed is dependent on the location of the real environment and the physical perspective of the user in the real world.
- The AR experience is interactive.

This definition is the one that this chapter will be based on, as it encompasses more dimensions to explain the usefulness of AR in a simple manner.

Augmented Reality Devices

The different AR devices that allow virtual and real environments to merge can be classified into the following categories.

Head-Worn Displays or Head-Mounted Displays

HMDs are devices, usually a helmet, which AR users can wear to provide virtual information or images through a small transparent medium in front of one or both eyes. HMDs are hands-free, wearable devices that were created for personal use and the offered information is unseen by others (https://www.roadtovr.com/meta-raises-50m-in-funding-to-accelerate-groundbreaking-ar-tech/). They hold unique features that non-HMDs do not submit such as the resolution, the field view and the sensing mechanisms (Kiyokawa, 2012). Such a device could be in two forms; Optical see-through and Video see-through. The first type, Optical see-through, grants the AR cover over a transparent display (Azuma et al., 2001). There is a diagonal semi-transparent mirror that permits the visualization of the real world and graphically displays imagery that will be reflected in the user's eyes (Kesim and Ozarslan, 2012). In contrast, the second type, Video see-through, is used for the AR user to experience something that is distant or to experience something when he/she is using an image enhancement system (Kesim and Ozarslan, 2012).

The first HMD made its appearance with Ivan Sutherland (Caudell et al., 1992). In order for HMDs to be light and comfortable, it is strongly recommended for them to be no larger than a pair of glasses (Azuma et al., 2001). The most popular type of AR devices are smart glasses (https://www.bankingtech.com/2014/02/caixabank-embraces-google-glass-augmented-reality/), like the Google Glasses or the Sony smart Eye Glass (Kiryakova et al., 2017).

Contact lenses

Contact lenses provide its users with a mixed image of virtual content planted into reality. The concept is to transform traditional lenses into a functional system by blending components such as circuits, LEDs, and miniature antennas to form an image directly in front of the eye Source: Silicon, 2011 (https://www.silicon.co.uk/workspace/augmented-reality-contact-lenses-could-display-emails-46804). However, AR contact lenses are still in the early phases of development (Kiryakova et al., 2017) and problems like how the lenses can be powered or how safety for the human eye can be ensured must still be overcome (DHL Trend Research, 2014).

Non-Head-Mounted Displays

There are some situations where HMDs are impractical, uncomfortable or can be a source of fatigue because of their unwieldy and heavy size. In light of this, the following subsections represent propositions of other types of AR devices (Fuchs and Ackerman, 1999).

Handheld devices

Nowadays, a great number of consumers are attracted to small form devices and displays including smartphones and tablet computers. This explains the massive boom of handheld devices that we are currently witnessing. Handheld devices are essentially types of AR devices (i.e. mobile phones, tablets) that operate with video see-through techniques to show virtual information such as texts and graphics in a physical environment (Wilson et al., 2018). Generally, they are small computing devices with a screen that the user can hold in his/her hand (Kesim and Ozarslan, 2012).

Stationary AR systems

These devices are the opposite of mobile AR devices since they can be equipped with better and more powerful camera systems; as a result, they will offer more accurate recognition of people and scenes. In addition, more realistic images are usually displayed by the large screen or display unit owing to their ability to be unaffected by environmental influences, such as cloudy and blurry lighting or sunlight (DHL Trend Research, 2014). This type of AR system is appropriate and useful in 'virtual trials and shopping', which are typically found in retail stores (Kiryakova et al., 2017). For example, the Uniqlo brand 'offered shoppers the unique ability to imagine themselves in a range of colour choices from a single silhouette without having to remove the garment' (Source: Uniqlo's magic mirror, 2013 (http://retail-innovation.com/uniqlos-magic-mirror).

Spatial augmented reality

Unlike other systems, spatial augmented reality directly displays virtual content on physical surfaces like walls, desks, and even the human body. SAR systems are often stationary, still standing devices that can offer a more precise reflection of reality owing to their ability to show virtual contents in their real size and dimensions. Moreover, it can be used on a wider scale, since its content can be simultaneously shared among many viewers (DHL Trend Research, 2014).

Museums

Museums can be divided into two categories (Table 1): traditional and modern digital museums (Li et al., 2019; Mei, 2004), where technology has been implemented by museums stakeholders to innovate and meet their visitors' needs (e.g. Smithsonian Museum of Natural History in Washington DC and Swedish National Museum of Science and Technology in Stockholm).

Traditional museums

Traditionally, museums were not public and were only accessible to a limited number of private individuals. When the first museum became public in the 18th century, it faced the challenge of telling stories and conveying meaning in a more effective way. Today, museums are being visited by many more individuals and are attracting more visitors each year owing to their classifications, such as Archaeology, Art, History, Science (Asghar and Nauman, 2010). According to Sparacino et al., (2000), for museums to handle the challenge of becoming public places, they have to turn to technology to help achieve a balance between leisure and learning.

Technology in Museums (Digital Museums)

New information technologies have changed the expectations of museum visitors, who have become more demanding. They are no longer satisfied with limited access to information about museum exhibits and collections, and they instead expect 24/7 access to these type of data (Marty, 2008). There is also a growing desire for more personalized information and experiences than the standard ones offered by traditional museums. The new age of technology has contributed to revolutionising the 'Interpretation Media Landscape', which gave cultural institutions alternative ways to offer experiences to their visitors (Damala et al., 2008). According to Rodrigues et al., (2016), 'We are used to technology in museums in the form of universal, inviting, and attractive websites with digital catalogues, providing the public with additional information to what is already displayed at the physical museum'. However, this kind of technology does not involve or engage the audience by offering them an enhanced experience, since the only functionality of these electronic catalogues is to offer extra information about an artwork. Now more than ever before, visiting a museum is considered an act of leisure (Arnaboldi et al., 2021), which explains some of the challenges being faced. As a result, many museums have been considering technology as a partner (rather than a simple tool) to help them be more effective communicators of stories and meaning (Kung, 2020).

Table 1. The difference between traditional and modern digital museums

Traditional museums	Modern digital museums
• Static museum experience • Passive visitors • Instructive • Neutral • Offering standard experiences	• Enhanced museum experience • Involving and engaging visitors • Instructive • Interactive • Offering personalized experiences

Mobile Augmented Reality

MAR applications consist of mixing virtual elements with real elements to provide an immersive view (Fan et al., 2020; Bach and Scapin, 2004). They help users to better understand their surroundings by adding digital information to a mobile user's physical environment. In this context, putting a mobile phone camera in front of an object and getting immediate information about it is considered revolutionary. This development effectively changed users' access to, knowledge of and enjoyment of virtual information. Moreover, newer opportunities have arisen for software designers and developers with the emergence of mobile versions of AR (Qin et al., 2021).

Multiple businesses have extended their offerings by introducing MAR in various fields like personalized shopping, tourism and cultural heritage (Dhir and Al-kahtani, 2013). However, it is important that they first understand users' needs and expectations before deploying such tools (Olsson et al., 2009).

MAR in museums

One of the reasons that museum visitors find the MAR to be useful is the insufficient and unavailable description labels. Hence, this drives visitors to search for more information on the internet. However, the available information is often untrustworthy, especially when the wanted information is unknown to the user. The MAR can fill this gap by offering more reliable information with less effort (Selvam et al., 2016). In fact, various museums around the world have implemented MAR, as in the case of the partnership between Gamar and Samsung to design 'A gift for Athena'. This MAR application offered British museum visitors, particularly children, the opportunity to learn more via games about Greek culture (Selvam et al., 2016). For example, a baroque vault that existed in 1682 was recently reconstructed in the cathedral of Valencia in Spain. Through this, visitors can appreciate the early Gothic style through a complete 3D reconstruction (Wilson et al., 2018), which was designed to precisely display its original morphology, character, and shape (Portalés et al., 2009).

Another study has also been conducted at the Greevor Tin Mine Museum in Cornwall, a UNESCO World Heritage Site, to assess the impact of AR on museum visitors, particularly in terms of its social presence (Jung et al., 2016).

User experience

Visitor experience

For years, various frameworks have been developed to survey users' needs and to guide museums through the challenges they are currently facing (Anderson, 2004). However, the majority of these frameworks have not been tested within museums or with visitors.

In addition, they do not take into consideration the role played by technology (Do et al., 2020; Pallud and Monod, 2010). To enhance museum experience, Kotler and Kotler (2000) suggested that those responsible for museum experience should focus more on the various types of visitor experience, the depth of visitor experience and the design of visitor experience. This is contrary to Anderson (2004), who suggested focusing on governance, institutional goals, management strategies and the style of communication. Some empirical frameworks do actually focus on visitor experience, but few of them address visitor experience while also considering technologies or UX (Peacock and Brownbill, 2007). According to the Institute of Museum and Library Services (2006, p124), in order for museums to assess user needs, they need training and guidance on topics including how to collect information about the characteristics of users, the way they use the technology and for what purpose.

Definition of User Experience

According to Hassenzahl's (2008, p12) definition, UX is 'a momentary, primarily evaluative feeling (good/bad) that one gets when interacting with a product or service…good UX is the consequence of fulfilling the human needs for autonomy, competency, stimulation…through interacting with a product or service'. However, an alternative definition has recently been developed by ISO 9241-11, which states that UX is 'a consequence of presentation, functionality, system performance, interactive behaviour and assistive capabilities of an interactive system, both hardware and software. It is also a consequence of the user's prior experience, attitudes, skills, habits and personality' (ISO, 2010, p7). This definition is going to be the one that this study will be based on. The enjoyment and well-being of users are considered by academia and industry in the field of human computer interaction owing to the fact that users are likely to face issues that can lower their enjoyment. Moreover, usability is about making the experience of interacting with a product or a system (Harrison, 2008) better for users. Since there is no commonly accepted definition of technology's UX (Bevan, 2009), it usually gets confused with Usability of devices (Rusu et al., 2015). However, these two expressions are not the same because UX is often illustrated as all the interaction facets between the user and the product (Marcus, 2006). This study will be based on Hassenzahl's (2003) UX model in cultural heritage contexts, which are divided into three features. The first is the presentation, including the general appearance, how the product feels and how this affects the UX; the second is the Information Content, which is considered the most important feature of MAR and varies from a simple visual notation to 3D modelling displays; and the third is functionality, which, in the case of MAR apps, is the ability to mix the real and virtual world in one environment.

Table 2. User experience model (Hassenzahl, 2003)

Product Features	Apparent Product Characters	Consequences
Content	**Pragmatic Attributes**	
• Easy to learn • Detailed • Richer	• Usefulness • Tutorial Suggestions	**Satisfied**
Presentation	**Hedonic Attributes**	
• Attractive colours	•	
• Bad choice of colours	Uncomfortable • Interactive • Innovative • Involving • Independent • Immersive • Entertaining	
Functionality		
• Language • Functional failure		**Not satisfied**
Interaction		
• Stability • Network connection		

RESEARCH SETTING AND SAMPLE

The choice of the museum where our research was conducted is the Sousse Archaeological museum which is considered one of the very rich and important Archaeological museums of Tunisia. In fact, it displays mosaic collections that reflect the lifestyle of romans centuries ago, the roman mosaic flooring one of 'the most important collections in the Mediterranean'… (Agency for the Development of National Heritage and Cultural Promotion, 2018). The museum was founded in 1951, and it has the second large mosaic collection in the world after the museum of Bardo which also situated in Tunis (Wikipedia, 2018).

Data Collection

We have chosen to conduct qualitative research for the purpose of this study to determine the dimensions that enhance the UX of MAR. According to our knowledge from conducting a literature review, studies that focus on the UX of MAR are scarce; particularly those conducted in the context of cultural heritage institutions.

The selected data collection instrument of this study is semi-structured interviews, since they mix freedom of speech with an interview protocol to guide the researcher (Romelaer, 2005). In order for the interviewees to describe their use of the app and what they felt in the process, they had to feel comfortable to express themselves freely without feeling judged, which made this type of instrument the most suitable approach for this study.

The interview guide is based on the themes extracted from the existing literature, as well as the researcher's intuition (Hirchman and Holbrook, 1992; Charfi, 2012). We started by introducing more general themes and then presented specific themes according to Hassenzahl's (2003) UX model.

Theme 1: Get to know the participant's familiarity with the target technology.

Theme 2: Features of MAR.

Theme 3: Apparent character of MAR.

Theme 4: Consequences of using MAR on visitor experience.

We applied themes that are related to the context of the study with the help of Hassenzahl's (2003) model, and we used both open-ended and closed-ended questions.

Sample

The sampling was based on convenience/availability, which is characterized by the diversity of its population (Evrard et al., 2009). Furthermore, the number of questioned visitors was not pre-fixed, it was fixed according to the thematic saturation. In other words, when additional interviews could no longer bring about new ideas and opinions, the collected data would reach 'repeat mode' (i.e. repeating what had already been said in previous interviews); reaching this point was taken as an indication that no further data collection was needed (Mucchielli, 1991).

We opted for 23 interviewees for this research. The interviews took place during the spring of 2018 in Sousse Tunisia. We took advantage of this period (when there were holidays and week-ends), because there were higher chances of Tunisians going to museums and higher chances of there being tourists. We gave attendees a tablet that contained the Mobile Augmented Reality application (MAR) prototype, and asked them to take their tour of the museum using the app. In fact, its's a Mobile Augmented Reality application prototype that brings the roman era back to life by simply allowing the visitors of museum of Sousse to interact with mosaic displays such as: roman sculptures, lapidaries, funerary objects of terraccota… via integrating digital visual and audio content into the museum real environment.

Once the tour was over from the museum, we started talking to the interviewees using two languages: French and English. These interviews were recorded via a tablet application called "Audio Recorder", after approval by the participants. Some screenshots, of the content provided by the MAR application on a certain exhibition, are included in the appendices.

The study sample was made up of people from different nationalities (14 local/ Tunisian visitors, 9 tourists), totalling 11 women and 12 men between 11 and 70 years old. This choice is justified by the fact that the cultural heritage museum of Sousse, where the study was conducted, receives visitors from all over the world. Therefore, it was important to take into consideration the UX of foreigners, as well as the local population.

Analysis Method

This study is based on thematic analysis, because it is the most relevant and suitable to the objective of the research, afterwards, all the interviews were transcribed and translated into English. A second person would then review the translation to ensure that it was correct and done objectively (i.e. that the correct meaning of each word was maintained). Next, the information collected was analyzed and simplified, taking into account all the details, and classified by themes and sub-themes. The illustration of this analyzed content has been summarized in the following table (Evrard et al., 2009).

RESULTS

The results are presented in the form of a summary table and each dimension is supplemented by associated verbatims. Three main points sum up the results of this study. The first deals with the technical and usage features of the application. The second point presents the characteristics of the application, and the third point analyses users' feelings of satisfaction or dissatisfaction.

DISCUSSION

This study was conducted to evaluate MAR's ability to enhance its UX in the context of the cultural heritage museum of Sousse in Tunisia. The results mostly show that AR is an excellent tool to implement in cultural institutions, particularly in museums. The findings also revealed that detailed content made visitors more engaged and interested with the help of storytelling and visuals, which enhanced the visitor experience and offered more enjoyment (Lee et al., 2020).

The findings have shown that the development of MAR applications should aim to offer high quality *content*, which explains why the content aspect received a lot of attention from participants. The content of the MAR app offered more rich and detailed information about the exhibits and allowed visitors to understand and to learn more easily than before (i.e. compared to not using the MAR app). It also provided visitors with additional knowledge on the exhibit's background and context via multiple tools: audio and visual (Pictures, 3D models), which made the presentation of the content more efficient, precise and accurate (Jaillot et al., 2020). Moreover, it was considered a better source of information than the labels displayed by the artworks. The content was easy to understand, and therefore enhanced the learning process, which supports Dieck et al.'s (2016) research. Additionally, in support of Zheng and Pulli's (2005) findings, participants believed that this information should be comprehensible, accessible and easy to recognize.

The majority of participants found the *presentation* (i.e. the interface of the app) very attractive (Hoyer et al., 2020). Previous studies focused not only on the visual aspect, but also on whether the design of the app was easy to use and understand to increase its adoption (Carmigniani et al., 2011).

According to Marimon et al., (2010) and Flavián et al., (2019), having the option to personalize the app depending on user preferences would enhance the experience. Other studies recommended arranging content into categories such as organisation by themes or era, and adding the keyword search (Jaillot et al., 2020) to facilitate interaction with the app and enhance the UX. On the other hand, the MAR app mostly received positive reviews concerning its performance and speed, even for participants who faced technical problems such as the app freezing or crashing. This kind of drawback could have led to participants' resistance, as such problems can leave visitors feeling frustrated and disappointed. For example, Glauser (2013) studied AR apps in the medical field and confirmed that participants were disappointed about the effects of an app freezing or having a short battery life. In addition, according to Pulli et al. (2007), one of the drawbacks of using an AR app is the issue of speed owing to the limited power of its processor. Therefore, the speed of performance is crucial in e-services and needs to be solved in future (Patricio et al., 2003).

Table 3. Results

Product features		
Content	Richer	**P6** *"...The information I've got about the letters was the best thing..."* **P21** *"The content is rich and detailed comparing to the label, and lack of a docent to mention some things about it .."*
	Detailed	**P17** *"Because I think the application provides further details, than those written..."* **P22** *"... it helps because they always explain history in general, they don't give the details..."*
	Easy to learn	**P15** *"... helps people to know and to understand more than the label shows.. understand better"* **P19** *"You get the meaning easier the app makes you learn more"*
Presentation	Attractive Colors	**P7** *"Colors are good, so close to reality... and to me that's what... and when i see it through the app..."* **P12** *"The visual impression is very good... they are attractive"*
	Bad choice of Colors	**P14** *"The colors don't change them with light effects..."* **P20** *"...a little too more fake..."*
Functionality	Language	**P1** *"I would like to have more languages, even though it's english, but I do understand arabic and french better:::"* **P4** *"The worst thing was the language, since it was English, it was complicated... .unfortunately there were some words that I didn't understand"*
	Functional Failure	**P3** *"It has stoped working a lot, I didn't like that part"* *P21* *"... I was a little dissapointed, because It failed to work sometimes, and interrupted what I was so focused on watching"*
Interaction	Network Connection	**P2** *"I appreciated the fact that the app can work without internet connection, it's great"*
	Stability	**P7** *"I found some difficulties in stability...I needed a little stability, in order to do that.... and not being at the same time trying to maintain it in your hand and clicking and working with it"*
Product characteristics		
Pragmatic attributes	Useful	**P14** *"..., the app is than useful, because when you just see it without the app, ..."* **P17** *"..., I appreciate it yes, I can understand what things mean, it will be very useful"*
	Tutorial Suggestion	**P8** *"..., and don't know what I am doing and what to expect... I start using the app, not like the existing help that shows when you are using it, so I find myself focusing on what the app is telling me to do, guiding me rather than on the content."*
Hedonic attributes	Entertaining	**P1** *"It was fun, so entertaining"* *P5* *"It's like playing a game to discover what's its story...amazing"* **P13** *"It's something playful"*
	Immersive	**P2** *"It gives you a big imagination... .I felt disconnected, like i was merged more than necessary, like I lived the app in reality"* **P21** *"I felt im part of that era"*
	Independent	**P11** *"...and the one that I want, and put it first, and it's faster"* **P21** *"It offered me a free will to choose...."*
	Involving	**P11** *"I felt im getting closer to that world, that they're living in before, I can understand what was like exactly.."* **P18** *"I felt more involved when i used the app"*
	Innovative	**P9** *"..., it's new level of innovation"* **P11** *"... it's something innovative and.. amazing"*
	Interactive	**P18** *"... I'll be more interactive, that's what I want when I come to a museum to interact with the past, history...I'll have my iphone first thing to give me live, interactive information"* **P21** *"It's amazing to be able to interact with a static art, and not just watch and that's it."*
	Uncomfortable	**P3** *"It's a problem for me the fact of staying like this.. it's uncomfortable"* **P6** *"To hold it for a quite a time, it can be heavy... use the app, and to see more pieces of art.. like this.. it could be quite heavy for your hand".*
Consequences		
Satisfied	**P3** *"It's a memorable experience, it was really appealing."* **P9** *"It's an extraordinary experience, I am satisfied and happy to be here today.. I didn't miss it".*	
Not satisfied	**P13** *" I am not used to experience something through technology, and I don't plan to change it now, it works for me im satisfied, if I didn't understand something, when i go back home I'll look for it on the internet or books, I have many, im passionate about history and art... so I see technology takes the charm of the original pieces, it targets maybe your generation more than mine... yes i enjoyed it because it was something new, different but not to use it"* **P16** *"It was good not to the point to be impressive that I will use it in the future, I'll stick to what I know, it's for young people not for me honnestly, i want a simple photo, mostly remember the whole experience in my head what I've read and seen, be able to talk to my friends about it not show them a phone or an application that will speak and show them the experience in my place"*	

A great number of participants considered it a big advantage to be able to use the MAR app without needing to connect to a network. According to Han et al., (2017), this feature effectively reduces accessing problems, since internet connection is not always required to get information. However, according to Van Krevelen and Poelman (2010), the accessibility to a 'stable WIFI infrastructure' will help visitors who want to get additional online information during their visits. Also, according to Gafni (2008), Wi-Fi will keep the content of the MAR app up to date.

Pragmatic attributes are mainly the efficiency and usefulness of the MAR app. Overall; the results have shown that the MAR app is efficient and useful since it fulfilled its development purposes. Furthermore, it was considered a helpful tool for visitors, as extraction of the information did not require a lot of effort.

Moreover, the absence of docents made the existence and use of the MAR app essential for understanding each artefact; it effectively allowed visitors to skip the confusion phase, where they interpret what they are seeing according to their own perspectives. In some cases, it is essential to have different interpretations from different visitors, but this is only accurate when the interpretation is based on information from an authentic source like a historian or researcher. However, in the case of our study, sufficient information from experts like docents and displayed labels, which usually offer much generalized knowledge about artefacts, was supplemented by the MAR app.

Results from participants' interviews showed that most users agreed that the MAR app could enhance their experience in the museum. Usually, museum visits are considered boring and predictable, especially by children. The visitor has to stroll around the museum and read labels, which generally do not provide enough information about artefacts, thereby disappointing visitors. So when participants tried the MAR app, most of them agreed that this interactive technology could radically change the tourism industry, because it did not only help them gather more information about displays, but also they had a fun time and a pleasant experience using it. According to visitors, the implementation of interactive technologies will provide a new dimension to cultural institutions, and the MAR app used herein is an example of that. This fun and playful side of the technology has a big impact on UX, and will entice visitors to go to more museums. This supports Dieck et al.'s (2017) finding that AR could change museums' static experience into a very enjoyable one by bringing old stories and history to life. Additionally, interactive games such as quizzes are an amazing tool to test visitors' learning outcomes and make them more active participants in the experience. As a consequence, this will lead to a sense of 'personal fulfilment' and will encourage more positive word of mouth, since visitors will share their experience with their family and friends (Gordillo et al., 2013).

As some participants were not familiar with AR and had never used it before, the novelty aspect was also cited as part of their motivation to participate. As such, their experience was made more pleasant, since they did not know the outcomes of the app in advance. This supports the studies of Dieck et al., (2017) and Jiang and Kim (2015), in that the availability of high technology enhanced visitors' experience, particularly younger visitors. Generally, young people tend to focus more when they are experiencing new ways of learning, which will make their visit more interesting and fun. However, for elderly people who are interested in exploring new ways to visit museums, it may be preferable to add a personal guide to make their experience more enjoyable.

The majority of this study's participants were very satisfied and concluded that their museum experience was indeed enhanced when the MAR application was included in the tour. They reported that they would definitely use AR in future visits and would mostly recommend the experience to their friends and family (Neuberger et al., 2018).

CONCLUSION

Even though AR was first developed more than 20 years ago, it has not yet reached its full potential, which explains why no one can foresee what its limitations will be. However, as AR is increasingly used in our contemporary everyday life, many people believe that it will be the '*Supreme Technology*' of the future, which makes it even more crucial and important to study its impact and potential applications in various contexts.

Museums are great places to learn about the history of human kind, the various cultures and numerous golden centuries that deserve to be preserved through knowledge. However, with the advancement of technology, generations of the 21st century no longer give museums their deserved value. Therefore, museums are facing multiple challenges to please their visitors, who have very high expectations of how they want to spend their leisure time and money. Not only do they want to have a pleasant time, they also expect to be engaged and have memorable experiences that will make them momentarily forget their everyday life and routine. By implementing interactive technologies inside museums, visitors' expectations can be met and their experiences can be enhanced.

This study is our theoretical contribution to understanding the difference between VR and AR. It also aimed to clarify the concept of UX and the difference between it and usability. Through this study, we attempted to enrich the academic field with contributions specific to a social context by testing Hassenzahl's (2003) model on the different attributes of UX in MAR technology.

Moreover, this study provides researchers, professionals (such as developers and designers) and tourism practitioners (like museum stakeholders) very detailed findings that can be taken into account when developing MAR apps. Also, these study outcomes can help stakeholders adopt more appropriate strategies to benefit museums, museum visitors and the whole cultural tourism sector. However, to enrich the findings of this study, the UX of MAR ought to be examined from museum visitors' perspectives in the long term (i.e. the expectations before the visit, during the visit and after the visit). This must also be studied in different periods to eliminate the impact of technological innovation. Finally, it would be interesting for future works to investigate the benefits of MAR apps in museums from the perspective of museum stakeholders.

REFERENCES

Anderson, G. (Ed.). (2004). *Reinventing the museum: Historical and contemporary perspectives on the paradigm shift*. Editions Rowman Altamira.

Arnaboldi, M., Melisa, L. & Lema, D. (2021). The participatory turn in museums: The online facet. *Poetics*. . doi:10.1016/j.poetic.2021.101536

Asghar, T., & Nauman, A. (2010). *Augmented Reality in Museum Environments: A Case Study at Lund University Historical Museum* [Unpublished Master Thesis]. Division of ergonomics and aerosol technology, Department of Design Sciences, Lund University, Sweden.

Azuma, R. T. (1997). A Survey of Augmented Reality. *Ed. Presence.*, *6*(4), 355–385. doi:10.1162/pres.1997.6.4.355

Azuma, R., Baillot, Y., Behringer, R., Feiner, S., Julier, S., & MacIntyre, B. (2001). *Recent advances in augmented reality*. Naval Research Lab. doi:10.1109/38.963459

Bach, C., & Scapin, D. L. (2004). Obstacles and Perspectives for Evaluating Mixed Reality Systems Usability. Acte du Dhir, W., A., & Al-kahtani, M. (2013). A case study on user experience (UX) evaluation of mobile augmented reality prototypes. *Journal of Universal Computer Science*, *19*, 1175–1196.

Bevan, N. (2009). What is the difference between the purpose of usability and user experience evaluation methods. *Proceedings of the Workshop UXEM '09*.

Carmigniani, J., Borko, F., Marco, A., Paolo, C., Ernesto, D., & Misa, I. (2011). Augmented Reality Technologies, Systems and Applications. *Multimedia Tools and Applications*, *51*(1), 341–377. doi:10.100711042-010-0660-6

Caudell, T. P., & Mizell, D. W. (1992). Augmented reality: An application of heads-up display technology to manual manufacturing processes. *System Sciences, 1992. Proceedings of the Twenty-Fifth Hawaii International Conference on, 2*(January), 659-669.

Chang, K., Chang, C., Hou, H., Sung, Y., Chao, H., & Lee, C. (2014). Development and behavioral pattern analysis of a mobile guide system with augmented reality for painting appreciation instruction in an art museum. *Computers & Education, 71*, 185–197. doi:10.1016/j.compedu.2013.09.022

Charfi, A. A. (2012). *L'expérience d'immersion en ligne dans les environnements marchands de réalité virtuelle.* Thèse de doctorat en gestion, Universite Paris-dauphine. France.

Chung, N., Hyunae, L., Jin-Young, K., & Chulmo, K. (2018). The Role of Augmented Reality for Experience-Influenced Environments: The Case of Cultural Heritage Tourism in Korea. *Journal of Travel Research, 57*(5), 627–643. doi:10.1177/0047287517708255

Chylinski, M., Heller, J., Hilken, T., Keeling, D. I., Mahr, D., & De Ruyter, K. (2020, November). Augmented reality marketing: A technology-enabled approach to situated customer experience. *Australasian Marketing Journal, 28*(4), 374–384. doi:10.1016/j.ausmj.2020.04.004

Craig, A. B. (2013). Understanding augmented reality: concepts and applications. Morgan Kaufmann.

Damala, A., Cubaud, P., Bationo, A., Houlier, P., & Marchal, I. (2008). Bridging the Gap between the Digital and the Physical: Design and Evaluation of a Mobile Augmented Reality Guide for the Museum Visit. In *Proceedings of the 3rd international conference on Digital Interactive Media in Entertainment and Arts - DIMEA '08*. Athens, Greece: ACM Press.

DHL Trend Research. (2014). Retrieved from: https://www.dhl.com/content/dam/downloads/g0/about_us/logistics_insights/csi _augmented_reality_report_290414.pdf

Dieck, T. M. C., & Jung, T. (2015). A theoretical model of mobile augmented reality acceptance in urban heritage tourism. *Current Issues in Tourism, 18*, 1–21.

Dieck, T. M. C., & Jung, T. (2016). Value of augmented reality to enhance the visitor experience: A case study of Manchester Jewish Museum. *Review of Tourism Research, 7*, 1–5.

Dieck, T. M. C., & Jung, T. H. (2017). Value of Augmented Reality at Cultural Heritage Sites: A Stakeholder Approach. *Journal of Destination Marketing & Management*, *6*(2), 110–117. doi:10.1016/j.jdmm.2017.03.002

Do, H.-N., Shih, W., & Ha, Q.-A. (2020, August). Effects of mobile augmented reality apps on impulse buying behavior: An investigation in the tourism field. *Heliyon*, *6*(8), e04667. Advance online publication. doi:10.1016/j.heliyon.2020.e04667 PMID:32923709

Evrard, Y., Pras, B., Roux, E., Desmet, P., Dussaix, A. M., & Lilien, G. L. (2009). *Market-Fondements et méthodes des recherches en marketing* (4th ed.). Dunod.

Fuchs, H., & Ackerman, J. (1999). *Displays for augmented reality: Historical remarks and future prospects. Mixed Reality Merging Real and Virtual Worlds*. Ohmsha Ltd.

Fan, X., Chai, Z., Deng, N., & Dong, X. (2020). Adoption of augmented reality in online retailing and consumers' product attitude: A cognitive perspective. *Journal of Retailing and Consumer Services*, *53*(March), 101986. Advance online publication. doi:10.1016/j.jretconser.2019.101986

Flavián, C., Ibáñez-Sánchez, S., & Orús, C. (2019). The impact of virtual, augmented and mixed reality technologies on the customer experience. *Journal of Business Research*, *100*(July), 547–560. doi:10.1016/j.jbusres.2018.10.050

Gafni, R. (2008). Framework for quality metrics in mobile-wireless information systems. *Interdisciplinary Journal of Information, Knowledge, and Management*, *3*, 23–38. doi:10.28945/88

Glauser, W. (2013). Doctors among early adopters of Google Glass. *Canadian Medical Association Journal*, *185*(16), 1385. doi:10.1503/cmaj.109-4607

Gordillo, A., Gallego, D., Barra, E., & Quemada, J. (2013). The City as a Learning Gamified Platform. *IEEE Frontiers in Education Conference (FIE)*. 10.1109/FIE.2013.6684850

Hammady, R., Ma, M., & Powell, A. (2018). User Experience of Markerless Augmented Reality Applications in Cultural Heritage Museums: 'Museum Eye' as a Case Study. Academic Press.

Han, D.-I., Dieck, M. C. T., & Jung, T. (2018). User Experience Model for Augmented Reality Applications in Urban Heritage Tourism. *Journal of Heritage Tourism*, *13*(1), 46–61. doi:10.1080/1743873X.2016.1251931

Harrison, C. M. (2008). *Exploring emotional web experience: More than just usability and good design* (Doctoral dissertation). University of York.

Hassenzahl, M. (2003). The thing and I: understanding the relationship between user and product. In M. A. Blythe, K. Overbeeke, A. F. Monk, & P. C. Wright (Eds.), *Funology* (pp. 31–42). Kluwer Academic Publishers. doi:10.1007/1-4020-2967-5_4

Hassenzahl, M. (2008). User experience (UX): towards an experiential perspective on product quality. *Proceedings of the 20th International Conference of the Association Francophone d'Interaction Homme-Machine on*. 10.1145/1512714.1512717

Hirschman, E., & Holbrook, M. B. (1992). *The semiotics of consumption*. Editions Mouton de Gruyter.

Hoyer, W., Kroschke, D.M., Schmitt, B., Kraume, K., & Shankar, V. (2020). Transforming the customer experience through new technologies. *Journal of Interaction Marketing*. . doi:10.1016/j.intmar.2020.04.001

Hunter, W. C., Chung, N., Gretzel, U., & Koo, C. (2015). Constructivist research in smart tourism. *Asia Pacific Journal of Information Systems*, *25*(1), 105–120. doi:10.14329/apjis.2015.25.1.105

ICOM. (2007, August 24). *The World Museum Community - Museum Definition*. Retrieved from ICOM International Council of Museums: https://icom.museum/the-vision/museumdefinition/

Institute of Museum and Library Services. (n.d.). *Status of technology and digitization in the nation's museums and libraries*. Retrieved from: https://www.imls.gov/resources/TechDig05/Technology% 2BDigitization.pdf

International Organization for Standardization. (2010a). *Ergonomics of human-system interaction — Part 11: Usability: Definitions and concepts* (ISO/DIS 9241-11.2, 3.1). Retrieved from: https://www.iso.org/obp/ui/#iso:std:iso:9241:-11:dis:ed-2:v2:en

International Organization for Standardization. (2010b). *Ergonomics of human-system interaction — Part 210: Human-centred design for interactive systems* (ISO 9241-210:2010, 2.15). Author.

Jiang, Y., & Kim, Y. (2015). Developing multi-dimensional green value: Extending social exchange theory to explore customers' purchase intention in green hotels–evidence from Korea. *International Journal of Contemporary Hospitality Management*, *27*(2), 308–334. doi:10.1108/IJCHM-08-2013-0383

Jung, T., Chung, N., & Leue, M. C. (2015). The determinants of recommendations to use augmented reality technologies: The case of a Korean theme park. *Tourism Management*, *49*, 75–86. doi:10.1016/j.tourman.2015.02.013

Jung, T., Dieck, M. C. T., Hyunae, L., & Chung, N. (2016). *Effects of Virtual Reality and Augmented Reality on Visitor Experiences in Museum. In Information and Communication Technologies in Tourism 2016*. Springer International Publishing.

Kesim, M., & Ozarslan, Y. (2012). Augmented Reality in Education: Current Technologies and the Potential for Education. *Procedia: Social and Behavioral Sciences*, *47*, 297–302. doi:10.1016/j.sbspro.2012.06.654

Kiryakova, G., Angelova, N., & Yordanova, L. (2017). The potential of augmented reality to change the business. *Trakia Journal of Sciences*, *15*(1, Suppl.1), 394–401. doi:10.15547/tjs.2017.s.01.066

Kiyokawa, K. (2012). Trends and Vision of Head Mounted Display in Augmented Reality. In *International Symposium on Ubiquitous Virtual Reality. IEEE*. 10.1109/ISUVR.2012.11

Kotler, N., & Kotler, P. (2000). Can museums be all things to all people? Missions, goals, and marketing's role. *Museum Management and Curatorship*, *18*(3), 271–287. doi:10.1080/09647770000301803

Kuniavsky, M. (2007). *User experience and HCI. In The Human-Computer Interaction Handbook: Fundamentals, Evolving Technologies, and Emerging Applications* (2nd ed.). Lawrence Erlbaum Associates Inc.

Kung, W.-S. (2020). Reconceptualizing the digital humanities in Asia: New representations of art, history and culture (2nd ed.). Springer. doi:10.1007/978-981-15-4642-6

Lee, H., Jung, T. H., Dieck, M. C. T., & Chung, N. (2020). Experiencing immersive virtual reality in museums. *Information & Management*, *57*(5), 103229. Advance online publication. doi:10.1016/j.im.2019.103229

Li, C., Guo, S., Wang, C., & Zhang, J. (2019). Veni, vidi, vici: The impact of social media on virtual acculturation in tourism context. *Technological Forecasting and Social Change*, *145*(August), 513–522. Advance online publication. doi:10.1016/j.techfore.2019.01.013

Loureiro, S. M. C., Guerreiro, J., Eloy, S., Langaro, D., & Panchapakesan, P. (2019). Understanding the use of Virtual Reality in Marketing: A text mining-based review. *Journal of Business Research*, *100*(July), 514–530. doi:10.1016/j.jbusres.2018.10.055

Marty, P. F. (2008). Museum websites and museum visitors: Digital museum resources and their use. *Museum Management and Curatorship*, *23*(1), 81–99. doi:10.1080/09647770701865410

Mei, Q. (2004). A knowledge processing oriented life cycle study from a digital museum system. *Proceedings of the 42nd annual Southeast regional conference*, 116-121. 10.1145/986537.986565

Milgram, P., & Kishino, F. (1994). A taxonomy of mixed reality visual displays. *IEICE Transactions on Information and Systems*, *77*, 1321–1329.

Neuburger, L., & Roman, E. (2017). *An Afternoon at the Museum: Through the Lens of Augmented Reality. In Information and Communication Technologies in Tourism 2017*. Springer International Publishing.

Neuburger, L., & Egger, R. (2018). Augmented reality: Providing a different dimension for museum visitors. In *Augmented Reality and Virtual Reality* (pp. 65–77). Springer. doi:10.1007/978-3-319-64027-3_5

Olsson, T., Ihamäki, P., Lagerstam, E., Ventä-Olkkonen, L., & Väänänen-Vainio-Mattila, K. (2009). User expectations for mobile mixed reality services: an initial user study. *Proceedings from the European Conference on Cognitive Ergonomics: Designing beyond the Product*, 177–184.

Pallud, J., & Monod, E. (2010). User experience of museum technologies: The phenomenological scales. *European Journal of Information Systems*, *19*(5), 562–580. doi:10.1057/ejis.2010.37

Patrício, L., Fisk, R. F., & Cunha, J. F. (2003). Improving satisfaction with bank service offerings: Measuring the contribution of each delivery channel. *Managing Service Quality*, *13*(6), 471–482. doi:10.1108/09604520310506531

Pauget, B., & Dammak, A (2019). The implementation of the Internet of Things: what impact on organizations? *Technol Forecast Soc Change*, *140*, 140-146. . doi:10.1016/j.techfore.2018.03.012

Peacock, D., & Brownbill, J. (2007). *Audiences, visitors, users: Reconceptualising users of museum on-line content and services*. Museums and the Web.

Pine, B. J. II, & Gilmore, J. H. (1999). *The Experience Economy: Work Is Theater and Every Business a Stage*. Harvard Business School Press.

Portalés, C., José, L. L., & Pérez C. (2009). Photogrammetry and Augmented Reality for Cultural Heritage Applications. *The Photogrammetric Record*, *24*(128), 316-331. doi:10.1111/j.1477-9730.2009.00549

Pulli, P., Zheng, X., Antoniac, P., Hickey, S., Manninen, T., Martikainen, O., & Kuroda, T. (2007). *Design and development of mobile services platform for senior citizens*. Paper presented at the 13th International Conference on Concurrent Enterprising, Nottingham, UK.

Qin, H., Peak, D. A., & Prybutok, V. (2021). A virtual market in your pocket: How does mobile augmented reality (MAR) influence consumer decision making? *Journal of Retailing and Consumer Services*, *58*(January), 102337. Advance online publication. doi:10.1016/j.jretconser.2020.102337

Rodrigues, J. M. F., Lessa, J., Gregório, M., Ramos, C., & Cardoso, P. J. S. (2016). An Initial Framework for a Museum Application for Senior Citizens. In *The 7th International Conference on Software Development and Technologies for Enhancing Accessibility and Fighting Info-exclusion - DSAI 2016*. ACM Press. 10.1145/3019943.3020002

Romelaer, P. (2005). L'entretien de recherche. Management des ressources humaines: méthodes de recherche en sciences humaines et sociales. *De Boeck*, 101-137.

Rusu, C., Rusu, V., Roncagliolo, S., Apablaza, J., & Rusu, V. Z. (2015). User Experience Evaluations: Challenges for Newcomers. In. Lecture Notes in Computer Science: Vol. 9186. *Design, User Experience, and Usability: Design Discourse*. Springer., doi:10.1007/978-3-319-20886-2_23.

Selvam, A., Yap, T. T.-V., Ng, H., Tong, H.-L., & Ho, C. C. (2016). Augmented Reality for Information Retrieval Aimed at Museum Exhibitions using Smartphones. *Journal of Engineering and Applied Sciences (Asian Research Publishing Network)*, *100*(3), 635–639.

Tscheu, F., & Buhalis, D. (2016). *Augmented Reality at Cultural Heritage Sites. Information and Communication Technologies in Tourism*. Cham Springer International Publishing.

Van Krevelen, D., & Poelman, R. (2010). A survey of augmented reality technologies, applications and limitations. *The International Journal of Virtual Reality: a Multimedia Publication for Professionals*, *9*(2), 1–20. doi:10.20870/IJVR.2010.9.2.2767

Wilson, P. F., Stott, J., Warnett, J. M., Attridge, A., Smith, M. P., & Williams, M. A. (2018). Museum visitor preference for the physical properties of 3D printed replicas. *Journal of Cultural Heritage*, *32*(July–August), 176–185. doi:10.1016/j.culher.2018.02.002

Zheng, X., & Pulli, P. (2005). Extending quality function deployment to enterprise mobile services design and development. *Journal of Control Engineering and Applied Informatics*, *7*(2), 42–49.

APPENDIX

These are some screenshots of the content provided by the MAR app about a certain exhibit

Figure 1.

Figure 2.

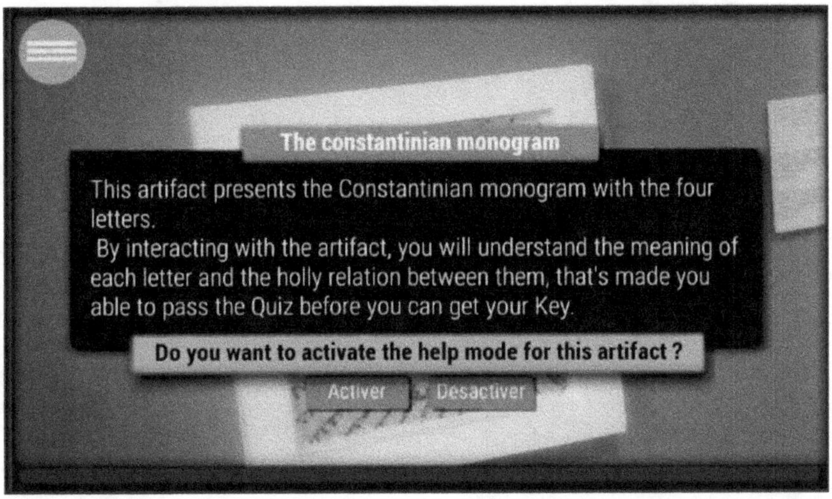

Figure 3.

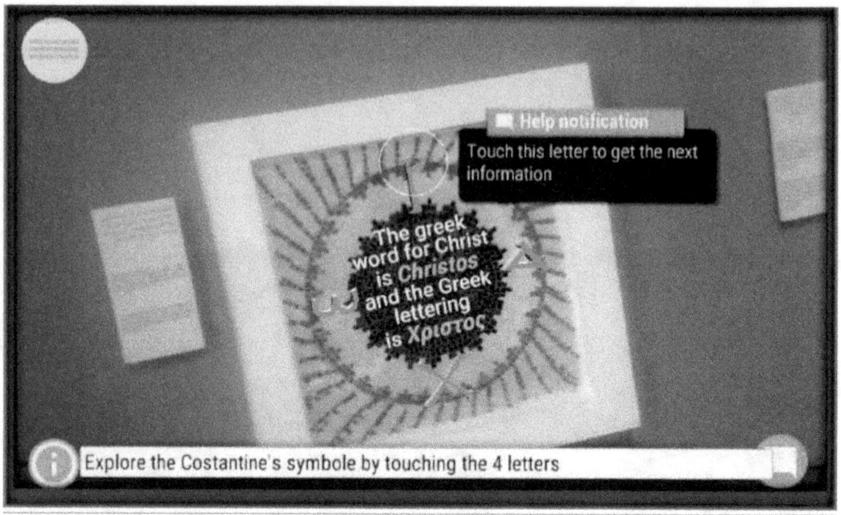
Help notification
Touch this letter to get the next information
The greek word for Christ is Christos and the Greek lettering is Χριστος
Explore the Costantine's symbole by touching the 4 letters

Chapter 7
Blockchain Technology: A Revolution in the Hospitality and Tourism Industry

Anusha Thakur
https://orcid.org/0000-0001-8761-2250
University of Petroleum and Energy Studies, India

ABSTRACT

Travel and technology tend to possess seamless amalgamation. Advancements in technologies are significantly changing the way people travel, with exciting and promising experiences. Incorporation of smart devices and solutions in businesses enables easier ways to analyze, collect, as well as share the real-time data with vendors. One such concept includes blockchain technology, inclusion of which brings about an extreme change in the way the data are being accessed and stored. This technology enables the businesses to fortify operations, transactions, as well as processes, thereby bolstering productivity and growth and discovering newer profit models. This chapter discusses the different competitive advantages offered by the tourism businesses on inclusion of blockchain technology in their strategic-decisions and emphasizes how the blockchain concepts help in simplifying the tourism and hospitality businesses, amidst different challenges being faced by these industries.

INTRODUCTION

In the era of digitalization, blockchain is one of the latest technologies, with much of its potentiality yet to be explored. This technology is on the edge of reforming trade finance sector, and hovers to leave aside the businesses and services which

DOI: 10.4018/978-1-7998-6904-7.ch007

do not cope up with the current scenario (PWC, 2018). Businesses nowadays are engaged in assigning more of its resources to research, with the incorporation of this technology in operational activities on a daily basis. It is considerably one of the foundational technologies rather than the disruptive ones, owing to the fact that the social and economic systems nowadays, can be based on the concepts of the same.

Blockchain technology owes the capability to completely turn over, the overall industrial structure. The technology was primarily associated with Bitcoins, however, in today's scenario, it is gaining significant traction across different verticals. Cryptographic, tamper-evident, trust-free, as well as proven chronological integrity and security structure of blockchain enables development of decentralized self-directed establishments, generating smart contracts, digitizing fiat currencies and several other applications (Dogru).

In the hospitality sector, blockchain emphasizes on opening new applications and potentials, which are not only promising, but also inevitable over the forecast period. The technology helps augment the guest experience, foster innovations, as well as enhance the efficiency and effectiveness of hotels (Srivastava A.). Features such as decentralization, programmability, transparency, and immutability, empowers organizational processes and structures, enables methods to strategize customer relations, and simplifies new forms of interorganizational associations. The technology is expected to bring in favorable changes, thereby leading to stress-free experiences from stressful experiences.

Incorporation of blockchain in the hospitality and tourism industry largely depends upon the take of new technologies and innovations within the business environment. However, there are few assumptions and concerns regarding integration of blockchain with current systems and its adoption. The chapter outlines the market determinants impacting the implementation of blockchain technology in the hospitality and tourism sector, with analysis of different challenges and trends in the market.

Research Questions

- Need for blockchain technology in the hospitality and tourism industry?
- What are the growth opportunities and trend analysis for the blockchain technology in the hospitality and tourism sector?
- What are the factors bolstering the demand for this technology?
- What are the competitive advantages offered by the concept of blockchain in hospitality industry?
- Impact of COVID-19 on the inclusion of the decentralized networks in businesses.

Purpose of the Article

This article illustrates the different factors boosting and hindering the market growth of blockchain technologies in the travel sector. The paper also focuses on the key growth opportunities for the application of these technologies in the hospitality and tourism space. In addition to these, future implementations of this technology, with a brief description of competitive strategies adopted by the businesses have also been illustrated.

Key Findings in the Research

Industrial development considerably depends on reliable associations; however, growing incidences of fraud and cybercrime is expected to restrain the market from expanding. Implementation of blockchain technology is expected to cater to such issues, and challenges.

Adoption of blockchain technology offers simplicity and transparency in business processes, hence, creating a surge in the industrial growth over the forecast period. Growing demand for digital payment and increasing need to control the issues of data breach and leakage in businesses are expected to propel the demand for the concept of "blockchain". Further, during the global crisis of COVID-19, this technology has further witnessed a surge in demand, owing to the promising features and solutions offered by the same.

LITERATURE REVIEW

Blockchain is significantly one of the emerging technologies of the "Fourth Industrial Revolution". Fundamentally, it is a decentralized, new and computational infrastructure which is expected to change several existing processes in society, governance, and business models. The technology is more than just a tool to empower digital currencies. Its features include advanced cryptography, with immutable and distributed ledgers, which empowers the transfer of a series of assets amongst parties inexpensively and securely without the third-party mediators (PWC, Fourth Industrial Revolution for the Earth Series: Building block(chain)s for a better planet, September 2018). Few characteristics of Blockchain can be illustrated as follows:

In the hospitality and travel & tourism industry, blockchain is projected to be one of the significant and influential technological forces, which offers several possibilities to revolutionize the industry's position. Blockchain permits the businesses to store and access several vital information directly and consistently, as it is stockpiled in a distributed ledger and shared across the systems (Alnemari, et al., 18 December 2018)

Figure 1. Characteristics of blockchain
Source: (Yaga, October 2018)

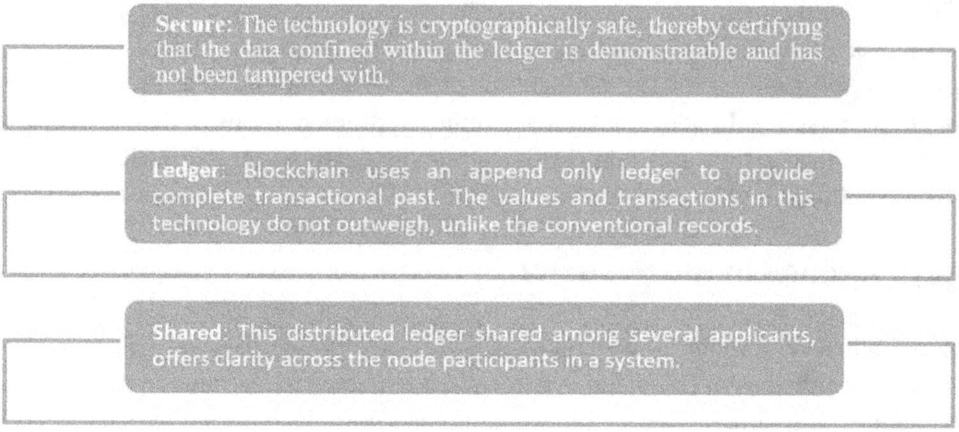

(Panarello, Tapas, Merlino, Longo, & Puliafito, 2018) (Pop, et al., 2018). Adoption of this latest technology in the tourism industry offers seamless experiences to the travelers and hoteliers for their direct interaction with service providers without the intervention of third-party intermediaries.

NEED FOR BLOCKCHAIN IN THE HOSPITALITY AND TOURISM INDUSTRY

In today's scenario, various travel mediators in travel businesses are engaged in making travel nearly 20 to 25 per cent expensive via different commission that they take (Magazine, 2017). Incorporation of technology such as "blockchain" is projected to make the experiences for both service providers as well as customers cost-effective and proficient without any commission.

Blockchain technology is a digitally encoded ledger wherein the transactions between different parties are stockpiled via smart contracts. It is a distributed system backing up reliable dealings. Blockchain records entries in the blocks which are permanent and immutable, thereby keeping the information securely accessible, transparent and safe. The essence of this technology lies in its capability to enable reliable transactions between varied parties via smart contracts which are accessible. To enhance the operational activies of businesses, it enables the hotels to easily record the transactions and also exchange the digital information and identities between the different parties in an effortless manner.

In addition to these, it also enables closer consumer relationships, faster integration with cloud and IT technology, quicker product innovations, and presence of more agile value chains. These factors pose as great opportunities for the hospitality and travel industry, owing to the fact that there are a lot of processes happening online such as payments, bookings, transportation, and few others within the industry (ChangeNOW, 2020).

APPLICATIONS OF BLOCKCHAIN IN HOSPITALITY AND TOURISM INDUSTRY

Usage of blockchain within the tourism and hospitality industry are considerably boundless, with some of the application segments gaining significant surge in demand with a transformative effect. Few of these application areas are illustrated below:

Loyalty and Rewards Management

In today's scenario, the hospitality and travel organizations are essentially focusing on the need for information systems, to manage the reward points. With the inclusion of blockchain technologies, it becomes easier for the partners to define their own digital assets and the related fiat values with the same. This further enables the guests to redeem their loyalty points with any merchant or partner with much ease and flexibility. The technology also enables easy gifting of the loyalty points to family members and friends, thereby making the rewards program more customer-friendly, seamless and flexible.

Supply Chain Management

Trust within the parties play a crucial role, in order to make the process of supply chain effective. Solutions such as "blockchain" is engaged in offering consecutive information persistence across a distributed structure wherein, the records can be cryptographically secured via reliable identities.

Implementation of this concept also helps to establish trust without the requirement of an intermediary, by ensuring that updates are attached with a suitable audit trail and that stored information is immutable. Hence, this helps in building an unconditional cooperation between the creator and the data via consent with indisputable proof of its origin.

Secure Payments

"Secure payments" is significantly one of the most obvious applications of blockchain technology in terms of provision of safe, bank payment, transparent, global ledger or the acceptance of cryptocurrencies in the hotel and travel industry. Incorporation of this technology helps to streamline and minimize the associated costs for the hotels.

ID and Security

One of the greatest possible usages of blockchain technology in the hospitality sector is related to the security and identification services. The technology enables for a shared digital database, thereby minimizing the waiting time for the customers.

For instance, systems such as SITA/ShoCard enables the ID details of customers to be stockpiled on blockchain, in an identical and decentralized format, thereby enabling the companies to easily retrieve and verify any user identity from anywhere at any time (Mazars, 2021).

Baggage Tracking

Applicability of blockchain within the hospitality industry outspread to things such as tracking of luggage/baggage's. During travelling, the luggage of any particular traveler considerably goes from one point to another, thereby causing several logistical issues at times. However, inclusion of blockchain in this process, enables easy and hassle-free accessibility to the baggage.

Food and Beverage (F&B)

Blockchain technology enables businesses along the supply chain processes in the food & beverage segment to easily trace and track every step during the storage, sourcing, and processing steps of the product. In addition to this, it also allows the information on the platform to be unchangeable, thereby increasing the transparency and visibility of the food & beverage supply chain processes.

Tax And Accounting

The processes of supply chain often pose to be challenging in the hospitality industry. This is expected to lead to the additional count of multiple sheets of taxation hence, creating trouble for both the suppliers as well as buyers. Blockchain helps easy integration of these taxation layers to simplify the payment procedures for both the payer as well as payee.

Smart Contracts

Adoption of smart contracts helps to simplify both major as well as minor transactions in the hospitality and tourism sector. For instance, the travel agencies and hotel businesses, owes the capability to streamline their alliances with smart contracts on the blockchain platforms. Just as in the case of legal contracts, a smart contract between the travel agencies and hotel businesses is expected to have contractual provisions which can be preordained by the transacting parties (Crosby, 2016) (Gupta, 2017) (Peters G. W., 2016).

Whenever a transaction occurs, it can easily be shared and recoded on the blockchain platform, thereby further enabling the payments to be processed immediately on the basis of contractual terms. The process not only simplifies the payment process, but is also further expected to optimize the sales via enhanced partnership between the travel agencies and hotels.

Smart contracts pose to be a crucial concept in terms of the guest experiences, as it helps to completely reduce the check-in process. The hotel rooms can easily be assigned to the guests via blockchain technology wherein, the digital IDs are already stored with an authorized account for payments and once the payment is received, the digital key can be recorded into the blockchain technology. Further, smart contracts also facilitate insurance related to the travel processes in case if the flight gets cancelled or delayed. With the help of smart contract, the contractual terms can be a part of the network and can easily be executed in cases of cancellation or delay.

COMPANIES IMPLEMENTING BLOCKCHAIN TECHNOLOGY

The concentrated shape of tourism and travel industry is expected to bear a weight on the industry and consumers itself, thereby eliminating the price competition and adding extra costs during the transactions among the travel providers. Several companies such as LockTrip, Winding Tree, Cool Cousin, Sandblock, Webjet, Hotel P2P, TUI, and Travelchain are engaged in the implementation of blockchain applications into different operational activities, in their businesses in order to gain a viable and competitive edge in the industry.

Implementation of blockchain technology in various segments of hotel and travel businesses helps minimize the hurdles by streamlining the processes and catering to several challenges while constructing an unbiased ecosystem.

Winding Tree

Nowadays, travel service companies such as Priceline and Expedia are increasingly dominating the airfare bookings segment with higher costs to both the airlines as well as travelers. However, startups such as "Winding Tree" are engaged in, to unseat these rooted giants by challenging their part as middlemen in the industry. Winding Tree focuses on creating the next-generation of decentralized travel display platform on Ethereum, thereby enabling a peer-to-peer economy to free up innovation and creativity in the travel industry (Izmaylov).

About the Company: Winding Tree, established in 2017, is a private Switzerland-based company. The company caters to different parts of the hospitality and tourism industry, and ensures a competitive travel market. In addition to this, the company also emphasizes on to minimize the costs for service providers and the fees for travelers. The company is engaged in the usage of ERC827 protocol and smart contracts which delivers better savings for the shareholders in the tourism industry. Its profit status ensures that there are no middlemen counting to the unwanted fees in the booking criteria and processes.

Webjet

Inaccurate hotel bookings are expected to significantly add anxiety and stress to the individual's travel experience. In case of such situations, the consumers need to take on the fees across different sections of the supply chain, with the providers constantly extracting the value. Companies such as "Webjet" have announced the immutable distributed ledger technology model to avoid such hassles in booking experiences. Inclusion of this technology offers improved security features with better protection for the ecosystem. Along with the benefits such as minimization of costs for hotels, and optimum supply chain efficiencies, the travelers are also expected to benefit from greater accountability and transparency during the booking process.

About the Company: Webjet is an Australia based company, established in 1998. The company is one of the leading online travel agencies with its offerings being focused at efficient online technology and tools. The company announced the launch of a blockchain platform in 2019, wherein, the system focused on enhancing the user experience by keeping a record of all the entries on the immutable ledger, with reduced probability of lost or mistaken bookings between the seller and the users. This is usually done by locating the real-time information problems occurring between the hotels, agents, and customers (Foxley, 2019).

GLOBAL MARKET OVERVIEW

Over the past few years, blockchain has emerged as one of the most able technologies, and is expected to further gain significant traction in the future as well. The technology is an immutable, open and a distributed public ledger which can be accessed by different parties engaged in the transactional processes. Further, its decentralization nature, helps to remove needless intermediates, thereby enabling seamless running of the business activities.

Adoption of blockchain is expected to empower the businesses to maintain and add loyalty partnerships without counting the complexity. A frictionless and robust partnership network offers several redemption possibilities outside the central travel product, thereby, creating a much-required valve for the rising balance-sheet pressures (Kowalewski, 2017).

Increasing need to avert the issues of counterfeiting is expected to boost the overall market growth. Blockchain, particularly in the hospitality and tourism sector, is gaining widespread adoption owing to its potentiality to minimize the overall transactional processing time, and complexities involved, thereby propelling its market demand. The technological concept provides abundant scenarios for the sector; but it is still not matured in today's scenario, nor a solution which can instantly applied. Hence, it is vital for the various economic, organizational, behavioural, and technical challenges to be catered to before the adoption of blockchain by businesses in the travel and hotel environment sector. In addition to these, in order to stay relevant and competitive in today's changing scenario, there is also a need to augment harmonization among different investors and accordingly, implement legal aspects in the processes to better cope up with the market.

MARKET DRIVERS

Increasing Demand for Digital Payment Systems

The digital payment systems rely on varied parties for transactional processes, comprising of retail banks, merchant banks, as well as payment software companies, thereby creating the blockchain technology demand for securing the transactions. In addition to this, reliability of users on trusted institutions in order to complete the electronic transactions is also expected to open avenues for the blockchain technology. Features of blockchain such as auditable, immutable, and transparency, holds the potentiality to augment the functionality, integrity, security, and data provenance of the business processes.

The digital payment providers offer opportunities to bolster the adoption and growth of these options by forcing the boundaries of innovation and creating awareness about the value these contributions bring in to several shareholders. Growing demand for the digital payment systems in the hospitality and tourism industry is expected to bolster the need for blockchain technology in different regions across the world.

MARKET RESTRAINTS

Regulatory Constraints

Regulatory concerns pose to be one of the major factors hindering the blockchain technology in the hospitality market. Like any cutting-edge technology, usage of blockchain in the hospitality sector lacks formal and regulatory guidance. In today's scenario, consideration of collaboration in between the industry and policy makers plays an important role to facilitate and understand the growth of the system within new policy objectives and framework.

Regulators in the different regions are anticipated to review these technologies in hospitality and tourism sector methodically, with security and privacy concerns being one of the key focus areas, over the next five years.

FUTURE SCENARIO OF BLOCKCHAIN IN HOSPITALITY AND TOURISM INDUSTRY

The blockchain technology has profound usage and applicability in the hospitality and tourism sector, with a focus to completely revolutionize the user experience from a stressful to stress-free. Applicability fields of blockchain is expected to increase manifolds, owing to its potential features and offerings. In the future, it is expected to change the way people make payments and interact.

IMPACT OF COVID-19 ON THE IMPLEMENTATION OF BLOCKCHAIN TECHNOLOGY IN THE HOSPITALITY AND TOURISM MARKET

The hospitality and tourism sector are one of the hardest hit sectors across different regions in the world. One of the latest technologies in the sector includes the blockchain technology, which possess unique characteristics such as decentralization, immutability and transparency, and have their profound usage across various domains such as credential, and identity management, content reservation, ticketing, baggage tracking, secure payments, loyalty programs and few more.

However, blockchain is significantly opening abundant avenues during this crisis situation, as it enables efficient monitoring and tracking solutions with a guarantee of transparent, safe and efficient processes in the business. This can be attributed to the fact that blockchain consists of a chronologically well-organized list of encoded signatures, a safe scattered ledger comprising of permanent transactional records which can be shared by all the associates in the system (Marbouh, October 12, 2020).

RESEARCH METHODOLOGY

This study entails a market research-based study with key analysis of market dynamics impacting the blockchain technology in the hospitality sector along with market sizing and forecasts over the projected period. The research focuses on different market drivers, trends, and opportunities, impacting market growth over the forecast period. The forecast analysis will be in terms of revenue at the global, and application levels, with the key trends varying for 2019-2028 (forecast period).

Research Objectives

Objectives of the research study can be illustrated as:

- To study the need of blockchain in the hospitality and tourism industry with the key focus on recent trends in the market.
- To emphasize on the future aspects, challenges and implications of the technology in the market.
- To recognize the factors propelling the growth of businesses across the travel and hotel industry verticals over the forecast period.
- Impact of COVID-19 on the blockchain technology in the hospitality market

Inclusions

- Financial services, government, healthcare, media & entertainment, retail, transportation & logistics, travel and others.
- North America, Europe, Asia Pacific, Latin America and Middle East & Africa

Scope of the Study

The article contemplates the current scenario of Global blockchain in the hospitality and tourism market for the forecast period of 2019 - 2028. The study focuses on regional as well as segment revenue for assessing the overall market scenario. It further emphasizes on the competitive strategies and market share analysis.

The study contemplates the study of the market size and estimates of the blockchain market in terms of revenue over the forecast period; as well as segmentation on the basis of region covering North America, Europe, Asia Pacific, and the Rest of the World (RoW). This helps in determining overall size of the market over the forecasted period in terms of revenue, get insights about the factors which are engaged in propelling the market growth and the factors which poses to be challenging in its development. Recognizing the competitors for incorporating the concepts of blockchain in hospitality and tourism market helps to get an idea about the products and strategies offered by these market players to sustain in the industry. The given information is derived from secondary sources collected and gathered via various investor documents, company annual reports, and statistics published by various market leaders. The data is analysed in view of the following factors:

- Demand and supply estimates
- Market developments and trends.
- Future aspects and opportunities offering insights on product commercialization as well as expansion in different regions.

The analysis is carried out on the basis of the historical market data and their respective trends. The data has been categorized based on varied parameters such as region, product, and application. Methodologies to study the gathered data can be illustrated as:

Top-Down Approach: The data are collected for the global scenario and are then separated into different entities (which include application/region, in this case it is based on regional analysis).

Bottom-Up Approach: The data are collected and estimated for regional segments and are then combined to obtain the global numbers. The data are then forecasted on the basis of different market initiatives and trends for over the forecast period of 2019 to 2025. This helps to gain holistic information and understand the current market scenario as well as future trends of the same.

Figure 2. Global blockchain technology marked 2019—2028 ($ billion)
Source: Secondary Sources

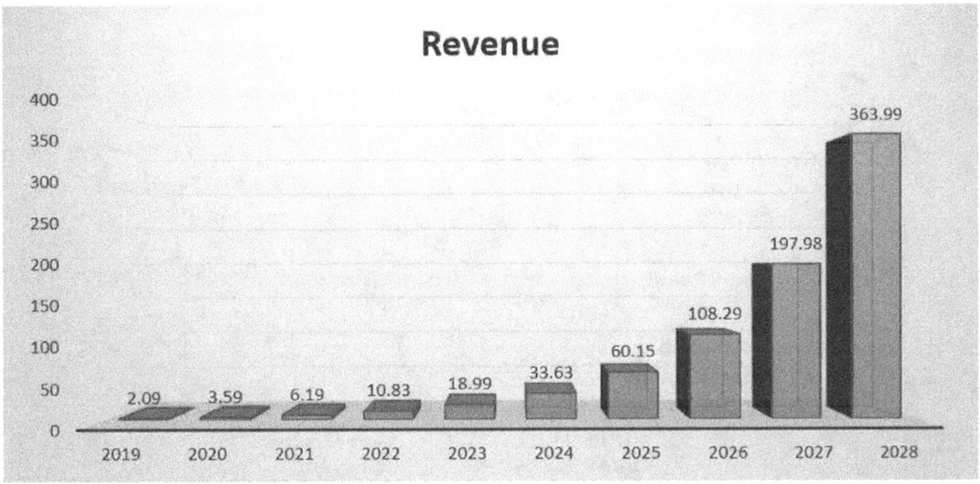

FINDINGS AND MARKET ANALYSIS

Global Analysis

Global blockchain technology market was valued at $2.09 billion in 2019, and is expected to witness a surge in its growth rate over the forecast period. The market was expected to be valued at $363.99 billion in 2028.

Application Analysis

On the basis of application segments, the market can be segmented into healthcare, government, financial services, media & entertainment, retail, transportation & logistics, hospitality & tourism, and others.

In 2019, in terms of blockchain technology, the hospitality and tourism sector accounted for nearly a share of 6.65 per cent, owing to the fact that still a lot of potentialities of blockchain is yet to be explored. However, the technology is expected to gain a huge surge in demand in the hospitality sector, owing to its rising applicability in the different end-use applications such as inventory management, tracking and maintenance, content reservations and ticketing, personalized marketing and loyalty programs, smart contracts and few other applications.

Figure 3. Application analysis of global blockchain technology market (in terms of market share)
Source: Secondary Sources

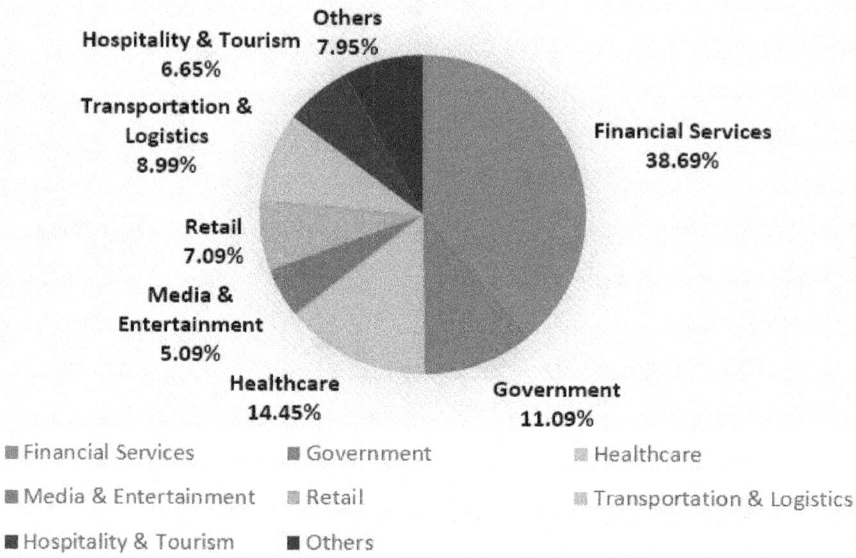

Regional Analysis

On the basis of geographies, the market can be segmented into regions of North America, Europe, Asia Pacific, Latin America, and Middle East & Africa. Europe accounted for the market share of nearly 46.79 per cent of the overall market. This can mainly be attributed to the increasing preferences among government and businesses to promote the usage of blockchain.

North America accounted for nearly 38.69 per cent of the total share in 2019, thereby witnessing a considerable growth along with the Asia Pacific regions as well. Presence of a large number of players and their strategic initiatives is expected to propel the market demand.

The Middle Eastern and Latin American regions are expected to witness considerable growth over the forecast period, owing to the increasing trends of identity detection solutions, and the development of new and efficient credential management system in the region.

Figure 4. Geographical analysis of global blockchain technology market in 2020 ($ billion)
Source: Secondary Sources

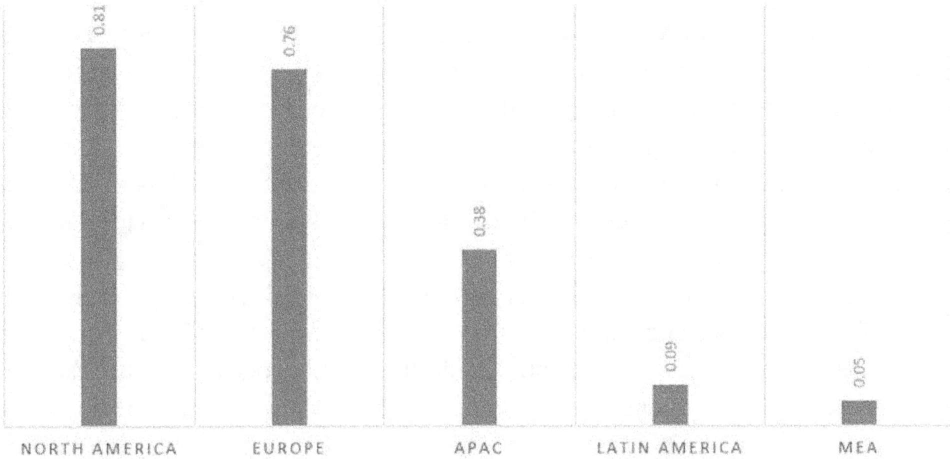

CONCLUSION

Blockchain technology is in its early stages when talking about its adoption in the travel industry. Researchers all across the world are working towards creating applications and frameworks involving the applicability of blockchain in the travel and tourism industry. The consequence of the rapid research being conducted in the past few years has led to an increase in acceptance of blockchain thus resulting in new and relevant use cases from within the travel industry. The integration of blockchain ensures enhanced customer data privacy and the facilitation of paperless travel experiences. The paper discusses significant insights to its readers regarding the importance of blockchain technology in the travel and hospitality industry. The aim of the research was to explore blockchain technology from the perspective of the hospitality industry.

The research conducted was exploratory in nature and talks through prominent works dealing with the implementation of blockchain in the hospitality industry. The work identifies the extent of research that has been conducted for the past five years in areas of blockchain technology and its integration with the travel and hospitality industry. The paper presents important case studies from the hospitality industry that help in enhancing the understanding of its readers. Furthermore, it allows all of us to think about further questions to investigate. What are the challenges of the adoption of blockchain technology in the hospitality industry? How integration with blockchain contributes towards revenue generation for the hospitality industry? However, the paper presents itself as an interesting contribution towards existing literature and will assist readers in exploring new avenues and adding advancements to the current scientific research

Blockchain technology is one of the latest technologies and has huge potential to conduct and change the way businesses are done. In the hospitality and tourism industry, blockchain technology, is still in its formative stage, but its offerings and values are exclusive and unique. Integration of blockchain guarantees improved privacy on user information with the facilitation of hassle-free experiences. Demand for this technology is expected to witness rapid expansion across a wide range of application segments and industries.

The technology ensures accurate transactions, correct business ratings, transparent flow of funds, solves frauds and helps for managing reward and loyalty schemes. In addition to this, it has significantly made the travel processes secure, simple, and transparent, thereby benefiting the industry and the travelers. Hence, inclusion of this latest concept is significantly expected to revolutionize the way operational activities are being conducted in the hospitality and travel industry.

REFERENCES

Alnemari, A., Arodi, S., Sosa, V., Pandey, S., Romanowski, C., Raj, R., & Mishra, S. (2018). *Protecting Infrastructure Data via Enhanced Access Control, Blockchain and Differential Privacy.* Springer.

ChangeNOW. (2020, May 14). Blockchain Technology in Hospitality & Travel industry. *ChangeNOW.*

Crosby, M. e. (2016). Blockchain technology: Beyond bitcoin. *Applied Innovation*, 6-10.

Dogru, T. M. (n.d.). *Blockchain Technology & its Implications for the Hospitality Industry.* Boston Hospitality Review.

Foxley, W. (2019, November 7). Digital Travel Firm Webjet Has Launched Its Booking Verification Blockchain. *CoinDesk.*

Gupta, M. (2017). *Blockchain for dummies.* John Wiley & Sons, Inc.

Izmaylov, M. A. (n.d.). *A Practical Application of Blockchain for the Travel Industry.* WindingTree.

Khanna, A. S. A. (2020). *Blockchain Technology for Hospitality Industry*. Springer. doi:10.1007/978-3-030-63396-7_7

Kowalewski, D. M. (2017, March 14). Blockchain Will Transform Customer Loyalty Programs. *Harvard Business Review*.

Magazine, X. (2017, September). Benefits of using Blockchain in the Hospitality Travel Industry. *XcelLab Magazine*.

Marbouh, D. A. (2020). *Blockchain for COVID-19: Review, Opportunities, and a Trusted Tracking System*. National Center for Biotechnology Information, U.S. National Library of Medicine.

Mazars. (2021). *Using Blockchain to Promote Trust in the Hospitality Industry*. Academic Press.

Panarello, A., Tapas, N., Merlino, G., Longo, F., & Puliafito, A. (2018). Blockchain and IoT Integration: A Systematic Survey. *MDPI*.

Peters, G. W. (2016). Understanding modern banking ledgers through blockchain technologies: Future of transaction processing and smart contracts on the internet of money. In *Banking Beyond Banks and Money* (pp. 239–278). Springer. doi:10.1007/978-3-319-42448-4_13

Peters, G. W. (2016). Understanding modern banking ledgers through blockchain technologies: Future of transaction processing and smart contracts on the internet of money. In *Banking beyond banks and money* (pp. 239–278). Springer. doi:10.1007/978-3-319-42448-4_13

Pop, C., Cioara, T., Antal, M., Anghel, I., Salomie, I., & Bertoncini, M. (2018). Blockchain Based Decentralized Management of Demand Response Programs in Smart Energy Grids. *MDPI*.

PWC. (2018). *Blockchain*. Digital Services.

PWC. (2018). *Fourth Industrial Revolution for the Earth Series: Building block(chain) s for a better planet*. Author.

Srivastava, A. (n.d.). *How blockchain is transforming the hospitality industry*. Today's Hotelier.

Yaga, D. (2018). *Blockchain Technology Overview*. U.S. Department of Commerce.

Chapter 8
Artificial Intelligence in the Delivery of Mobile Tourism Services

Zafer Türkmendağ
https://orcid.org/0000-0002-7712-1500
Atatürk Üniversitesi, Turkey

ABSTRACT

Tourists are more attached to mobile phones which oblige tourism businesses to give importance to the automated services in mobile phones for better services. This chapter discussed the artificial intelligence-integrated mobile technologies in tourism in terms of technological, marketing, and managerial perspectives. The integration of artificial intelligence-based open access application programming interfaces such as language translation, recommendation making, mapping, etc. to the mobile technologies created great changes in tourist behavior and business processes in tourism. Moreover, businesses can provide automated and contactless services by using mobile big data and analyze better the needs of the tourists, especially in the post-pandemic era. Finally, a framework has been proposed for artificial intelligence integration to mobile tourism services, and theoretical-managerial implications were developed.

INTRODUCTION

Mobile phone usage is one of the most important issues that have come to the fore recently. Studies and practitioners in tourism developed a better experience for tourists with new technologies such as data analytics, machine learning, and artificial

DOI: 10.4018/978-1-7998-6904-7.ch008

intelligence (AI) (Gretzel et al., 2015; Tussyadiah, 2020). The use of technology in tourism literature is generally discussed in terms of technological, managerial, and marketing perspectives. The technological perspective has focused on technical and computational developments in tourism, mostly on hardware infrastructure and algorithm-oriented developments. The managerial perspective is concerned with the changes that technologies created in tourism organizations and business processes. These changes reformed the competition due to the increase in the use of technology and the technological developments in the market. The common point in all perspectives is that new technologies destroy the old ones and thus force businesses to become innovative. However, one of the issues that have shown itself more recently with the inclusion of large companies in business processes is AI. Both the augmented use of mobile phones and the developments in AI make tourism more complex to handle processes. The traditional tourist who was obliged to travel with a camera, clock, folding map, tourist guide and who was looking for a payphone to communicate has left its place to the modern tourist who can do all these and more just with a mobile phone. On the other hand, self-guided tours have emerged that provide intelligent and comprehensive tour support by using IoT technologies. They can provide tourists with a choice of popular and theme routes so that users can plan their itineraries based on the recommendations (W. Wang et al., 2020). Besides, tourism is an industry that is quite open to information production and digitalization due to its intangible structure. Especially, by using smartphones, travelers can generate more information about their trip, accommodation, and transportation even while they are away. One of the biggest benefits of the mobile internet is that tourists can exchange information when they want to access the information they need during the trip. Also, with the integration of AI into mobile computing, tourists and tourism businesses have gained advantages in all processes of traveling. In the light of the aforementioned issues, in this book section, the literature review for research on mobile computing and AI in the field of tourism is systematically presented, and at the end of the chapter, a process framework for AI integration into mobile applications and some recommendations for future researches are presented.

TECHNOLOGICAL PERSPECTIVE

Studies on AI date back to ancient times and have become very popular in the information age. The "Bombe", which was worked on by the British physicist Alan Turing to solve the Enigma used by the German army in secret communication during the First World War, is the first example of this. The article "Computing Machines and Intelligence", published in 1950 and is still in use of the determination of the cognitive ability of systems, stated that machines could be smart when they

Figure 1. Perspectives for AI integration in mobile tourism services

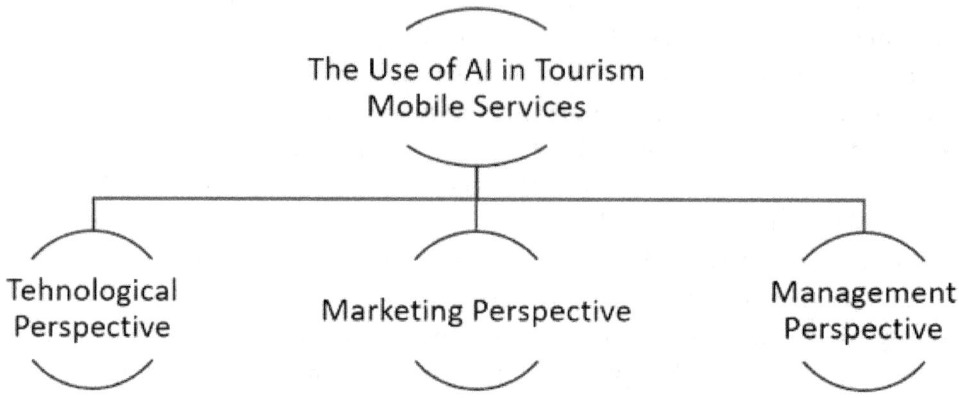

interact like humans. In 1956, Marvin Minsky and John McCarthy started to work, using the term AI (Haenlein & Kaplan, 2019), on the development of different prediction models to reach clearer and accurate results. Since then, studies in the field of information technology have increased considerably and have led to the emergence of AI techniques and big data analysis. Especially in the last decade, with the development of mobile computing and wireless internet networks, great progress has been made in the production and transfer of big data (Jordan & Mitchell, 2015). In the coming period, obtaining the big data needed to train AI models will be possible with the spread of mobile 5G technology that can provide huge data flows through basic daily life activities like emotional sharing, walking, eating, traveling, etc. (Pires et al., 2018; W. Wang et al., 2020). Besides, different models using econometrics and AI algorithms have emerged in the estimation of tourist numbers with the data obtained over the internet (X. Li et al., 2021). Machine learning algorithms can be seen as searching through candidate programs that best optimize the performance (Jordan & Mitchell, 2015). The most comment examples of generally used applications in the field of AI are (Fadlullah et al., 2017):

1. *Robotics*, having the basis of the mechanical functioning of robots in human form or non-human form by the integration of the AI, allows robots to train the data autonomously with machine learning algorithms in the physical environment.
2. *Object recognition applications*, as one of the most important applications, include AI models to distinguish the characteristic differences of objects and entities such as humans.

3. *Speech recognition*, like object recognition, can distinguish sound characteristics by analyzing them over frequencies.
4. *Natural language processing* includes analyzing the spoken or written language of communication using machine learning and deep learning algorithms through categorization and labeling and establishing relationships between them. While supervised data training can be used in this method, an unsupervised learning path can be chosen by using frequently available texts on the internet.
5. *AI games*, on the other hand, are applications based on algorithms such as deep learning networks and reinforcing learning that enables players to make decisions in an abstract set of rules with (supervised) or without (unsupervised) them.

Machine learning is used in the clustering, classification, and prediction of the data, especially in large enterprises. Clustering is an algorithm that can determine homogeneous objects and group patterns calculating their variances with unsupervised learning and it can be used in business operations. The classification, on the other hand, can categorize a new object with a supervised or unsupervised learning model that is saturated with a certain data set and can be used to personalize customers according to their characteristics. Finally, prediction is about estimating from data patterns the outcomes of future problems. The prediction is important in improving performance by forecasting possible developments or features in the market (Lee & Shin, 2020). Besides, regression algorithms using sequential and continuous data, and classification algorithms using unordered categorical data constitute the two approaches used in the AI models. These can be done by both supervised learning, where an algorithm maps inputs and outputs through a dataset, and unsupervised learning where patterns in data are determined by identifying commonalities and anomalies (Kakatkar et al., 2020). After machine learning models have been installed and deployed, they need to be constantly monitored and optimized for misleading. Machine learning algorithms become underfitting when there is high bias, but when the variance is high, they can be overfitting, which means they are very well fitted (Desouza et al., 2020). Therefore, learning for machines should be viewed with a holistic approach that includes pre and post-deployment.

Providing sufficient data for machine learning requires attracting people to use the mobile app, which therefore should be designed user-friendly and have functional quality. Studies conducted in the literature of mobile technologies in tourism generally examined the perception of the ease of use and the usefulness of the applications. However, the quality perceived by the user from the system affects the system functionality and the perceived ease of use together with the information quality predicts the adoption of the system (C.-C. Chen & Tsai, 2019). Tourists generally do not prefer unstable applications which are not integrated with the web environment, have incorrect run-time, and are not personalized (D. Wang, Xiang, & Fesenmaier, 2016). Particularly, the first-time adoption is determined by the transmission of the appropriate message (Calvignac & Smolinski, 2020). Further, the use of mobile phones by tourists during the trip is not only affected by personal thoughts or momentary events but rather they take and edit photos and videos that include more creative activities (D. Wang et al., 2014). Thus, detailed information about the travel patterns of tourists can be obtained by analyzing the experiences, photos, texts (even in different languages) visual / text processing, or c-in information using data mining techniques over mobile phones (Vu et al., 2018). Beyond the user-generated contents, mobile phones can provide enormous data about tourist's movements, feelings, and thoughts with technologies like accelerometers, gyroscopes, magnetometers, Bluetooth, Wi-Fi, microphones, proximity and light sensors, GPS, heart rate sensors. The data obtained from sensors of mobile and wearable devices can be used in the training and classification of AI and deep learning models (Nweke et al., 2018). For example, Park et al. (2020), as a result of their mobile big data analysis in South Korea, obtained findings that destinations can greatly benefit from these technologies. They used the orbital mining method to model travel patterns between different cities. Compared to other studies, this study was able to obtain clearer findings of tourists' travel behavior through mobile sensor data. Likewise, measurement and analysis using the GPS data offer opportunities to destinations, with lower costs, such as providing information about the locations, creating travel routes (Raun et al., 2016), investigating crowd levels (Lu et al., 2020), and determining the destination networks by using big data analytics techniques (Y. Xu et al., 2021). Besides, the data independence should be ensured by planning well the combination process of the data obtained from different environments (Reif & Schmücker, 2020; Vecchio et al., 2019). Bustamante et al. (2020), in their study, created a database automatically by collecting data from a social network, open map platform, and two accommodation services on a platform called BITOUR. Using this database and data mining method, they could have determined the relationships between the frequency of posting on social media by tourists in specific locations, the average length of stays, and the activities they participated in. Beyond that, cloud-based smart tourism systems include machine learning algorithms that integrate positioning data with image tags, allowing tourists to share past experiences through the map interface and make personal recommendations (Kontogianni & Alepis, 2019). Consequently, by creating labels in specific positions on the map, tourists can greatly affect the movements of other users (Högberg et al., 2020). Besides, viewing positioning data and storing personal data may cause some important problems related to security (Raun et al., 2016). But, Saravanan & Sadhu Ramakrishnan (2016) argue that travelers' data can be exchanged with the service provider using the hidden zone and cryptographic GPS data to prevent it from being passed on to third parties.

MARKETING PERSPECTIVE

In tourism marketing, mobile applications can serve as intermediate devices that give tourist information and attract (D. Wang, Xiang, Law, et al., 2016) or offer important solutions in improving the experiences of new tourists and create effective communication between tourists and the local people. Besides, visits with the use of mobile phones have greatly changed the traditional visit by speeding up, choking on information, and reducing the focus of the tourist by making fewer pauses and moving faster. They tend to learn more about the place visited, places to discover, and stores to experience (Calvignac & Smolinski, 2020). Recent research has provided empirical evidence that online reviews and suggestions made via mobile phones affect tourists on a large scale and provide reliable information (Ghaderi et al., 2019). In this way, hospitality businesses can collect personal and environmental data through numbers of downloads in different countries, keywords used, application activities, payment points, clicks, visit statistics, view times, forms, etc. (Morosan & DeFranco, 2016a; Pierdicca et al., 2019). Further, mobile applications should be designed with content that can attract the user with personalized context and sensitive advertisements to encourage the users' spending and increase the consumption tendency (Rivera et al., 2016) and integrate AI algorithms to do it autonomously and in a much productive way. The hospitality businesses, gathering data on customers' needs, develop predictive business models to provide qualitative goods and services (T. Davenport et al., 2020). They can also deliver novel experiences to their consumers using AI technologies such as facial recognition, virtual reality, chatbots, robots, maps, language translators, audio tours, etc. (Samala et al., 2020). To reach potential consumers effectively, it is important to provide qualitative information on travel, food & beverage, destinations, interactive maps, and promotions through mobile applications and mobile compatible websites (Stienmetz et al., 2013). Further, all these technologies require an infrastructure that enables tourism stakeholders to work harmonizingly and be in communication with each other (Buhalis, 2019).

The mobile phone can be seen as a very important channel for the distribution of services offered in the tourism and hospitality industry. It increases efficiency by reducing costs in marketing efforts by providing reliable and real-time quality information for hospitality businesses and destinations (No & Kim, 2014). Using mobile applications, hospitality businesses can differentiate themselves independently from intermediary companies. Applications including mobile c-in, room entrance, and many other services can create customer loyalty (Gibbs et al., 2016) with the support of AI technology. The AI models that learn from the reservation data obtained from mobile applications can give suggestions or warning notifications to users. Beyond that, mobile applications can offer many opportunities such as making reservations, self-c-in and c-out, notify users that the room is ready, and open the door of the room (M.-M. Chen et al., 2016). Further, these functions or creative solutions are very useful in providing more data that can be used by businesses to better know and direct tourists with AI models. For instance, the tourism advice systems can develop recommendations by analyzing travelers' interests and preferences. Using various AI techniques, they provide consultancy and support to tourists during the decision-making process (Borràs et al., 2014). They can increase engagement by offering rewards or with gamification by developing interaction designs that appeal to consumers' emotions and provide brand loyalty (Morosan & DeFranco, 2016b). Through gamification, the usage of the application can be increased at all stages of the travel with virtual rewards such as free tickets (F. Xu et al., 2016).

Mobile Application and AI Adoption

The acceptance and use of mobile applications is a subject that has been studied extensively in the literature. Researches have shown that the functional features of mobile applications should be developed, and especially the user-friendly interfaces of the map-related features significantly affect the behavioral intentions of the tourists (Y. C. Huang et al., 2019). The application interface design quality is the most important factor affecting the loyalty of tourist even more than the content and system quality (J. Y. Kim et al., 2019). To further enhance the tourist experience, advanced images must be well defined and graphical interfaces must be well designed (Borràs et al., 2014). Thus, destination marketers demand front-end developers that put the functions of AI in more visible and creative ways in their interface designs, and the use of AI in the selection of elements, measurements, and colors can provide great benefits. For example, in a mobile application designed for Kuala Lumpur, a one-time login feature has been introduced for users to view comfortably and to save time/effort (Yusoff et al., 2019). Besides, the differences in information between the designs of intermediary businesses (GDS, OTA, recommendation sites, reservation sites, etc.) and the hospitality businesses that provide services also affect the consumer (D. Wang, Xiang, Law, et al., 2016). Therefore, it is necessary to know well how, where and under what conditions these differences and information were obtained during the production and analysis of information. Mobile designs are very effective in creating value by using co-creation approaches with tourists. Algorithms that learn the consumption preferences of tourists can increase marketing communication with tourists (Morosan & DeFranco, 2016a). Thereby, businesses can create value by developing efficient communication with their users' through the created media content and instant messaging towards mobile technologies (Lei et al., 2020).

Mobile applications can autonomously create value by increasing the use intentions of personalized services for tourists (Morosan & DeFranco, 2016b) and providing such technologies as NFC (Liebana-Cabanillas et al., 2020). For this reason, this type of application should be followed closely by businesses and efficiently meet the needs of new tourists (Do et al., 2020). Likewise, wearable technologies and applications, connected to mobile phones, can provide information about the experiences and behaviors of tourists in smart destinations (Dorcic et al., 2019; Neuhofer et al., 2014). Increasing the use of applications requires technological infrastructure, so smart destinations can provide free internet access in all areas where tourists can visit (Falcao et al., 2019). Further, applications that personalize the experiences of tourists by using location services such as GPS, both facilitate the travel of tourists and provide information to businesses (Dorcic et al., 2019). Mobile positioning can provide continuous real-time data, so its reliability is higher. In their study, Saluveer et al. (2020) obtained tourism statistics across the country by tracking mobile positioning data and had the opportunity to obtain more detailed information such as the duration of the visits to certain places, statistics for transport use, and travel patterns. Although there were some obstacles such as legal regulations and the unwillingness of network operators due to privacy, they could show how important data can be produced by mobile technologies, especially for AI models. Further, businesses need to develop their applications based on trust and use (Falcao et al., 2019). The enjoyment, confidence, and ease of use of the application is an antecedent of the downloading behavior (Kwon et al., 2013). To build trust, hospitality businesses should create an environment where tourists can share personal and financial information, and ally with banks (Park & Huang, 2017).

Tourism Recommendation Systems

AI technologies produce knowledge reasoning and logic in the field of tourism. Tourism recommendation systems help tourists make personalized decisions about the places to visit. Mobile devices offer great advantages to tourists thanks to their portable screens and the ability to be used everywhere. Almost all of them connect to mobile tourism advice systems with positioning services such as GPS, allowing them to discover new places or plan travel and share the user's current location and recommendations or suggestions (Borràs et al., 2014). The importance of these systems in marketing is hidden in creating loyalty by developing personal recommendations to tourists, enriching their experiences, and increasing their satisfaction. Therefore, businesses and destinations need to integrate AI models into their mobile applications to stay strong in the market (Zheng et al., 2020). In tourism, big data can be obtained from user-generated contents (47%), devices (36%), user transactions (17%) data and can be used in tourist sentiment, behavior, tourism marketing, tourist recommendation analysis (J. Li et al., 2018). Moreover, ways of seeking information may also vary from culture to culture. For example, social media in Bangladesh and China, and content sites created by users in Australia are more preferred (Fardous et al., 2019). When it comes to tourism, cultural assets have a great impact on consumer decisions. The recommendation strategy of an application that defines cultural assets as objects and develops cultural tour suggestions for tourists consists of the following three stages (Su et al., 2019):

Stage 1: Before filtering (Developed ML algorithm creates clusters for cultural objects by learning information about the person),

Stage 2: Ranking (The clusters formed are ranked according to criteria such as interest and popularity),

Stage 3: After filtering (the result of the suggestion is presented by inferring according to the location of the user).

MANAGEMENT PERSPECTIVE

The world has become a place where much more can be done than counting data. Mobile computing can produce its data beyond getting it from or giving it to conventional users. Self-learning computers shed light on the world of the future. Since the tourism industry is under the intense pressure of information technologies due to its information-intensive structure; mobile phones and wearable devices, that can learn and make decisions, going to be the most important part of the industry. Further, tourism businesses that have to minimize their costs have started to discover robots and other AI technologies. Robotics and AI both provide the industry with a cost advantage and help the services to be performed more accurately. The hospitality industry offers a lot of opportunities for AI in mobile computing with its advanced operations and knowledge. The AI applications, which are also very efficient in providing customer satisfaction, have started to be studied by hospitality researchers and it is expected to be increased even more in the coming period (Gursoy, 2018). Studies in the literature of tourism and AI indicate that it provides efficiency and profitability in the industry, as it can provide customers with personalized and useful experiences. Due to its abstract structure, tourism has the potential to grow faster with AI (Samara et al., 2020). Since innovative businesses are preferred by new generation consumers, they can respond to them by obtaining useful information about their purchasing patterns (Morosan & DeFranco, 2016b). Thus, innovation managers or teams can reduce the complexity of the algorithms to parse and analyze unstructured data by effectively leveraging big data and AI in businesses. Because, AI can craft better innovation scenarios than humans by validating creative insights and minimizing creative blind spots (Kakatkar et al., 2020).

AI approaches to production are handled in two ways. Predictive AI includes advanced algorithms such as predicting future failure in more concrete production.

Normative AI, on the other hand, determines optimum decisions with statistical methods in operational processes such as supply chain decision support systems (Olsen & Tomlin, 2020). Hospitality businesses need to produce fast and effective solutions to visitor complaints to increase their service quality. Thus, in organizations with flexibility, such solutions can be achieved by developing alert or decision support systems through data mining, machine learning, and AI (Buhalis & Sinarta, 2019). AI-supported decision support systems based on photo recognition let destinations self-manageable (R. Wang et al., 2020) since mining big data and learning through appropriate models can be used for decision-making, forecasting, and problem-solving (Jordan & Mitchell, 2015). Furthermore, machine learning has improved technologically, helping hotels to both evaluate and improve their performance. Thanks to machine learning-based tools, the hospitality profession has become flexible and motivates the management to use it by helping to increase customer satisfaction and support the tasks of the employees (Parvez, 2020).

The use of AI in hospitality businesses can have different effects on employees. The fact that AI models automate many business processes can affect especially customer-facing employees. AI-based services and robotic applications have started to be used more in hospitality businesses due to the reduction of costs and efficiency by automating the works (Ivanov et al., 2019; J. (Justin) Li et al., 2019). The deployment of the AI technologies that can automate routine tasks that require less talent in hospitality businesses can cause a variety of problems, such as an increased intention to leave the job. Although AI applications are generally considered to pose a risk for employees, just as Jun Li et al. (2019) found that AI applications have a negative effect on employees' intention to leave their jobs. According to the authors, hotel employees may be aware of the inevitable change that new technologies create in business processes and may think it can be useful as it can help control things. Furthermore, hospitality businesses can regularly organize training programs to reduce the employee's turnover intention and to make them positive towards innovations or new technologies. The AI implementation reduces the dependency and participation of employees within organizations. Therefore, employees should be informed about the benefits of this technology and how it can change business processes (Braganza et al., 2020). Moreover, organizations can also benefit from AI in finding and recruiting their employees. Recruitment with the help of AI can be based on data and analysis rather than instinct and intuition. Besides, it can have the ability to recognize and distinguish many candidates faster than humans (van Esch & Black, 2019).

Service Automation

Tourists using mobile applications for businesses always stay connected during and after their trip (M.-M. Chen et al., 2016). Thus, technological compliance and clarity are important in making hotel reservations through mobile applications. Due to the financial power of large hotels, they can invest in the adoption of such technologies. For this reason, software companies mostly see large hospitality businesses as primary customers (Y. S. Wang et al., 2016). Hospitality businesses can use mobile phone data for autonomous services and strategic decision support systems. In this respect, applications that meet the work-related needs of employees and are easy to learn/use give more efficient results for the organization (García et al., 2019). Service delivery may need different intelligence in providing it with machines as a process involving employees and customers. These are - sequential among themselves in developing successful AI applications – mechanical which means learning and adapting at the lowest level; analytical which means learning and adapting systematically based on data; intuitive which learn and adapt by understanding; and empathetic which means learning and adapting based on experience (M. H. Huang & Rust, 2018). For example, museum managers can define the information, symbols, and pictures that visitors need, and share them instantly through sign applications (Carvajal-Trujillo et al., 2020).

AI produces large amounts of data for service production and evaluation with the help of sensors, and this data can be used in strategic decisions with techniques such as data mining (Olsen & Tomlin, 2020). By using different AI capabilities together, solutions can be produced in the service area. For example, models such as image recognition and NLP are included in the AI models so that robots can make decisions autonomously. Consequently, the guest can receive the order directly from the robot and benefit from the automated service (Blöcher & Alt, 2020). The data generated through these models can be analyzed and optimization can be made in business processes. Therefore, robotics and AI solutions create related changes in different dimensions in the ecosystem of hospitality businesses. These changes are generally manifested by the developments in servitization and industry 4.0. These two concepts contain many functions that complement each other (Frank et al., 2019). When AI applications are examined in terms of business capabilities, they support three important business activities (T. H. Davenport, 2018; T. H. Davenport & Ronanki, 2018):

1. Process automation (self-completion of the tasks through robotics),
2. Cognitive insight (creating smart solutions as a result of data analysis with machine learning),

3. Cognitive engagement (creating interactions with consumers and employees through natural language processing).

There are many issues that managers should pay attention to, especially in the integration of robotics (AI models that have mechanical abilities) with mobile computing. The use of machines in jobs where communication with guests requires empathy and creates changes in the form of service. In this respect, human-robot collaboration through mobile computing is a positive approach for the efficient execution of activities (Reis et al., 2020). For example, research conducted in a restaurant business has shown that business processes require mechanical intelligence and many of them can be met with AI. However, there are some areas where AI cannot affect emotions. Moreover, emotional tasks were encountered in 24% of the solutions (Blöcher & Alt, 2020). Therefore, human interaction needs to be considered when automating jobs. Because some consumers have a high anthropomorphic tendency, appropriate designs should be used to reach this audience. In this respect, anthropomorphism is important for understanding tourist responses in AI-supported service processes, where mobile phone users give more credit to the information suggested by AI (Martin et al., 2020). Besides, in the studies of Vatan and Dogan (2021), hotel employees worried about the robots stating that they might be nonemphatic, weak interactive, and less communicative, and might cause costs such as service errors and technical expenses due to operational and technical malfunctions.

Automated Environmental Management

The AI-supported mobile computing can also be used for disaster preparedness and response to the international crisis. Knowing the expectations, perceived benefits, and barriers from the use of such applications provide opportunities for governments as well as disaster managers (Aliperti & Cruz, 2020). Besides, data analytics offers potential scales for determining tourist behavior. It can produce smart solutions for inequality, food, and water security, health, wellbeing, and carbon emission issues, which are required for sustainable tourism management (F. Xu et al., 2020). Thus, machine learning models can be integrated into mobile applications to protect the environment and residents' health by preventing pollutant gas emissions from cars using GPS or speed sensor data (Vecchio et al., 2019). Further, for carrying capacity, which is one of the most important pillars of sustainable tourism management, AI can provide mathematical benefit maximization by using data such as visitor flow tracking to determine the number of visits (Camatti et al., 2020). Further, with the help of AI, virtual assistants can provide social feedback to change the behavior of tourists. By using such technologies, tourism can be regulated for its negative effects on the environment. For example, by using voice virtual assistants, behavior modifying incentives for the efficient use of resources by the visitor in hotel rooms can be suggested (Tussyadiah & Miller, 2019).

Social Distance Management and Contactless Services in Post-Pandemic Era

During the Covid-19 pandemic, the use of mobile data has increased. The number of interregional personal contact and duration time of people who stay in a specific place can be determined by mobile GPS tracks. Besides, the touristic places in the destination can be categorized and the contact density of mobile users with these places can be determined (Kishore et al., 2020). This situation is likely to continue after the pandemic as customers are accustomed to having contactless service for a long time during the pandemic period. For this reason, many hotel chains made investments to produce digital solutions such as robots, kiosks, face detection, mobile payment, and self-check-in / check-out to maintain their post-pandemic competitiveness (Hao et al., 2020). Using AI and deploying robots in touristic venues can be an effective way to reduce crowd and provide important measures such as social distance (Hou et al., 2021), and determine their directions and plans in difficult situations (Douglas et al., 2018). In the new normal after the pandemic era, IoT technologies supported with 5G networks can provide advanced solutions for tourism. With the synchronization of the health and travel data obtained through mobile phones, big data can be utilized in machine learning models for virus spread and contact prevention solutions. Besides, these data can be used to replace traditional service understanding with smart operations in the tourism industry. For example, by using AI algorithms in robots or drones, food and beverage delivery can be made without human contact and further, using the data obtained from IoT devices, social distance solutions in transportation can be developed (Abubakar et al., 2020).

SOLUTIONS AND RECOMMENDATIONS

Hospitality and travel businesses invest heavily in the area of AI, as its applications have begun to automate all industrial processes in terms of tourism marketing. Automation in tourism means rendering faster, error-free, and reliable services and having big data about customers. AI applications offer tourists the opportunity to receive useful information and recommendation before, during, and after travel. Theories in the field of mobile technologies and AI in tourism mostly focused on the adoption, booking preferences, self-c-in, market analysis, recommendation systems, etc. However, considering these studies, it is undoubtedly expected that AI-based models will replace many of the traditional business models. Besides, since tourists constantly use mobile phones, with the integration of AI algorithms, they may benefit from autonomous mobile services by providing information which in turn can be learned by AI models. The data required for machine learning can be obtained faster and simpler with mobile applications since they are easy and enjoyable for tourists to use. Mobile technologies, with their sensors and daily usage information, can produce big data for businesses to gain an advantage over their competitors. Mobile phones, as a metaphor, are remote controls of tourists for businesses to manage them through AI models. Because, these technologies can change the ideas and experiences of tourists before or during the trip (D. Wang, Xiang, & Fesenmaier, 2016). Moreover, these technologies can provide services such as contactless check-in / check-out, face recognition, electronic key, and cleaning robots (Seyitoğlu & Ivanov, 2020; Shin & Kang, 2020). Therefore, when taken into account that the customers' attitudes have changed in the pandemic era and expected to continue in the post-pandemic, hotels should use autonomous and robotic technologies that provide social distance to reduce customers' health preoccupations.

In managerial terms, AI can support employees as well as reduce processes in workplaces with the help of mobile technologies. At first glance, it seems to be taking away the jobs of the employees, in fact, according to the disruptive innovation theory, existing jobs are getting older and new jobs are formed. The integration of autonomous services and robots which are the inevitable future of hospitality businesses with mobile applications offers opportunities to businesses such as reducing their costs, providing accurate and quality services, ensuring customer satisfaction and loyalty, increasing revenues, etc. Managerially, the changes caused by the integration of AI into mobile tourism services can be discussed in two dimensions:

1. *By providing full automation (excluding humans)*: The AI models provide the service without being dependent on humans by using the data obtained through mobile applications. For example, a hotel robot bellboy can welcome the guest who made a reservation through the mobile phone and escort him/her to the room by asking if he/she needs any help.
2. *By supporting with semi-automation (including human)*: This provides information for employees to make decisions in the service process by using mobile data. For example, the AI algorithm can analyze the data pattern of the food and drinks information searched on the guest's mobile application, who comes to the hotel restaurant and recognized with the face at the entrance, and develop menu item recommendations on waiter's mobile device.

An integration decision is a very important and burdensome process that requires a large investment. A model proposal has been developed to clarify this process (Figure 2). Accordingly, the hospitality business should start by defining exactly the problem and determining the goals. Then, they should determine what data can be obtained through mobile technologies and what are the conditions and requirements. The next processes require a little more technical and tight programming. The AI model should be developed, trained, and tested for validity. The model, after passing the test, should be integrated into the mobile application with lifelong maintenance and updates. Qomariyah et al. (2020) developed an AI-supported mobile application in Indonesia that can be given as an example of this framework. Indonesia Online Tourism (SONIA) develops personalized tourist tour suggestions by providing sufficient information about the city or region. Due to the COVID-19 situation, people in Indonesia have stayed in their homes and international tourism movements have also decreased. In this "new normal", the idea of developing an AI-powered mobile application has emerged to safely increase tourist activity. Developers have identified the problem as "how can safe and clean places can be recommended to people?". Later, it was thought that user experience data on taste, interest, cleanliness, and security levels according to the demographic characteristics could be used over mobile or web applications to solve this problem. In the next stage, the requirements for the development of SONIA are determined. According to these requirements, a quick plan was made with the stakeholders and the application model was designed by creating a UML use case diagram. AI-based personal recommendation system, which also includes third-party systems, has been integrated into the application system as a complement. The AI model was designed to offer users recommendations on specific spots, food places, tour plans. For this, APARELL (active Pairwise Relation Learner) framework is combined with the decision support system to develop the best alternative recommendation for the user. After this stage, the application was deployed according to the prototyping method and developed by receiving feedback. The authors did not give detailed information about execution, but according to the model suggested in this book section, the AI model should be trained with data after this stage, it should be analyzed and tested to determine whether it gives the best recommendation (validation). After passing the test stage, the SONIA developers may deploy the application to tourists by making necessary investments to the infrastructure. Ultimately, the AI model integrated into SONIA requires continuous maintenance and updating.

FUTURE RESEARCH DIRECTIONS

This book section has been endeavored to discuss how tourism services and their consumption was changed with the integration of AI to mobile tourism services. Future studies can be conducted that examine the technical and managerial challenges in integrating AI models into mobile tourism services. Besides, the adoption and use

Figure 2. Processes for integrating AI to mobile tourism services

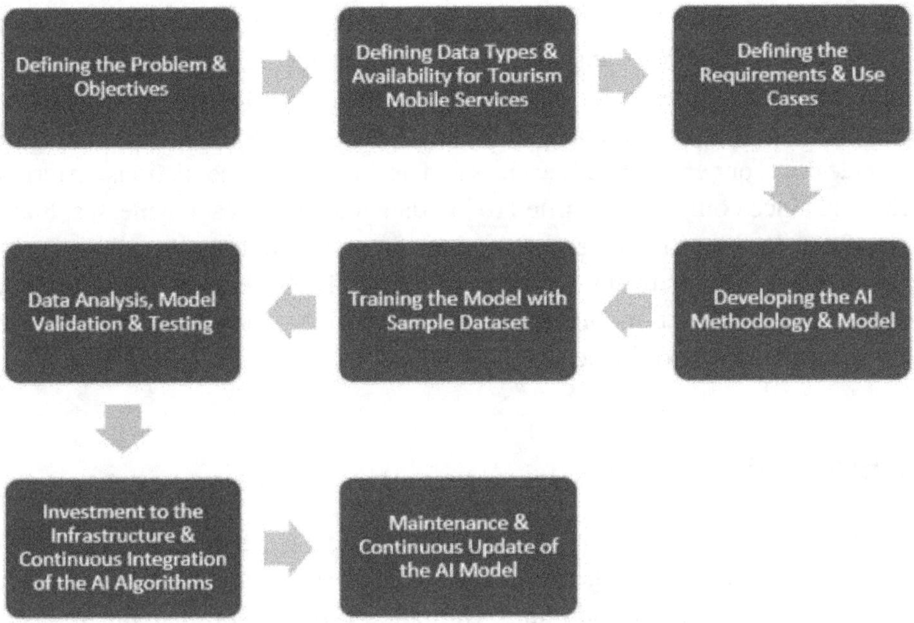

of AI in tourism enterprises are seen as an important knowledge gap. Further, the tourism service has phenomena that are difficult to measure due to its abstract and complex nature. It is possible to digitize these abstract structures with psychometric measurements and to analyze them using advanced statistical methods. Especially approaches such as NLP can provide useful solutions for tourist behavior. AI-powered solutions integrated into mobile computing to determine these behavioral patterns of tourists can provide a competitive advantage to businesses.

CONCLUSION

Tourism offers great opportunities for digitalization with its abstract and service-based structure. Many processes or products in this area can be transferred to the virtual environment through digitization. This reveals the importance of data use and analytics in the field of tourism which has been one of the industries most affected by mobile computing since there is a positive linear relationship between mobile internet use and international tourist numbers after a certain level of penetration (Adeola & Evans, 2019). The AI integrated applications can autonomously provide better information and services to tourists and hospitality businesses, as tourists use their smartphones in their whole travel stages. In the foreground, tourists benefit from many mobile applications such as making reservations, viewing maps, searching for information on the internet, messaging, sharing experiences on social media, while, in the background, transactions, network connections, sensors, and sharing data is continuously produced. These data provide can support AI models to learn supervised and unsupervised or with the combination of both (J. Kim et al., 2003) to provide novel or personalized services and develop recommendations. In terms of marketing, since competition shifted to big data and analytics, businesses, that are aware of this, have seriously started to collect data from tourists, especially through mobile devices. That is to say, to succeed, hospitality businesses should integrate AI applications into their mobile applications, enriching tourist's continuous, well-predicted and instant experiences.

REFERENCES

Abubakar, A. I., Omeke, K. G., Ozturk, M., Hussain, S., & Imran, M. A. (2020). The Role of Artificial Intelligence Driven 5G Networks in COVID-19 Outbreak: Opportunities, Challenges, and Future Outlook. *Frontiers in Communications and Networks*, *1*, 575065. Advance online publication. doi:10.3389/frcmn.2020.575065

Adeola, O., & Evans, O. (2019). Digital tourism: Mobile phones, internet and tourism in Africa. *Tourism Recreation Research*, *44*(2), 190–202. doi:10.1080/02508281.2018.1562662

Aliperti, G., & Cruz, A. M. (2020). Promoting built-for-disaster-purpose mobile applications: An interdisciplinary literature review to increase their penetration rate among tourists. *Journal of Hospitality and Tourism Management*, *44*, 193–210. doi:10.1016/j.jhtm.2020.06.006

Blöcher, K., & Alt, R. (2020). AI and robotics in the European restaurant sector: Assessing potentials for process innovation in a high-contact service industry. *Electronic Markets*. Advance online publication. doi:10.100712525-020-00443-2

Borràs, J., Moreno, A., & Valls, A. (2014). Intelligent tourism recommender systems: A survey. *Expert Systems with Applications*, *41*(16), 7370–7389. doi:10.1016/j.eswa.2014.06.007

Braganza, A., Chen, W., Canhoto, A., & Sap, S. (2020). Productive employment and decent work: The impact of AI adoption on psychological contracts, job engagement and employee trust. *Journal of Business Research*. Advance online publication. doi:10.1016/j.jbusres.2020.08.018 PMID:32836565

Buhalis, D. (2019). Technology in tourism-from information communication technologies to eTourism and smart tourism towards ambient intelligence tourism: A perspective article. *Tourism Review*, *75*(1), 267–272. doi:10.1108/TR-06-2019-0258

Buhalis, D., & Sinarta, Y. (2019). Real-time co-creation and nowness service: Lessons from tourism and hospitality. *Journal of Travel & Tourism Marketing*, *36*(5), 563–582. doi:10.1080/10548408.2019.1592059

Bustamante, A., Sebastia, L., & Onaindia, E. (2020). BITOUR: A Business Intelligence Platform for Tourism Analysis. *ISPRS International Journal of Geo-Information*, *9*(11), 671. doi:10.3390/ijgi9110671

Calvignac, C., & Smolinski, J. (2020). How can the use of a mobile application change the course of a sightseeing tour? A question of pace, gaze and information processing. *Tourist Studies*, *20*(1), 49–74. doi:10.1177/1468797619885955

Camatti, N., Bertocchi, D., Carić, H., & van der Borg, J. (2020). A digital response system to mitigate overtourism. The case of Dubrovnik. *Journal of Travel & Tourism Marketing*, *37*(8–9), 887–901. doi:10.1080/10548408.2020.1828230

Carvajal-Trujillo, E., Molinillo, S., & Liébana-Cabanillas, F. (2020). Determinants and risks of intentions to use mobile applications in museums: An application of fsQCA. *Current Issues in Tourism*, 1–20. doi:10.1080/13683500.2020.1780200

Chen, C.-C., & Tsai, J.-L. (2019). Determinants of behavioral intention to use the Personalized Location-based Mobile Tourism Application: An empirical study by integrating TAM with ISSM. *Future Generation Computer Systems*, *96*, 628–638. doi:10.1016/j.future.2017.02.028

Chen, M.-M., Murphy, H. C., & Knecht, S. (2016). An Importance Performance Analysis of smartphone applications for hotel chains. *Journal of Hospitality and Tourism Management*, *29*, 69–79. doi:10.1016/j.jhtm.2016.05.001

Davenport, T., Guha, A., Grewal, D., & Bressgott, T. (2020). How artificial intelligence will change the future of marketing. *Journal of the Academy of Marketing Science*, *48*(1), 24–42. doi:10.100711747-019-00696-0

Davenport, T. H. (2018). *The AI Advantage How to Put the Artificial Intelligence Revolution to Work*. MIT Press. doi:10.7551/mitpress/11781.001.0001

Davenport, T. H., & Ronanki, R. (2018). Artificial intelligence for the real world. *Harvard Business Review*, (January-February), 2–10.

Del Vecchio, P., Secundo, G., Maruccia, Y., & Passiante, G. (2019). A system dynamic approach for the smart mobility of people: Implications in the age of big data. *Technological Forecasting and Social Change*, *149*, 119771. doi:10.1016/j.techfore.2019.119771

Desouza, K. C., Dawson, G. S., & Chenok, D. (2020). Designing, developing, and deploying artificial intelligence systems: Lessons from and for the public sector. *Business Horizons*, *63*(2), 205–213. doi:10.1016/j.bushor.2019.11.004

Do, H. N., Shih, W., & Ha, Q. A. (2020). Effects of mobile augmented reality apps on impulse buying behavior: An investigation in the tourism field. *Heliyon*, *6*(8), e04667. Advance online publication. doi:10.1016/j.heliyon.2020.e04667 PMID:32923709

Dorcic, J., Komsic, J., & Markovic, S. (2019). Mobile technologies and applications towards smart tourism – state of the art. *Tourism Review*, *74*(1), 82–103. doi:10.1108/TR-07-2017-0121

Douglas, A., Lubbe, B., & van Rooyen, A. (2018). Business travellers' use of mobile travel applications: A generational analysis. *Information Technology & Tourism*, *18*(1–4), 113–132. doi:10.100740558-017-0103-6

Fadlullah, Z. M., Tang, F., Mao, B., Kato, N., Akashi, O., Inoue, T., & Mizutani, K. (2017). State-of-the-Art Deep Learning: Evolving Machine Intelligence Toward Tomorrow's Intelligent Network Traffic Control Systems. *IEEE Communications Surveys and Tutorials*, *19*(4), 2432–2455. doi:10.1109/COMST.2017.2707140

Falcao, R. P. Q., Ferreira, J. B., & Carrazedo Marques da Costa Filho, M. (2019). The influence of ubiquitous connectivity, trust, personality and generational effects on mobile tourism purchases. *Information Technology & Tourism*, *21*(4), 483–514. doi:10.100740558-019-00154-1

Fardous, J., Du, J. T., Hansen, P., Choo, K.-K. R., & Huang, S. (Sam). (. (2019). Group trip planning and information seeking behaviours by mobile social media users: A study of tourists in Australia, Bangladesh and China. *Journal of Information Science*. Advance online publication. doi:10.1177/0165551519890515

Frank, A. G., Mendes, G. H. S., Ayala, N. F., & Ghezzi, A. (2019). Servitization and Industry 4.0 convergence in the digital transformation of product firms: A business model innovation perspective. *Technological Forecasting and Social Change*, *141*, 341–351. doi:10.1016/j.techfore.2019.01.014

García, M. V., Blasco López, M. F., & Sastre Castillo, M. Á. (2019). Determinants of the acceptance of mobile learning as an element of human capital training in organisations. *Technological Forecasting and Social Change*, *149*, 119783. Advance online publication. doi:10.1016/j.techfore.2019.119783

Ghaderi, Z., Hatamifar, P., & Ghahramani, L. (2019). How smartphones enhance local tourism experiences? *Asia Pacific Journal of Tourism Research*, *24*(8), 778–788. doi:10.1080/10941665.2019.1630456

Gibbs, C., Gretzel, U., & Saltzman, J. (2016). An experience-based taxonomy of branded hotel mobile application features. *Information Technology & Tourism*, *16*(2), 175–199. doi:10.100740558-016-0052-5

Gretzel, U., Sigala, M., Xiang, Z., & Koo, C. (2015). Smart tourism: Foundations and developments. *Electronic Markets*, *25*(3), 179–188. doi:10.100712525-015-0196-8

Gursoy, D. (2018). Future of hospitality marketing and management research. *Tourism Management Perspectives*, *25*, 185–188. doi:10.1016/j.tmp.2017.11.008

Haenlein, M., & Kaplan, A. (2019). A brief history of artificial intelligence: On the past, present, and future of artificial intelligence. *California Management Review*, *61*(4), 5–14. doi:10.1177/0008125619864925

Hao, F., Xiao, Q., & Chon, K. (2020). COVID-19 and China's Hotel Industry: Impacts, a Disaster Management Framework, and Post-Pandemic Agenda. *International Journal of Hospitality Management*, *90*, 102636. doi:10.1016/j.ijhm.2020.102636 PMID:32834356

Högberg, J., Wästlund, E., Aas, T. H., Hjemdahl, K., & Nordgård, D. (2020). Herding the Hordes: Using Location-Based Services and Mobile Messaging to Affect Visitor Behavior. *Journal of Hospitality & Tourism Research (Washington, D.C.)*, *44*(5), 870–878. doi:10.1177/1096348020912449

Hou, Y., Zhang, K., & Li, G. (2021). Service robots or human staff: How social crowding shapes tourist preferences. *Tourism Management*, *83*, 104242. Advance online publication. doi:10.1016/j.tourman.2020.104242

Huang, M. H., & Rust, R. T. (2018). Artificial Intelligence in Service. *Journal of Service Research*, *21*(2), 155–172. doi:10.1177/1094670517752459

Huang, Y. C., Chang, L. L., Yu, C. P., & Chen, J. (2019). Examining an extended technology acceptance model with experience construct on hotel consumers' adoption of mobile applications. *Journal of Hospitality Marketing & Management*, *28*(8), 957–980. doi:10.1080/19368623.2019.1580172

Ivanov, S., Gretzel, U., Berezina, K., Sigala, M., & Webster, C. (2019). Progress on robotics in hospitality and tourism: a review of the literature. *Journal of Hospitality and Tourism Technology, 10*(4). doi:10.1108/JHTT-08-2018-0087

Jordan, M. I., & Mitchell, T. M. (2015). Machine learning: Trends, perspectives, and prospects. *Science, 349*(6245), 255–260. doi:10.1126cience.aaa8415 PMID:26185243

Kakatkar, C., Bilgram, V., & Füller, J. (2020). Innovation analytics: Leveraging artificial intelligence in the innovation process. *Business Horizons, 63*(2), 171–181. doi:10.1016/j.bushor.2019.10.006

Kim, J., Wei, S., & Ruys, H. (2003). Segmenting the market of West Australian senior tourists using an artificial neural network. *Tourism Management, 24*(1), 25–34. doi:10.1016/S0261-5177(02)00050-X

Kim, J. Y., Chung, N., & Ahn, K. M. (2019). The impact of mobile tour information services on destination travel intention. *Information Development, 35*(1), 107–120. doi:10.1177/0266666917730437

Kishore, N., Kiang, M. V., Engø-Monsen, K., Vembar, N., Schroeder, A., Balsari, S., & Buckee, C. O. (2020). Measuring mobility to monitor travel and physical distancing interventions: A common framework for mobile phone data analysis. *The Lancet. Digital Health, 2*(11), e622–e628. doi:10.1016/S2589-7500(20)30193-X PMID:32905027

Kontogianni, A., & Alepis, E. (2019). Moments of interest: A novel cloud-based crowdsourcing application enhancing smart tourism recommendations. In *11th Computer Science and Electronic Engineering Conference (CEEC)* (Vol. 11, pp. 144-149). Colchester, UK: University of Essex. 10.1109/CEEC47804.2019.8974337

Kwon, J. M., & Bae, J. (2013). Mobile applications in the hospitality industry. *Journal of Hospitality and Tourism Technology, 4*(1), 81–92. doi:10.1108/17579881311302365

Lee, I., & Shin, Y. J. (2020). Machine learning for enterprises: Applications, algorithm selection, and challenges. *Business Horizons, 63*(2), 157–170. doi:10.1016/j.bushor.2019.10.005

Lei, S. I., Ye, S., Wang, D., & Law, R. (2020). Engaging Customers in Value Co-Creation Through Mobile Instant Messaging in the Tourism and Hospitality Industry. *Journal of Hospitality & Tourism Research (Washington, D.C.), 44*(2), 229–251. doi:10.1177/1096348019893066

Li, J., Bonn, M. A., & Ye, B. H. (2019). Hotel employee's artificial intelligence and robotics awareness and its impact on turnover intention: The moderating roles of perceived organizational support and competitive psychological climate. *Tourism Management*, *73*, 172–181. doi:10.1016/j.tourman.2019.02.006

Li, J., Xu, L., Tang, L., Wang, S., & Li, L. (2018). Big data in tourism research: A literature review. *Tourism Management*, *68*, 301–323. doi:10.1016/j.tourman.2018.03.009

Li, X., Law, R., Xie, G., & Wang, S. (2021). Review of tourism forecasting research with internet data. *Tourism Management*, *83*, 104245. Advance online publication. doi:10.1016/j.tourman.2020.104245

Liebana-Cabanillas, F., Carvajal-Trujillo, E., Villarejo-Ramos, Á. F., & Higueras-Castillo, E. (2020). Antecedents of the intention to use NFC mobile applications in tourism. *Journal of Hospitality and Tourism Technology*, *11*(2), 369–383. doi:10.1108/JHTT-03-2019-0048

Lu, Y., Yuan, F., Lin, J., & Yuan, K. (2020). TouristGo: A location-based mobile game to improve tourist experience by visiting path optimisation. *Personal and Ubiquitous Computing*, *24*(3), 405–418. doi:10.100700779-019-01327-9

Martin, B. A. S., Jin, H. S., Wang, D., Nguyen, H., Zhan, K., & Wang, Y. X. (2020). The influence of consumer anthropomorphism on attitudes towards artificial intelligence trip advisors. *Journal of Hospitality and Tourism Management*, *44*, 108–111. doi:10.1016/j.jhtm.2020.06.004

Morosan, C., & DeFranco, A. (2016a). Co-creating value in hotels using mobile devices: A conceptual model with empirical validation. *International Journal of Hospitality Management*, *52*, 131–142. doi:10.1016/j.ijhm.2015.10.004

Morosan, C., & DeFranco, A. (2016b). Modeling guests' intentions to use mobile apps in hotels: The roles of personalization, privacy, and involvement. *International Journal of Contemporary Hospitality Management*, *28*(9), 1968–1991. doi:10.1108/IJCHM-07-2015-0349

Neuhofer, B., Buhalis, D., & Ladkin, A. (2014). A Typology of Technology-Enhanced Tourism Experiences. *International Journal of Tourism Research*, *16*(4), 340–350. doi:10.1002/jtr.1958

No, E., & Kim, J. K. (2014). Determinants of the adoption for travel information on smartphone. *International Journal of Tourism Research*, *16*(6), 534–545. doi:10.1002/jtr.1945

Nweke, H. F., Teh, Y. W., Al-garadi, M. A., & Alo, U. R. (2018). Deep learning algorithms for human activity recognition using mobile and wearable sensor networks: State of the art and research challenges. *Expert Systems with Applications*, *105*, 233–261. doi:10.1016/j.eswa.2018.03.056

Olsen, T. L., & Tomlin, B. (2020). Industry 4.0: Opportunities and challenges for operations management. *Manufacturing & Service Operations Management*, *22*(1), 113–122. doi:10.1287/msom.2019.0796

Park, S., & Huang, Y. (2017). Motivators and inhibitors in booking a hotel via smartphones. *International Journal of Contemporary Hospitality Management*, *29*(1), 161–178. doi:10.1108/IJCHM-03-2015-0103

Park, S., Xu, Y., Jiang, L., Chen, Z., & Huang, S. (2020). Spatial structures of tourism destinations: A trajectory data mining approach leveraging mobile big data. *Annals of Tourism Research*, *84*, 102973. Advance online publication. doi:10.1016/j.annals.2020.102973

Parvez, M. O. (2020). Use of machine learning technology for tourist and organizational services: high-tech innovation in the hospitality industry. *Journal of Tourism Futures*. doi:10.1108/JTF-09-2019-0083

Pierdicca, R., Paolanti, M., & Frontoni, E. (2019). eTourism: ICT and its role for tourism management. *Journal of Hospitality and Tourism Technology*, *10*(1), 90–106. doi:10.1108/JHTT-07-2017-0043

Pires, I. M., Garcia, N. M., Pombo, N., Flórez-Revuelta, F., Spinsante, S., & Teixeira, M. C. (2018). Identification of activities of daily living through data fusion on motion and magnetic sensors embedded on mobile devices. *Pervasive and Mobile Computing*, *47*, 78–93. doi:10.1016/j.pmcj.2018.05.005

Qomariyah, N. N., Sari, S. A., & Fajar, A. N. (2020). Sonia: An integrated indonesia online tourism system in new normal era. *International Journal of Innovative Computing, Information, & Control*, *16*(6), 1829–1843. doi:10.24507/ijicic.16.06.1829

Raun, J., Ahas, R., & Tiru, M. (2016). Measuring tourism destinations using mobile tracking data. *Tourism Management*, *57*, 202–212. doi:10.1016/j.tourman.2016.06.006

Reif, J., & Schmücker, D. (2020). Exploring new ways of visitor tracking using big data sources: Opportunities and limits of passive mobile data for tourism. *Journal of Destination Marketing & Management*, *18*, 100481. Advance online publication. doi:10.1016/j.jdmm.2020.100481

Reis, J., Melão, N., Salvadorinho, J., Soares, B., & Rosete, A. (2020). Service robots in the hospitality industry: The case of Henn-na hotel, Japan. *Technology in Society, 63*, 101423. Advance online publication. doi:10.1016/j.techsoc.2020.101423

Rivera, M., Croes, R., & Zhong, Y. Y. (2016). Developing mobile services: A look at first-time and repeat visitors in a small island destination. *International Journal of Contemporary Hospitality Management, 28*(12), 2721–2747. doi:10.1108/IJCHM-02-2015-0052

Saluveer, E., Raun, J., Tiru, M., Altin, L., Kroon, J., Snitsarenko, T., Aasa, A., & Silm, S. (2020). Methodological framework for producing national tourism statistics from mobile positioning data. *Annals of Tourism Research, 81*, 102895. Advance online publication. doi:10.1016/j.annals.2020.102895

Samala, N., Katkam, B. S., Bellamkonda, R. S., & Rodriguez, R. V. (2020). Impact of AI and robotics in the tourism sector: a critical insight. *Journal of Tourism Futures*. doi:10.1108/JTF-07-2019-0065

Samara, D., Magnisalis, I., & Peristeras, V. (2020). Artificial intelligence and big data in tourism: A systematic literature review. *Journal of Hospitality and Tourism Technology, 11*(2), 343–367. doi:10.1108/JHTT-12-2018-0118

Saravanan, S., & Sadhu Ramakrishnan, B. (2016). Preserving privacy in the context of location based services through location hider in mobile-tourism. *Information Technology & Tourism, 16*(2), 229–248. doi:10.100740558-016-0056-1

Seyitoğlu, F., & Ivanov, S. (2020). Service robots as a tool for physical distancing in tourism. *Current Issues in Tourism, 1–4*. Advance online publication. doi:10.1080/13683500.2020.1774518

Shin, H., & Kang, J. (2020). Reducing perceived health risk to attract hotel customers in the COVID-19 pandemic era: Focused on technology innovation for social distancing and cleanliness. *International Journal of Hospitality Management, 91*, 102664. doi:10.1016/j.ijhm.2020.102664 PMID:32921871

Stienmetz, J. L., Levy, S. E., & Boo, S. (2013). Factors Influencing the Usability of Mobile Destination Management Organization Websites. *Journal of Travel Research, 52*(4), 453–464. doi:10.1177/0047287512467698

Su, X., Sperli, G., Moscato, V., Picariello, A., Esposito, C., & Choi, C. (2019). An Edge Intelligence Empowered Recommender System Enabling Cultural Heritage Applications. *IEEE Transactions on Industrial Informatics, 15*(7), 4266–4275. doi:10.1109/TII.2019.2908056

Tussyadiah, I. (2020). A review of research into automation in tourism: Launching the Annals of Tourism Research Curated Collection on Artificial Intelligence and Robotics in Tourism. *Annals of Tourism Research*, *81*, 102883. Advance online publication. doi:10.1016/j.annals.2020.102883

Tussyadiah, I., & Miller, G. (2019). Nudged by a robot: Responses to agency and feedback. *Annals of Tourism Research*, *78*, 102752. Advance online publication. doi:10.1016/j.annals.2019.102752

van Esch, P., & Black, J. S. (2019). Factors that influence new generation candidates to engage with and complete digital, AI-enabled recruiting. *Business Horizons*, *62*(6), 729–739. doi:10.1016/j.bushor.2019.07.004

Vatan, A., & Dogan, S. (2021). What do hotel employees think about service robots? A qualitative study in Turkey. *Tourism Management Perspectives*, *37*, 100775. Advance online publication. doi:10.1016/j.tmp.2020.100775

Vu, H. Q., Li, G., Law, R., & Zhang, Y. (2018). Tourist Activity Analysis by Leveraging Mobile Social Media Data. *Journal of Travel Research*, *57*(7), 883–898. doi:10.1177/0047287517722232

Wang, D., Xiang, Z., & Fesenmaier, D. R. (2014). Adapting to the mobile world: A model of smartphone use. *Annals of Tourism Research*, *48*, 11–26. doi:10.1016/j.annals.2014.04.008

Wang, D., Xiang, Z., & Fesenmaier, D. R. (2016). Smartphone Use in Everyday Life and Travel. *Journal of Travel Research*, *55*(1), 52–63. doi:10.1177/0047287514535847

Wang, D., Xiang, Z., Law, R., & Ki, T. P. (2016). Assessing Hotel-Related Smartphone Apps Using Online Reviews. *Journal of Hospitality Marketing & Management*, *25*(3), 291–313. doi:10.1080/19368623.2015.1012282

Wang, R., Luo, J., & Huang, S. (2020). Developing an artificial intelligence framework for online destination image photos identification. *Journal of Destination Marketing & Management*, *18*, 100512. Advance online publication. doi:10.1016/j.jdmm.2020.100512

Wang, W., Kumar, N., Chen, J., Gong, Z., Kong, X., Wei, W., & Gao, H. (2020). Realizing the Potential of Internet of Things for Smart Tourism with 5G and AI. *IEEE Network*, *34*(6), 295–301. doi:10.1109/MNET.011.2000250

Wang, Y. S., Li, H. T., Li, C. R., & Zhang, D. Z. (2016). Factors affecting hotels' adoption of mobile reservation systems: A technology-organization-environment framework. *Tourism Management*, *53*, 163–172. doi:10.1016/j.tourman.2015.09.021

Xu, F., Nash, N., & Whitmarsh, L. (2020). Big data or small data? A methodological review of sustainable tourism. *Journal of Sustainable Tourism*, *28*(2), 144–163. doi:10.1080/09669582.2019.1631318

Xu, F., Tian, F., Buhalis, D., Weber, J., & Zhang, H. (2016). Tourists as Mobile Gamers: Gamification for Tourism Marketing. *Journal of Travel & Tourism Marketing*, *33*(8), 1124–1142. doi:10.1080/10548408.2015.1093999

Xu, Y., Li, J., Belyi, A., & Park, S. (2021). Characterizing destination networks through mobility traces of international tourists— A case study using a nationwide mobile positioning dataset. *Tourism Management*, *82*, 104195. Advance online publication. doi:10.1016/j.tourman.2020.104195

Yusoff, N. H. M., Isvaramurty, A., & Razalli, H. (2019). Mobile application for tourist's personal travelling management in Kuala Lumpur. In *IEEE 9th International Conference on System Engineering and Technology (ICSET)* (Vol. 9, pp. 354-359). Shah Alam, Malaysia: Institut Teknologi Bandung. 10.1109/ICSEngT.2019.8906443

Zheng, W., Liao, Z., & Lin, Z. (2020). Navigating through the complex transport system: A heuristic approach for city tourism recommendation. *Tourism Management*, *81*, 104162. Advance online publication. doi:10.1016/j.tourman.2020.104162

KEY TERMS AND DEFINITIONS

Anthropomorphism: The interpretation of nonhuman things or events in terms of human characteristics.

Application Programming Interface (API): A set of standardized requests that allow different computer programs to communicate with each other.

Decision Support System: A computerized program used to support determinations, judgments, and courses of action in an organization or a business.

Machine Learning: The discipline concerned with the implementation of computer software that can learn autonomously.

Natural Language Processing (NLP): The analysis of digitally recorded natural-language information from the semantic viewpoint is a matter of considerable complexity.

Robotics: The design, construction, and use of machines (robots) to perform tasks done traditionally by human beings.

Supervised Learning: The use of labeled datasets to train algorithms that to classify data or predict outcomes accurately.

Use Case Diagram: The primary form of system/software requirements for a new software program underdeveloped.

Chapter 9
A Conceptual Model of Emerging Mobile Travel Apps for Smart Tourism Among Gen X, Gen Y, and Gen Z

Phoebe Yueng Hee Sia
https://orcid.org/0000-0003-3796-4244
Graduate School of Business, Universiti Sains Malaysia, Malaysia

Yulita Hanum P. Iskandar
https://orcid.org/0000-0002-8037-5800
Graduate School of Business, Universiti Sains Malaysia, Malaysia

Siti Salina Saidin
Faculty of Hospitality, Tourism, and Wellness, Universiti Malaysia Kelantan, Malaysia

ABSTRACT

Mobile travel apps are essential tools in trip planning; they provide local insights and recommendation on destinations. Smart tourism features the extensive use of information and communication technology (ICT) which is a new evolution of old-style tourism and e-tourism, emphasised on two approaches: augmented reality (AR) and big data (BD). Several tourism studies have discussed the positive and negative impacts of adopting smart mobile travel apps in the tourism industry. Different factors may affect the app's adoption and acceptance of new technology. However, the level of adoption of smart mobile travel apps depends on the traveller's characteristics as each generation has different characteristics in the adaptability of smart technology. Therefore, this research model is based on the integration of the DeLone and McLean IS success (IS) model and consumer acceptance and use of information technology (UTAUT2) model to determine the factors influencing behavioural intention to use mobile travel apps for smart tourism among Generations X, Y, and Z.

DOI: 10.4018/978-1-7998-6904-7.ch009

INTRODUCTION

Travelling has a great impact on the evolution of our society and species. Less than a century old, tourism has become a worldwide phenomenon. After 60 years, the travel industry has become one of the biggest industries worldwide (Shoutem, 2016). Mobile travel apps are essential tools in trip planning and gaining local insights and recommendation on destinations. It ensures a smoother trip and spares travellers extensive long-term planning itineraries from booking flights to navigating a new destination. It provides a one-stop solution for users to perform travelling-related activities such as booking tickets, booking hotels, car rentals, insurance, restaurants and interesting places to visit; all at a single place without the need to go through different portals.

The advancement of the internet and technology has evolved the smart tourism industry. Recently, the smart tourism industry put a lot of efforts in two areas: Augmented Reality (AR) and Big Data (BD) (Pradhan, Oh, & Lee, 2018). AR augments the surroundings of the real world with digital objects or information, thereby helping the traveller acquire information throughout the experience. BD is a data-driven approach which is important in tourism applications for improving operations, provide better services, create personalized marketing campaigns based on specific user preferences and, ultimately increase better decision making and creates value for stakeholders (Mariani, Baggio, Fuhs, & Hoepken, 2018). BD analytic is accustomed to auto-provide appropriate suggestions for supporting the decisions of travellers.

Smart tourism has the potential to suggest more appropriate information in BD with more privacy of information disclosed by the traveller, however, it increases the risk of privacy and personal information (Pradhan, Oh, & Lee, 2018) accessed irresponsibly for inappropriate purposes. The level adoption of smart apps may depend on traveller's characteristics. As each generation has different characteristics on the adaptability of smart technology, the factors affect the adoption and acceptance of new technology may differ. Therefore, the main objective of this research is to determine the factors that influence the behavioural intention to use mobile travel apps for smart tourism among Generation X, Y and Z.

BACKGROUND

The emergence of mobile apps for smart tourism has raised the trends of personalization and privacy paradox (Litsa, 2018). In terms of convenience, digital media usage favours mobility and it leads to the rapid development of mobile apps. From app download statistics *(Figure 1)*, 197 billion in 2017 and 149 billion in 2016 have been

Figure 1. Number of Mobile App downloads worldwide

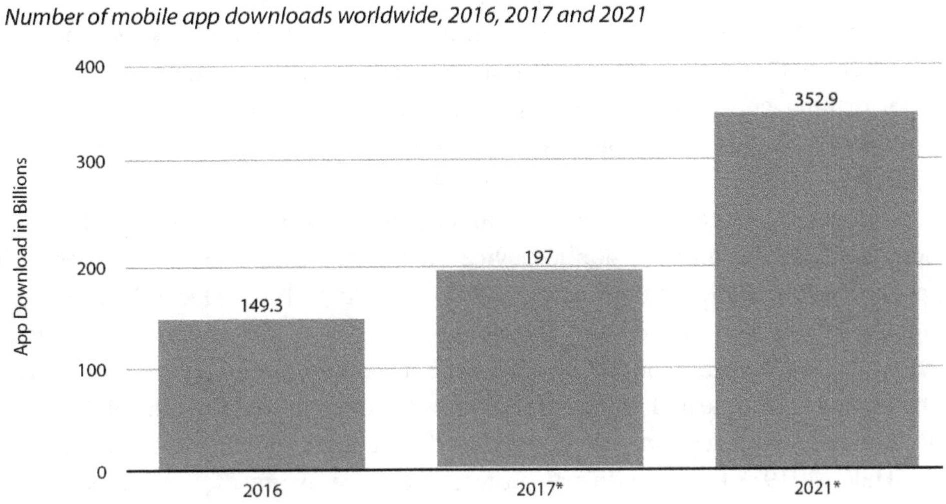

downloaded respectively and it was projected that this will increase to 352 billion by the year 2021 (Dogtiev, 2019). Thus, mobile apps are essential promotion tools for the company to advertise their services or products.

From another statistic, the estimated worldwide mobile apps users are 1.2 billion at the end of 2012 while it was expected to reach 4.4 billion by the end of 2017 (Kuan, Ann, Mohd Badri, & Tang, 2014).

The travel marketplace is growing in popularity and there are many opportunities available in this segment. About $10B has been invested in travel technology and it was predicted will generate $1091B revenue in the travel marketplace by the year 2022 (Sharma, 2018). A consolidated view for the travel market was carried out by Travelport Digital in 2019. The survey was compiled from the insights of 100+ professionals and 1,100+ travelers across the world (MacHale, 2019). From the survey, 92% of travel brands mentioned that digital strategy is very crucial for their organisational success in the future. Travelers spend 90% of their time on mobile apps compared to only 10% of the time on mobile web browsing. 58% of them use mobile apps to search for flights and 53% use mobile apps for finding accommodation according to a global study by Travelport Digital (MacHale, 2019). Mobile phones are now an indispensable part of the travel experience where 98% of travelers bring their smartphone with them throughout their journey and they have high expectations for the apps (MacHale, 2019). They expect lightning speed, on-demand information, voice or visual search, etc. 62% of Generation Z prefer visual search capabilities over other types of technology and about 50% of searches will be via voice search by 2020 (MacHale, 2019).

By embedding AI and machine learning features in the apps, users can search for a flight with a screenshot of a location. The app will auto detect the location by performing geo-lookup to find a destination and suggest flights for booking subsequently. Datafication could personalise a trip based on a traveler's preference or personal data shared (MacHale, 2019). Another trend in new travel apps is the use of Augmented Reality (AR) and Virtual Reality (VR) features (MacHale, 2019). There are over 500 million mobile devices with AR capabilities while over 2,000 apps with AR and VR technology available in the App Store (MacHale, 2019). For instance, KLM, Aeromexico and Kayak use an AR bag sizer app to check luggage size to ensure it is within the airline's baggage size requirement. Besides, big players in technology, Google and Apple, also invested and released its AR developer tools, ARCore and ARkit respectively, as well as dedicated large amounts of time to AR (MacHale, 2019). Travel industry professionals are aware of the importance of mobile travel apps where 90% plan to shift their investment into mobile and 60% are looking at the enhancement of mobile apps in 2018 (MacHale, 2019).

Smart Tourism refers to the application of innovative technology tools such as Internet of Thing (IoT), Artificial Intelligent (AI), Mobile Communication, Cloud Computing and Information and Communication Technology (ICT) in tourism apps (Pradhan, Oh, & Lee, 2018). Recently, the most effort is to utilize two approaches on an app: AR and BD (Pradhan, Oh, & Lee, 2018). Mobile augmented reality (MAR) was first introduced in the mid-1990s. It enhances the overall tourism experience and augments reality by overlaying digital information on top of the real world (Barten, 2019). Users can view the world surrounding them interactively augmented with digital annotations, graphics and other information. BD is a data-driven approach which provides relevant information for tourism companies to adopt a strategy for their marketing efforts—targeting the right people in the right way (Mariani, Baggio, Fuhs, & Hoepken, 2018).

Generations are defined by birth year rather than age (Kasasa, 2019). As Gen X, Y and Z have grown up in different generations and have different influences, therefore their behavior is slightly different. Generation-X was born during the emergence of the internet, so they are to some extent comfortable with technology and online presence (Charlotte, 2019). They are more concerned with proper instructions to complete their online tasks (Charlotte, 2019). Generation-Y, also known as the millennial generation, are digital natives who grew up with rapid improvements in technology (Prensky, 2001). The most significant characteristic of Generation-Y is that they are technology-savvy and highly dependent on complex technology (Prensky, 2001). Generation-Z grew up with internet access and most of them are familiar with technology (Charlotte, 2019). They have a short attention span and prefer to multi-task across multiple devices simultaneously (Charlotte, 2019).

LITERATURE REVIEW

Mobile apps have a great impact on business because the use rate of smartphones has grown rapidly within the last few years. Almost 80% of internet users possess a smartphone for internet browsing and they spend 89% of mobile media time on apps (Sukhraj, 2017); 42% of travellers utilise their smartphone for trip planning. Over 65% of companies targeted mobile marketing as their marketing strategy and over 70% believe mobile marketing is essential for their business.

Smart tourism integrates the hardware, software and network technologies to provide a real-time intelligence data for enabling stakeholders making more intelligence decision making (Gretzel, Reino, Kopera, & Koo, 2016). Apart from "e-tourism" provide travel information in pre/post-travel phase, smart tourism is extended to offer smart technology mediated tourism experiences with greater mobility and better decision support travel information (Yoo, Kwon, Na, & Chang, 2017). Visualization technology is highly adopted in the development of mobile applications in tourism, while Augmented Reality (AR) and personalization services through Big Data Analytics (BDA) are the two major approaches in tourism industry to boost their services (Stfalcon.com, 2018) (Vakhnenko, 2019) (Barten, 2019) (Singh, 2018).

AR technology offers tourists to experience real travel in remote destinations with their own smartphones or digital cameras using see-through displays (Maria, Loureiro, Guerreiro, & Ali, 2019). Mobile Augmented Reality (MAR) has the potential to influence the travelling market and reach more customers by enhancing travel experience (Kourouthanassisa, Boletsisb, Bardakia, & Chasanidou, 2015). MAR systems provide guidance to tourists for unfamiliar environments enjoyably and educationally, reduce the mental effort required for navigation and discover unknown areas via visual, audit and 3D location-based information (Kourouthanassisa, Boletsisb, Bardakia, & Chasanidou, 2015).

BD refers to the large and complex data sets available on the internet which comprises of both unstructured and structured data generated by technology development. The exponentially increasing adoption of network devices such as tablets and smartphones has brought the explosion of data availability (Mariani, Baggio, Fuhs, & Hoepken, 2018). The large volume of BD is difficult to process by traditional statistical methods and software techniques, therefore it arises in Business Intelligence (BI) technology by incorporating machine learning (ML) with embedded learning algorithms to process large data sets by filtering based on several variables and clustering large objects into a small number of classes (Mariani, Baggio, Fuhs, & Hoepken, 2018). Thus, the willingness for privacy and information disclosure for personalisation plays a critical role to streamline the processing of large volumes of data. Most of the advanced smart technologies require accessing to tourists' data for instance: personal information on social media, real-time positions, etc for improving tourist services or prediction of future needs (Buhalis & Leung, 2017) for getting better insights to offer services tailored to the tourist's preferences.

The Generation-X segmentation was born between 1966 and 1976 with a population of 1.44 billion (Venngage Inc., 2017). They are digital immigrants as most of them grew up without a computer in school and the introduction to computers came in middle school or high school (Fourhooks, 2015). Generation-Y was born between 1977 and 1994 and has a population of 1.72 billion (Venngage Inc., 2017). Simultaneously, they are digital natives, the earliest group to adopt new technologies and have less brand loyalty (Quintal, Phau, Sims, & Cheah, 2016). Generation-Z was born between 1995 and 2012 and has a population of 2.52 billion (Venngage Inc., 2017). This generation never knew a world without computers and cell phones and have used technology heavily since a young age (Fourhooks, 2015). They are digital integrationists and are also referred to as "Technoholics" as they are entirely dependent on IT (Fourhooks, 2015). Because individuals at different phases of life are impacted by economic changes, political ideologies, technological innovations and social upheavals. Each generation that experienced the same social, economic, political and cultural throughout their lifetime (Fernández-Durán, 2016) create the generational identity.

ISSUES OF THE STUDY

Through the analysis, the following are the problem statements that are deeply considered for ensuring the mobile app enhancements or development are in line with travelers' expectations.

1. A mobile travel app is high in download but low in usage.
2. A mobile travel app is high in usage but low in sustainability.
3. A mobile travel app is high in adoption but low in royalty or re-download rate.

Moreover, app personalization is happened everywhere, anytime and it became a marketing strategy to customize the needs of each individual which is equivalent to provide a personal travel agent individually to each user or customer for meeting higher expectation along the journey (Hande, 2020). Nowadays, customers are not brand loyalty but experience loyalty therefore the deep levels of user experience personalization is the way to keep customers' loyalty. Besides, the other challenge on app personalization is hard to maintain unified view of customer mainly due to the collected customer data and feedbacks online are derived from multiple devices and non-linear customer journey. It extremely increases the customer acquisition cost with approximately $7 on each customer in the travel sector (Hande, 2020).

App rule-based personalization based on user demographics such as age, gender, income, location, web-browsing history and past product purchases is essential and endless for cluster or segment similar customers together in the process (Hande, 2020). Rely on existing demographic data alone to personalize is insufficient to target personalization in particular. Travel brands need to supplement their personalization strategies based on customer's behavioural intention, rather than just their demographics (Savage, 2020). Demographic-based personalization have been proven to be unreliable which leading ads not bound the large number of consumers. Since generation cohort is one way to group age cohorts by birth year and shared the same social, economic, political and cultural events throughout their lifetime. Each generation with similar characteristics would act similarly in behavioural intention and making decisions over their lifetimes. A further or deeper personalization based on generation cohort is proposed to conquer the shortage of existing demographic-based personalization.

There is numerous research which seeks to determine the factors affecting the intention to use mobile apps but no research embeds AI technology in smart tourism with augmented reality and privacy & information disclosure. For example, research by Yoo et al. determine factors influencing the adoption of Gamified Smart Tourism Applications (GSTA) that incorporate game elements focused on hedonic utility (Yoo, Kwon, Na, & Chang, 2017). Research by Pradhan et al. seeks to understand travelers' behavior for sustainable smart tourism which emphasis the negative perceptions of smart tourism (Pradhan, Oh, & Lee, 2018). Research by Lai identifies antecedents and determinants affecting travelers' intention to use app-based mobile tour guides (AMTG) (Lai, 2015). Kourouthanassis et al. studies tourists' response to mobile augmented reality travel guides (Kourouthanassisa, Boletsisb, Bardakia, & Chasanidou, 2015). Yet, no research makes comparisons among Generation X, Generation Y and Generation Z.

Besides, previous researches are not emphasized on mobile travel apps where the findings may inadequate to determine the usage. For example, research by Smit et al. is restricted to self-service technology in the airline industry (Smit, Roberts-Lombard, & Mpinganjira, 2018); Research on the Taiwan Medical Travel App by Chang et al. is restricted to TMT providers (Chang, Chou, Yeh, & Tseng, 2016); Research by Hsiao et al. focused on mobile social apps rather than mobile travel apps (Hsiao, Chang, & Tang, 2016); Research by Alotaibi et al. is restricted to user's intention to use Mobile Government Applications (Alotaibi, Houghton, & Sandhu,

2017); Research by Olsson et al. on expected user experience of mobile augmented reality services in the context of shopping centres (Olsson, Lagerstam, Kärkkäinen, & Väänänen-Vainio-Mattila, 2013).

PROPOSED SOLUTIONS AND RECOMMENDATIONS

The DeLone and McLean IS Success model is integrated with the extension of the UTAUT2 model by incorporating the constructs from Technology Readiness Index (TRI) model, to evaluate the factors influencing the behavioural intention to use mobile travel apps towards smart tourism among Generation X, Y and Z. The DeLone and McLean IS Success model is used to evaluate the factors influencing the app's adoption on app perspective while the UTAUT2 model is used to evaluate the factors influencing the adoption of the app from user perspective. The TRI model is incorporated by introducing two new constructs from smart perspective: Personal Innovativeness and Privacy & Information Disclosure into the UTAUT2 model. It was aimed to measure individuals' adoption of mobile travel apps for smart tourism among Generation X, Y and Z.

The ability of the smart apps making predictions and proposing useful information to the travellers depends on how much privacy and travel information the travellers are willing to disclose. An empirical study examines the negative impact of travellers' usage intentions of smart devices through perceived risks and benefits. The results showed that the perceived benefits had a significant effect on usage intentions while travelling, and also confirmed that the influence of perceived risks is dependent on the characteristics of travellers. Therefore, a new construct, Privacy & Information Disclosure on BDA (which derived from the negative view towards new technologies/ undesirable dimensions: 'discomfort' and 'insecurity') is added to determine the willingness of travellers to share their privacy and travel information for perceived benefits mainly in-app personalisation. Besides that, Personal Innovativeness on AR in smart tourism (which derived from the positive view towards new technologies/ desirable dimensions: 'optimism' and 'innovativeness') was also included to determine the percentage of users who perceive reality which acts as a direction for app developers or organisations to design and develop travel apps in an AR perspective.

Nevertheless, the level adoption of mobile travel apps may defer depending on the travellers' characteristics. Different generation poses different characteristics may influence the behavioural intention to use mobile travel apps. Both comparison on factors influencing behavioural intention to use mobile travel apps among Generation X, Y & Z and generation cohort moderating effects are carried out to analysis the generational differences and provide empirical evidence of the main differences. The differences behaviour finding in generation group will serve as guidance for tourist providers to deliver specific marketing strategies for different generations on mobile travel apps toward smart tourism.

Figure 2. DeLone and McLean Original IS Success Model, 1992

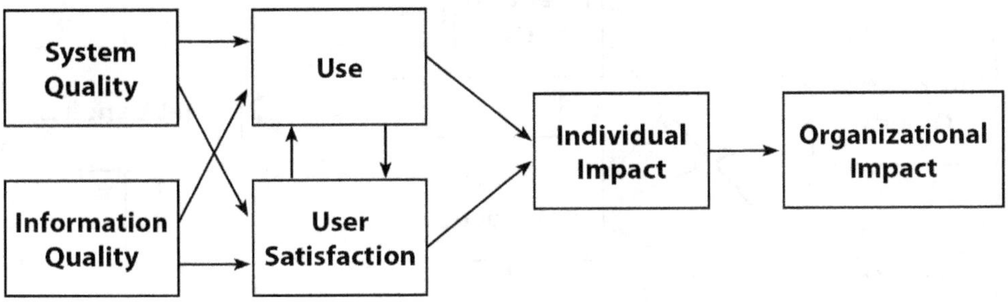

UNDERLYING THEORIES

DeLone and McLean IS Success Model

In 1992, DeLone and McLean introduced their first information systems (IS) success model *(Figure 2)* to measure a different aspect of IS success covering different evaluation perspectives (DeLone & McLean, 1992).

Ten years after publication, DeLone and McLean proposed an updated IS success Model in the year 2003 by further enhancing the original model based on the evaluation and validation of the model (DeLone & McLean, 2003). The three reinstated perspectives are 1) "service quality" based on SERVQUAL (Parasuraman & Zeithaml, 1988) added to reflect the importance of IS service support for successful e-commerce systems; 2) "intention to use" added as an alternative measure of use to measure user attitude towards a system; 3) and "net benefits" added to collapse the two variables, individual impact and organisational impact (Urbach & Mueller, September 2011). As a result, the refined model comprises six interrelated variables such as system quality, information quality, service quality, intention to use, user satisfaction and net benefits *(Figure 3)*.

However, this model poses a difficulty for measuring between intention to use and actual use as the two constructs are inseparable. The intention to use refers to the intention to use a system (intention before use) while used refers to the behaviour of actual use (action after use). Research results may be biased if these two categories are conflated as people might overestimate their future use by giving positive attribution to intention to use. Therefore, Peter and Mclean revisited and

Figure 3. DeLone and McLean Updated IS Success Model, 2003

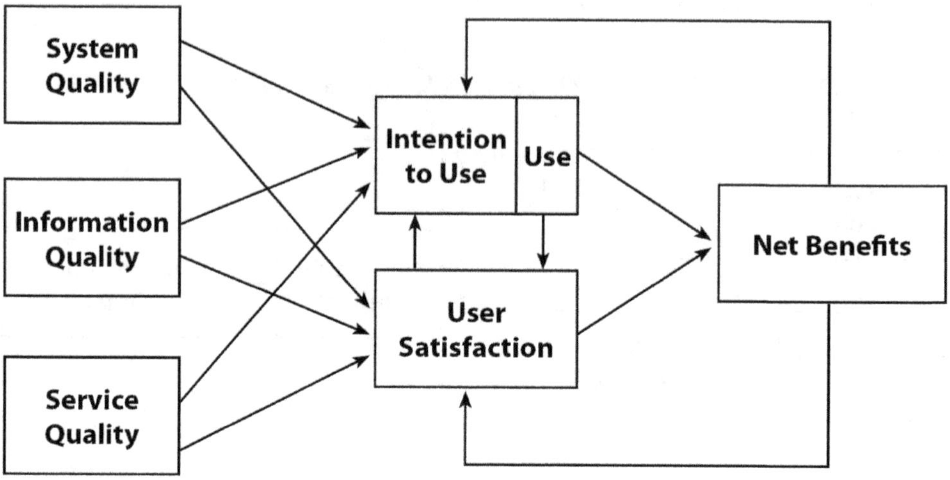

suggested this model through another meta-analysis study and the results showed the need to separate the two constructs *(Figure 4)* to avoid the complexity of the model (Peter & McLean, 2009).

Consumer Acceptance and Use of Information Technology (UTAUT2)

Benbasat and Barki (2007) highlight that the Technology Acceptance Model (TAM) theory is the most influential theory in IS with 18,932 citations among theories introduced (Benbasat & Barki, 2007). In 2003, UTAUT *(Figure 5)* was introduced to overcome the limitations of TAM to recognise technology acceptance in social aspects. Social influence (SI) is included in UTAUT to capture the influence of important others, especially to make a work-related decision in an organisational environment. UTAUT consists of four core determinants: performance expectancy, effort expectancy, social influence, which influences the intention to use; and facilitating conditions that affect the behavioural intention to use a technology or technology use (Venkatesh, Morris, Davis, & Davis, 2003). These relationships are moderated by four moderators: gender, age, experience and voluntariness of use.

Subsequently, UTAUT was extended to UTAUT2 by adding three core determinants: hedonic motivation, price value and habit (Venkatesh, Thong, & Xu, 2012). The UTAUT2 model *(Figure 6)* is better to justify technology acceptance than the UTAUT model as the percentage of variance explained in both behavioural intention (56% to 74%) and technology use (40% to 52%) are improved in the

UTAUT2 model (Maurits, 2012). Moderator voluntariness use is no longer retained.

Figure 4. The Separation of Intention to Use and Use of DeLone and McLean Model, 2009

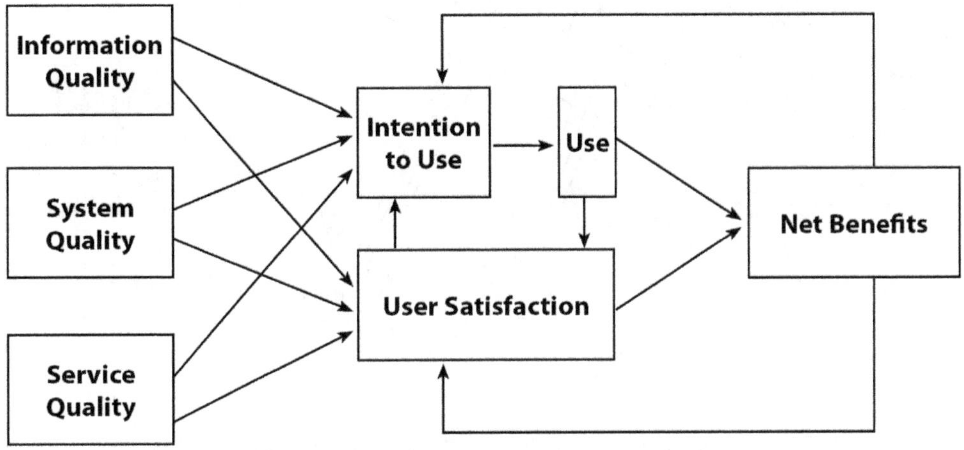

Figure 5. Unified Theory of Acceptance and Use of Technology 2003

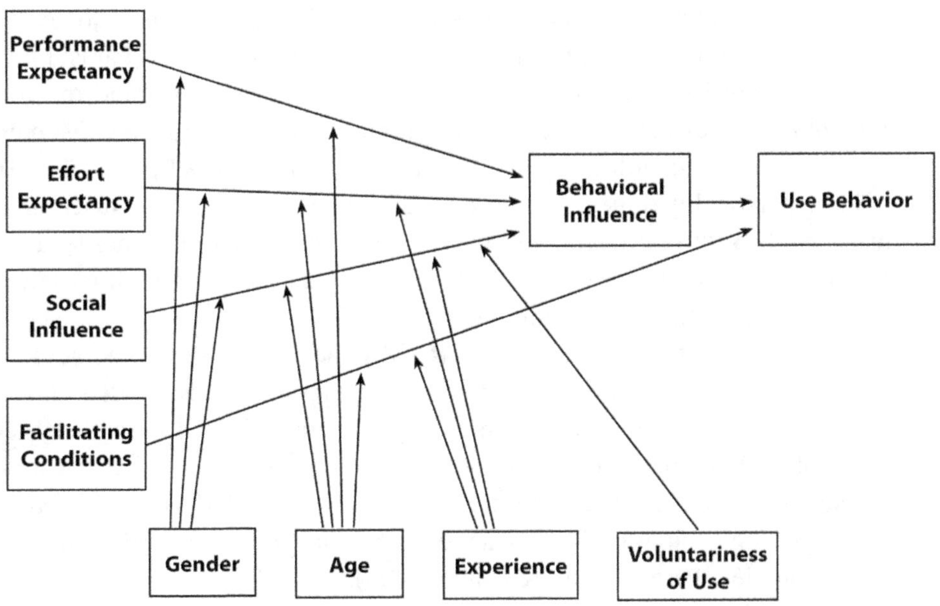

Figure 6. Consumer Acceptance and Use of Information Technology 2012

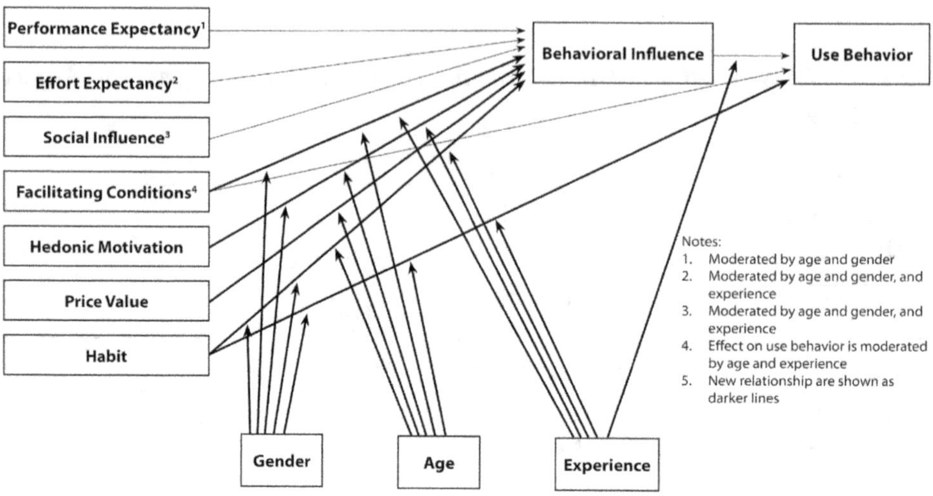

Additionally, a new relationship between facilitating conditions and behavioural intention is established in this enhanced model.

Technology Readiness Index (TRI)

The Technology Readiness Index is defined as 'people's propensity to embrace and use new technologies for accomplishing goals in home life and at work (Parasuraman A., 2000). The developed constructs measure 'an overall state of mind resulting from a combination of mental enablers and inhibitors to determine a consumer's preference to use new technologies' (Parasuraman A., 2000). The belief comprised of four different personality dimensions such as 'optimism', 'innovation', 'discomfort' and 'insecurity'. Both optimism and innovativeness as mental enablers lead to a positive view while both discomfort and insecurity as mental inhibitors have a negative view in adopting new technologies.

In 2005, an integrated model named TRAM (The Technology Readiness and Acceptance Model) was established by integrating TRI with TAM models *(Figure 7)*. It aimed to merge the general personality dimensions of TRI with system acceptance of TAM to measure how personality dimensions could affect the actual usage of new technologies. The personality dimensions of TRI act as antecedents to the cognitive dimensions of TAM in a measure of how people perceive and react to new technologies (Parasuraman & Colby, 2001).

Figure 7. The Technology Readiness and Acceptance Model 2005

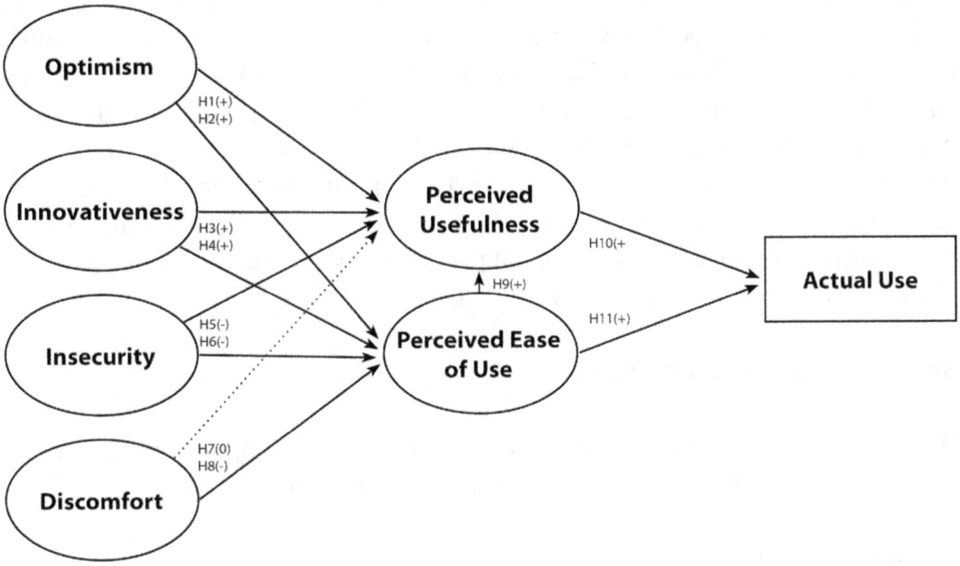

CONCEPTUAL FRAMEWORK AND HYPOTHESIS DEVELOPMENT

The theoretical framework of this study is based on the integration of the IS Success Model with the extension of the UTAUT2 model to determine the factors influencing behavioural intention to use mobile travel apps for smart tourism among Generation X, Y and Z. The UTAUT2 model is extended by adding two new variables named Personal Innovativeness and Privacy & Information Disclosure which is derived from the TRI model.

The IS Success Model is adopted to measure the system or application process to transmit information to the recipients effectively by attaining 3 problems: technical, semantic and effectiveness (DeLone & McLean, 1992). It also was the highest cited for numerous researches on information technology value which achieved over 10,000 in 2019. The UTAUT2 model is chosen in this study because it was the most widely used for assessing behavioural intention to use and also the variable available in the framework is the best predictor for behavioural intention to use. This could be proved by the highest number of citations, 8,731 cited by numerous researches (Mardiana, Tjakraatmadja, & Aprianingsih, 2015).

The dependent variable of this study is the behavioural intention to use mobile travel apps, and subsequently, it will be used to determine the usage of mobile travel apps and the results are used for further analysis and evaluation. The independent variables comprise of 12 variables which are System Quality, Information Quality and Service Quality derived from IS model on apps perspective; Performance Expectancy, Effort Expectancy, Social Influence, Facilitating Conditions, Hedonic Motivation, Price Value and Habit derived from the UTAUT2 model on user perspective; and Personal Innovativeness on AR and Privacy & Information Disclosure on BDA derived from the TRI model on smart perspective. In addition, the moderator Generation Cohort is introduced in this model for determining the moderating effects among Generation X, Y and Z on behavioural intention to use mobile travel apps in smart tourism. All the variables are clearly defined under the following conceptual framework *(Figure 8)*.

Hypothesis Development

The following section represents the empirical evidence that supports the relationships of the research model and the developed research hypothesis.

System quality

System Quality measures the desired characteristics of an e-commerce system to identify the reliability, usability, flexibility, integration, accessibility and timeliness of the system or application (Wixom & Todd, 2005). A mobile business app is required to be useful, understandable, and to provide information timely for attracting user adoption of the app (Koo, Wati, & Chung, 2013).

Thus, the study hypothesised that:

H1: System quality on apps has a positive effect on behavioural intention to use mobile travel apps for smart tourism.

Information quality

Information Quality measures the content of an e-commerce system based on several dimensions such as personalisation, currency, relevance, reliability, completeness, understandability, accuracy and security (Wixom & Todd, 2005). Empirical findings indicated that information quality has a strong significant influence on IS success (Zaied, 2012). If an information system does not deliver information timely or do not conform to the user needs (Gorla, Somers, & Wong, 2010), then the user will feel dissatisfaction and refuse to utilize the system subsequently. Therefore, Information Quality is another construct to measure mobile travel app could deliver information to the user efficiently and effectively at semantic perspective.

Thus, the study hypothesised that:

Figure 8. Conceptual Framework

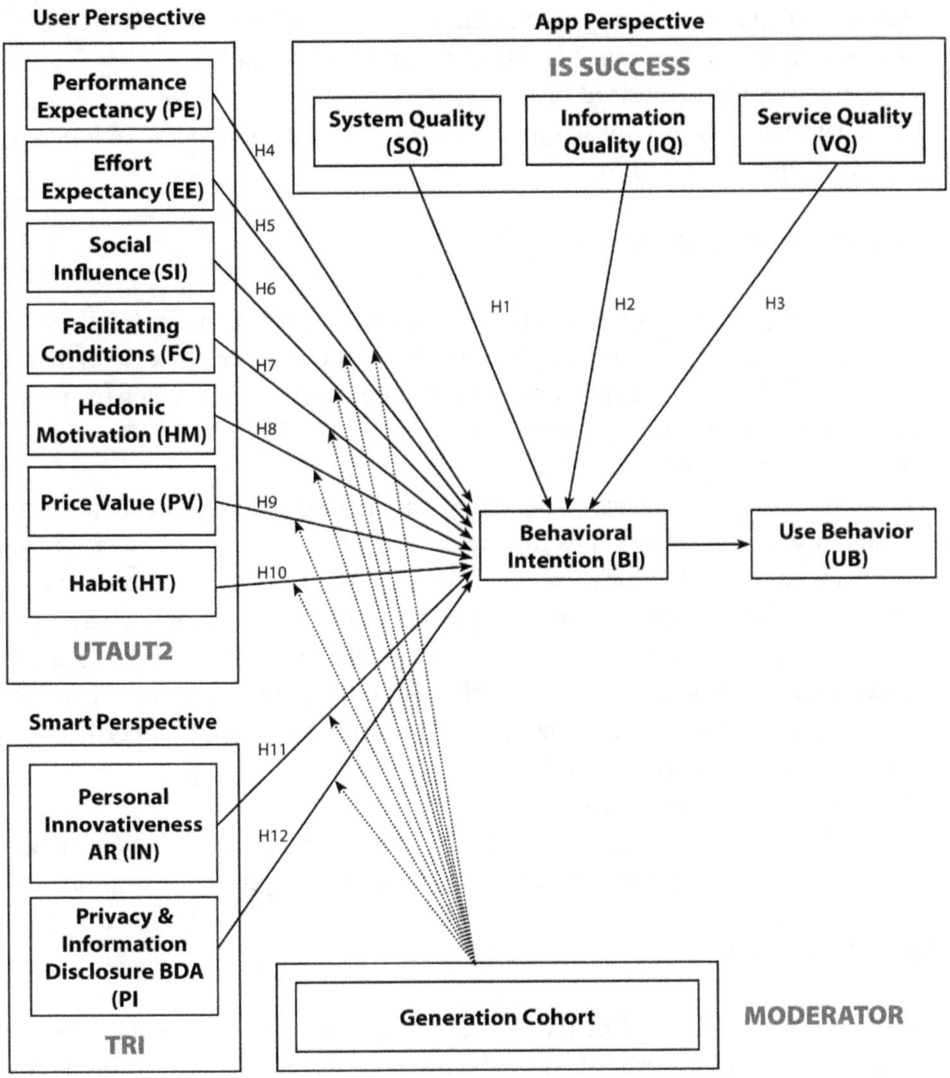

H2: Information quality on apps has a positive effect on behavioural intention to use mobile travel apps for smart tourism.

Service quality

Service Quality measures the overall system support delivered by the service provider and applies regardless of whether the support is delivered by the information system personnel. The dimensions of the variables are tangibles, reliability, responsiveness, assurance and empathy (Parasuraman & Zeithaml, 1988). Empirical findings indicated that service quality has a strong significant influence on IS success (Zaied, 2012) which can enhance system usage and foster user involvement.

Thus, the study hypothesised that:

H3: Service quality on apps has a positive effect on behavioural intention to use mobile travel apps for smart tourism.

Performance expectancy

Performance Expectancy is defined as the degree to which using technology will improve performance and bring benefits in performing certain activities (Venkatesh, Thong, & Xu, 2012). The user perception on mobile apps depends on app responses, the speed of the app and the aesthetic design (Trice, 2014).

Thus, the study hypothesised that:

H4: Perceived performance expectancy has a positive effect on behavioural intention to use mobile travel apps for smart tourism.

Owing to the lower adoption of mobile devices among Generation X, performance expectations to the mobile smart travel apps could be lower among Generation X compared to Generation Y and Z. Mainly Generation Z who considered highly educated, creative, innovative and capable of multi-tasking in an ever-changing environment (Robinson & Schanzel, 2019) would place higher performance expectations on mobile smart travel apps.

Thus, the study hypothesised that:

H4.1: Generation cohort moderates the degree of perceived performance expectancy on behavioural intention to use mobile travel apps for smart tourism.

Effort expectancy

Effort expectancy is defined as the degree of ease associated with the use of a particular system (Venkatesh, Morris, Davis, & Davis, 2003) which is similar to perceived ease of use in the TAM. An app with a clean, easy-to-use interface allows users to engage and solve problems faster tend to be more popular (Rios, 2017). Therefore, a good user experience (UX) is an essential component of product strategy for producing a successful app (Babich, 2018).

Thus, the study hypothesised that:

H5: Perceived effort expectancy has a positive effect on behavioural intention to use mobile travel apps for smart tourism.

The younger generation such as Generation Y and Z who are technology-savvy are more averse to internet usage, comfortable with technology and online activities

(Lee, 2020). If compare with Generation X who are lacking familiarity to the internet usage, Generation Y and Z may have lower effort expectations on mobile smart travel apps due to their familiarity to the apps.

Thus, the study hypothesised that:

H5.1: Generation cohort moderates the degree of perceived effort expectancy effect on behavioural intention to use mobile travel apps for smart tourism.

Social influence

Social Influence refers to the extent to which individual perceive important others such as family members, friends or colleagues and believe that they should use a particular technology (Venkatesh, Thong, & Xu, 2012). The choice of particular mobile apps is also largely influenced by others who intend to keep close relationships with each other for ease of information sharing, trip planning and ease of communication.

Thus, the study hypothesised that:

H6: Social influence has a positive effect on behavioural intention to use mobile travel apps for smart tourism.

Generation Y's decision of purchase or consumption of products are more rely on parents' advice or highly influenced by their friends or peers as they actively look for recognition from their friends or peers. The influence of family or peers on intention usage is low in Generation X which could be due to they are alienated from their parents as well as with their rebellious and nonconformist behaviour (Ivanova, Flores-Zamora, Khelladi, & Ivanaj, 2018). Generation Z has significant influence on family holidays, as parents would usually check with them prior to booking a trip (Robinson & Schanzel, 2019).

Thus, the study hypothesised that:

H6.1: Generation cohort moderates the degree of social influence on behavioural intention to use mobile travel apps for smart tourism.

Facilitating conditions

Facilitating Conditions refers to individuals' perceptions of the technical infrastructure that exists and the support available to users of the system (Venkatesh, Thong, & Xu, 2012). It mainly focuses on the environmental factors that could affect the usability of technology. Supporting facilities such as internet connectivity, sufficient smartphone memory and information needed to use the mobile app would lead the user intention to use the app (Gupta, Dogra, & George, 2017).

Thus, the study hypothesised that:

H7: Individual perception concerning facilitating condition has a positive effect on behavioural intention to use mobile travel apps for smart tourism.

Generation Z is characterized likely to expose remote places through fun or life experiences (Robinson & Schanzel, 2019) and even engage in multiple travels activities simultaneously. The incredible technical infrastructure and technical support is highly expected to support their needs. Although older generation such as Generation X and baby boomer do find mobile multi-channel is challenge in usage (Dorie & Loranger, 2020) but they do see mobile as useful tool toward online usage.

Thus, the study hypothesised that:

H7.1: Generation cohort moderates the degree of individual perception concerning facilitating conditions on behavioural intention to use mobile travel apps for smart tourism.

Hedonic motivation

Hedonic motivation is defined as the fun, enjoyment or pleasure that is derived from using technology (Venkatesh, Thong, & Xu, 2012). Individuals feel enjoyment, fun and entertainment from adopting the app which increases the satisfaction and positively affects the behavioural intention toward the usage of mobile travel apps (Hsiao, Chang, & Tang, 2016).

Thus, the study hypothesised that:

H8: Hedonic motivation has a positive effect on behavioural intention to use mobile travel apps for smart tourism.

Generation Y mostly use and get engaged with technology for entertainment purpose while Generation X who more concerned with the practical aspects of technology mainly use technology for utilitarian purposes and information search (Calvo-Porral & Pesqueira-Sanchez, 2019). Generation Z tourist who always seek for rich digital and gamified tourism experience is even prefer more on "fun experiences" (Robinson & Schanzel, 2019) where they expected to seek out remote places virtually through using their smartphone with AR capability.

Thus, the study hypothesised that:

H8.1: Generation cohort moderates the degree of hedonic motivation on behavioural intention to use mobile travel apps for smart tourism.

Price value

Price value refers to the consumers' cognitive trade-off between the cost of using the technology and the perceived benefits (Venkatesh, Thong, & Xu, 2012). A previous study found that the adoption of technologies allows consumers to acquire a product or service at a lower price (Escobar-Rodrígueza & Carvajal-Trujillo, 2014).

Thus, the study hypothesised that:

H9: Price value has a positive effect on behavioural intention to use mobile travel apps for smart tourism.

Generation X allocate more budget to luxuries hotel and upgrade airfare while Generation Y and Z are characterized as budget-conscious travellers as they usually start their journey without specific destination in mind (Robinson & Schanzel, 2019). Therefore, MAR technology is a tactic for them to explore a new destination within budget cost.

Thus, the study hypothesised that:

H9.1: Generation cohort moderates the degree of price value on behavioural intention to use mobile travel apps for smart tourism.

Habit

Habit refers to the individual tendency to perform automatic behaviours—based on learning (Venkatesh, Thong, & Xu, 2012). Several researchers found that habit was a strong predictor influencing the intention to use mobile apps such as user acceptance of mobile apps for a restaurant, MARSR (Ramon, Santiago, Javier, & Emilio, 2019). Habits influence tourists' intention to use travel apps installed on smartphones (Gupta, Dogra, & George, 2017).

Thus, the study hypothesised that:

H10: Habit has a positive effect on behavioural intention to use mobile travel apps for smart tourism.

Smartphone usage becomes a habit, and therefore influences the use of mobile technologies during travel (Dorcic, Komsic, & Markovic, 2018), especially Generation Y and Z who adopt their smartphone in daily use. Generation X individuals show lack engagement with technology due to the lack positive evaluation of credibility, trustworthiness and sufficiency of the information obtained by technology (Calvo-Porral & Pesqueira-Sanchez, 2019).

Thus, the study hypothesised that:

H10.1: Generation cohort moderates the degree of habit on behavioural intention to use mobile travel apps for smart tourism.

Personal Innovativeness on AR

Individual with optimism and innovativeness characteristics are enablers of new technology as they believe adoption of new technology lead to flexible control, improve efficiency and effectiveness as well as have tendency to be the leader of new technologies. It is defined as individual's willingness to try out new information technology (Agarwal & Prasad, 1998) which will affect individual's intention to use new technology for accomplishing goals (Parasuraman A., 2000). AR refers to the degree of an individual's perception of the reality of a physical environment thru the usage of digital facilities. It was the artificial environment created by combining the digital world with the user's physical world (Manuel, 2018). This construct is derived from the desirable dimensions: 'Optimism' and 'Innovativeness' of TRI (Demirci & Ersoy, 2008).

Thus, the study hypothesised that:

H11: Personal Innovativeness: AR has a positive effect on behavioural intention to use mobile travel apps for smart tourism.

The latest generations such as Generation Y and Z grew up in a media-saturated and brand-conscious context preferred communication online via internet or social media, while Generation X primary media channels are newspaper, TV or radio may reluctant to adopt internet or social media for communication.

Thus, the study hypothesised that:

H11.1: Generation cohort moderates the degree of personal innovativeness: AR on behavioural intention to use mobile travel apps for smart tourism.

Privacy and Information Disclosure on BDA

Privacy and information disclosure refer to an individual's willingness to disclose their private information for obtaining benefits mainly on app personalisation in smart apps environments. Big Data Analytics (BDA) increase the efficiency through app personalization based on disclosure of privacy information to the mobile travel app. The study on this perception is aimed to find the degree of privacy and information individuals will disclose personalization based on user's profile and preference in more specific needs individually. This construct is derived from the undesirable dimensions: 'Discomfort' and 'Insecurity' of TRI (Demirci & Ersoy, 2008) (Parasuraman A., 2000).

Thus, the study hypothesised that:

H12: Privacy & Information Disclosure: BD has a negative effect on behavioural intention to use mobile travel apps for smart tourism.

Commonly, old generation such as baby boomer and Generation X who is risk avoidance (Lissitsa & Kol, 2016) is more sensitive to their privacy information while younger generation such as Generation Y and Z are more "open minded' who may not reluctant to disclose their current location while travelling.

Thus, the study hypothesised that:

H12.1: Generation cohort moderates the degree of privacy and information disclosure: BD on behavioural intention to use mobile travel apps for smart tourism.

Behavioural Intention and Use Behaviour

Behavioural Intention is a psychological process of a user's intention to use a system. It acts as an antecedent to use behaviour and is suited to predict whether the system is acceptable to use (Mardiana, Tjakraatmadja, & Aprianingsih, 2015). Use behaviour is an important indication of a success system. It does not guarantee a system success but behaviour not to use does guarantee a system failure (Mardiana, Tjakraatmadja, & Aprianingsih, 2015).

Thus, the study hypothesised that:

H13: The degree of behavioural intention directly influences the use of behaviour of mobile travel apps.

GUIDELINES FOR FUTURE EMPIRICAL TESTING

Measurement Issues

The quantitative approach was adopted in this study mainly to gather a large amount of data through survey questionnaires and analyse statistically further to measure the specific variable such as behavioural intention with less bias. Besides, data collection via quantitative approach is relatively fast and data analysis is relatively less time consuming via statistical tools. Quantitative approach would justify the causal relationships identified between variables and identify the predictors of outcomes. It quantifies the result of hypothesis relationships through running statistical software. Further make predictions, establish facts, and examine developed hypothesis based on the constructed theories to determine the degree of developed hypothesis relationship is support or does not support.

This study intended to measure twelve independent variables namely system quality, information quality, service quality, performance expectancy, effort expectancy, social influence, facilitating conditions, hedonic motivation, price value, habit, personal innovativeness on AR and privacy & information disclosure on BDA, which are predicted to have a direct effect on behavioural intention to use mobile travel apps toward smart tourism. Additionally, the relationship of those predictors and criterions are further measure to determine the moderating effect by generation cohort context.

Unit of Analysis

The selected populations in this study are those born between the years 1965 to 2010 in Malaysia. The population of Malaysia is estimated at 32.6 million, reported

Figure 9. Percentage Population by Age Group and Median Age, Malaysia, 2018 and 2019

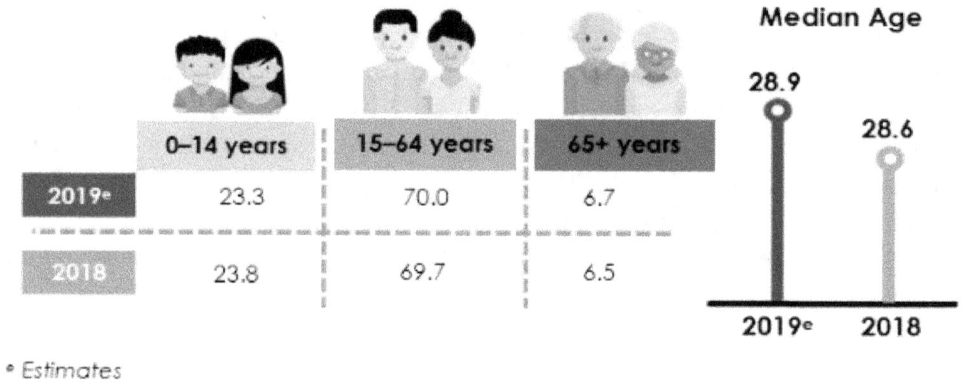

by Department of Statistics, Malaysia on 15, July 2019. Based on the percentage population by age group as Figure 9, the three generations X, Y and Z who were born between the years 1965 to 2010, are from the major population aged 15 - 64 years which occupied average 70% population in year 2019. Thus, their respondents are essential to determine the factors influencing the individual's behavioural intention to use and drive their attitude towards the usage of mobile travel apps toward smart tourism.

Each individual will be the unit of analysis (UOA) of this study, the individual is expected the respondents among Generation X, Y and Z who are smart phone users/owners aged 9 to 54 in year 2019 from all states in Malaysia. Apart respondents from Generation X and Generation Y, only those adults with age 18 and above in year 2020 (those born between year 1996 and 2002) are included as respondents from Generation Z to eliminate bias in the responses. As adults who are reaching mature age are responsible to make decision by their own and bear the consequences. They usually not control or influence by others on their decision making. As a result, the individual among Generation X, Y and Z who born between 1965 to 2002 with age 18 and above in year 2000 are selected to be the respondents of this study.

Convenience sampling is adopted to collect large sampling data to meet the sample size for hypothesis relationships analysis, participants are chosen based on availability and willingness to participate. Considering it was less time consuming and the most convenient to gather data (Sekaran & Bougie, 2016), as well as the randomly selected respondents are less bias. Quota sampling technique is further applied in this study where the total respondents from each generation must at least achieving the quota number from sample size based on the percentage of each generation in the population of Malaysia for balancing the comparison in subsequent data analysis.

Statistical Data Analysis

Several statistical tools are applied under this study such as IBM SPSS software version 22.0 and Smart Partial Least Square (SmartPLS) software version 3.0. The first-generation data analysis technique, IBM SPSS is used to analyse the collected data on the first part, where the demographic summary is generated. Due to this current study comprise multiple variables include independent variables ("exogenous" in SEM), dependent variable and moderating variable ("endogenous" in SEM) to form various interrelated relationships. Thus, the collected data on the second part is considered complex, with multiple latent constructs, mediator and moderator are more advisable to further analysed using PLS Algorithm in SmartPLS (Urbach & Ahlemann, 2010).

Under the SPSS model, descriptive statistics summarizes and visualizes the demographic information such as gender, age, educational level and intent of travel obtained in questionnaires. The frequency and percentage comparisons are used to clearly define the distribution of the data and are further computed in terms of average, standard deviation in order to verify the accuracy of the data obtained, which helps to determine the degree of users' feedbacks. Furthermore, both structural model (an inner model) and measurement model (an outer model) are adopted in this study under the SmartPLS model. The structural model represents the correlations obtained from theoretical deliberation between latent variables. Meanwhile, the measurement model demonstrated the correlation between empirically observable items and each of the latent variables based on auxiliary theories of measurement, such as reflective and formative models (Urbach & Ahlemann, 2010). All the latent variables in this study were categorized as reflective measurement models as all measure the same underlying concept, interchangeable and can be eliminated (Urbach & Ahlemann, 2010).

Minimum sample size required to estimate the PLS path model is equal to the maximum number of arrows pointing in the PLS path model (Hair, Hult, Ringle, & Sarstedt, 2017). In this study, there are total of 22 arrows which comprise of 12 independent variables, 1 dependent variable and 9 moderators. Therefore, according to the 10 times rule, the minimum sample size for model estimation is 220. While G*Power software is adopted to calculate required sample size (Cohen, 1992) based on F tests - Linear multiple regression: Fixed model, R^2 deviation from zero with population medium effect size (f^2) at 0.15, statistical power (1-β) of 95%, error margin (α) at 5% and number of predictors at 22. The total sample size to generate adequate statistical power of 0.80 is 230 sample size. Due to sample size 230 generated by

G*Power is greater than 10 times rule with sample size of 220, thus the minimum sample size at least 230 is required for PLS model analysis.

Both measurement model and structural model are used to evaluate the constructs and relationships. Three standards namely Internal Consistency, Convergent Validity and Discriminant Validity used to determine the validity and reliability of the construct variables. Internal Consistency comprise of Cronbach's Alpha and Composite Reliability evaluate the internal consistency based on the upper and lower bound value and the reliability value of 0.60 – 0.70 is acceptable in exploratory studies. Convergent Validity comprise of Outer Loader and Average Variance Extracted (AVE) which evaluate the outer loader with a minimum value of 0.60 as significant and AVE with 0.5 value and higher to meet the threshold value (Hair, Hult, Ringle, & Sarstedt, 2017). Discriminant Validity is access via Fornell-Larcker criterion, Cross-loading and heterotraitmonotrait ratio (HTMT). Fornell-Larcker asserts that the AVE of each latent variable should be higher than the squared correlations with all other latent variables. The loading of each indicator must greater than the rest of its cross-loadings under the cross-loadings assessment. Heterotrait-monotrait ratio (HTMT) evaluate the correlation between two constructs with a threshold value of 0.85 or lower (Henseler, Ringle, & Sarstedt, 2015). For the structural model, the primary assessment is the ability to predict one or more target constructs as well as the relationships between constructs (Hair, Hult, Ringle, & Sarstedt, 2017). The model evaluation included Variance Inflation Factor (VIF) Collinearity Values on predictor constructs is less than 0.5; Standard Beta Path Coefficient near to +1 and vice versa for negative values; t-value is exceeding the threshold value at a certain significant level (p-value); predictive capabilities (R^2) in variance of endogenous construct; effect size estimation (f^2) are 0.005, 0.01 and 0.025 and predictive relevance (Q^2) value greater than zero (Hair, Hult, Ringle, & Sarstedt, 2017) (Cohen, 1988).

SIGNIFICANCE OF STUDY

This research aims to identify the factors influencing the behavioral intentions of using mobile travel apps among Generation X, Generation Y and Generation Z for obtaining usage benefit and leading continuance usage. The significance of this research study is the completeness of finding results on the most current mobile travel apps for smart tourism. The findings identify the factors influencing the behavioral intention to use mobile travel apps on both app and user perspective, with the integration of the IS Success Model and UTAUT2 theories. Besides, the finding is also the most current with the extension of UTAUT2 by incorporating the constructs from TRI theory to identify both positive and negative impacts on smart tourism adoption.

AR and BDA two recent concerns in the new trend of AI Technology are taken into account to evaluate its impact on the behavioral intention to use mobile travel apps toward smart tourism. It helps to determine the traveler's perception of augmented reality and the willingness to disclose their private information to obtain benefits via app personalization. Also, this research compares Generation X, Generation Y and Generation Z to determine behavioral intention to use mobile travel apps for smart tourism.

As a result, this study is significant for the tourism industry for mobile app development, smart-device manufacturing companies, travelling organizations and tourism stakeholders to incorporate suitable or new features into smart devices, introduce new services to the apps for promoting their services and products in more effective and sustainable methods. Apart from that, the findings also significantly to optimize app personalization service by generation segment based on the traveler's preference in data-driven approach for turning the digital information into valuable data that can be minded and used (MacHale, 2019).

CONCLUSION

This conceptual framework is built based on the integration of DeLone and McLean IS Success (IS) with the extension of Consumer Acceptance and Use of Information Technology (UTAUT2) by adding two new constructs derived from the Technology Readiness Index (TRI) to investigate the factors influencing behavioural intention to use mobile travel apps for smart tourism among Generation X, Y and Z. The study aims for findings based on three perspectives: app perspective, user perspective and smart perspective. The DeLone and McLean IS Success (IS) model is adopted to evaluate factors influencing behavioural intention to use mobile travel apps on app perspective. The Consumer Acceptance and Use of Information technology (UTAUT2) model is adopted to evaluate factors influencing behavioural intention to use mobile travel apps on user perspective.

In addition, the Technology Readiness Index (TRI) model is incorporated to evaluate factors influencing behavioural intention to use mobile travel apps toward smart tourism. However, the degree of factors influencing app adoption may defer depend on the travellers' characteristics. Therefore, this research will segment the respondents into generational categories to collect feedbacks among Generation X, Y and Z. Generational respondents aimed to collect feedback from specific generations of people based on their preferences, attitudes, characteristics and upbringings that distinguish them from other groups. It could help collect mass feedback to tailor and customise feedbacks by specific age groups. By understanding the tendencies of each generation, it will help to determine factors influencing behavioural intention to use in specific contexts and also in getting to know what will work best to increase the adoption of apps by generation.

There are numerous researches on mobile travel apps but little or none of it has targeted three generations simultaneously. Therefore, this research also aimed at finding out what each generation prefers and what they value by gathering the different reactions needed to succeed. As a result, this study is significant for tourism industries for mobile app developers, smart-device manufacturing companies, travelling organisation and tourism stakeholders to incorporate suitable or new features into smart devices and introducing new services to the apps for promoting services and products in more effective and sustainable methods. Besides, by segmenting the market by generation, it will also help the marketers utilise sales tactics to reach existing and potential or new customers effectively, which could save both time and money for the organisation.

ACKNOWLEDGMENT

This work was supported by The Islamic Tourism Research Grant (ITRG) 2020/2021 (2020-ITRG-131).

REFERENCES

Agarwal, R., & Prasad, J. (1998). A Conceptual and Operational Definition of Personal Innovativeness in the Domain of Information Technology. *Information Systems Research*, *9*(2), 204–215. doi:10.1287/isre.9.2.204

Alotaibi, R., Houghton, L., & Sandhu, K. (2017). Factors Influencing Users' Intentions to Use Mobile Government Applications in Saudi Arabia: TAM Applicability. *International Journal of Advanced Computer Science and Applications*, *8*(7), 200–211. doi:10.14569/IJACSA.2017.080727

Babich, N. (2018, February 12). A Comprehensive Guide To Mobile App Design. *Design & Development*.

Barten, M. (2019). *How Augmented Reality is Revolutionising the Travel Industry*. Retrieved from Revfine Web Site: https://www.revfine.com/augmented-reality-travel-industry/

Benbasat, I., & Barki, H. (2007). Quo vadis, TAM? *Journal of the Association for Information Systems*, *8*(4), 211–218. doi:10.17705/1jais.00126

Buhalis, D., & Leung, R. (2017). Smart Hospitality—Interconnectivity and Interoperability Towards an Ecosystem. *International Journal of Hospitality Management*, 41–50.

Calvo-Porral, C., & Pesqueira-Sanchez, R. (2019). Generational differences in technology behavior: Comparing millennials and Generation X. *Kybernetes*, *49*(11), 2755–2772. doi:10.1108/K-09-2019-0598

Chang, I.-C., Chou, P.-C., Yeh, R. K.-J., & Tseng, H.-T. (2016). Factors influencing Chinese tourists' intentions to use the Taiwan Medical Travel App. *Telematics and Informatics*, *33*(2), 401–409. doi:10.1016/j.tele.2015.09.007

Charlotte, P.-D. (2019). *Best Practices for Researching Generation X, Y and Z.* Retrieved from FlexMR Web Site: https://blog.flexmr.net/researching-gen-x-y-and-z

Cohen, J. (1988). *Statistical Power Analysis for the Behavioral Sciences* (2nd ed.). Routledge.

Cohen, J. (1992). A Power Primer. *Psychological Bulletin*, *112*(1), 155–159. doi:10.1037/0033-2909.112.1.155 PMID:19565683

Current Population Estimates, Malaysia, 2020. (2020, July 15). Retrieved from Department of Statistics Malaysia: https://www.dosm.gov.my/v1/index.php?r=column/cthemeByCat&cat=155&bul_id=OVByWjg5YkQ3MWFZRTN5bDJiaEVhZz09&menu_id=L0pheU43NWJwRWVSZklWdzQ4TlhUUT09

DeLone, W. H., & McLean, E. R. (1992). Information Systems Success: The Quest for the Dependent Variable. *Information Systems Research*, *3*(1), 60–95. doi:10.1287/isre.3.1.60

DeLone, W. H., & McLean, E. R. (2003). The DeLone and McLean Model of Information Systems Success: A Ten-Year Update. *Journal of Management Information Systems*, *19*(4), 9–30. doi:10.1080/07421222.2003.11045748

Demirci, A. E., & Ersoy, N. F. (2008). *Technology Readiness for Innovative High-Tech Products: How Consumers Perceive and Adopt New Technologies*. Academic Press.

Dogtiev, A. (2019, February 16). *App Download and Usage Statistics (2018).* Retrieved from BusinessofApps Web Site: https://www.businessofapps.com/data/app-statistics/

Dorcic, J., Komsic, J., & Markovic, S. (2018). Mobile Technologies and Applications Towards Smart Tourism. *Tourism Review*.

Dorie, A., & Loranger, D. (2020). The multi-generation Generational differences in channel activity. *International Journal of Retail & Distribution Management, 48*(4), 395–416. doi:10.1108/IJRDM-06-2019-0196

Escobar-Rodríguez, T., & Carvajal-Trujillo, E. (2014). Online purchasing tickets for low cost carriers: An application of the unified theory of acceptance and use of technology (UTAUT) model. *Tourism Management, 43*, 70–88. doi:10.1016/j.tourman.2014.01.017

Fernández-Durán, J. (2016). Defining generational cohorts for marketing in Mexico. *Journal of Business Research, 69*(2), 435–444. doi:10.1016/j.jbusres.2015.06.049

Fourhooks. (2015, April 28). *The Generation Guide - Millennials, Gen X, Y, Z and Baby Boomers*. Retrieved from fourhooks: http://fourhooks.com/marketing/the-generation-guide-millennials-gen-x-y-z-and-baby-boomers-art5910718593/

Gretzel, U., Reino, S., Kopera, S., & Koo, C. (2016). Smart Tourism Challenges. *Journal of Tourism*, 41-47.

Gupta, A., Dogra, N., & George, B. (2017). What determines tourist adoption of smartphone apps?: An analysis based on the UTAUT-2 framework. *Journal of Hospitality and Tourism Technology*, 50–64.

Hair, J., Hult, G. T., Ringle, C. M., & Sarstedt, M. (2017). *A Primer on Partial Least Squares Structural Equation Modeling (PLS-SEM)* (2nd ed.). SAGE Publications.

Hande, P. (2020, May 5). *Personalization in Travel – AI-powered Recommendations To Help Your Brand Take Off*. Retrieved from Smartech: https://netcoresmartech.com/blog/personalization-in-travel

Henseler, J., Ringle, C. M., & Sarstedt, M. (2015). A new criterion for assessing discriminant validity in variance-based structural equation modeling. *Journal of the Academy of Marketing Science, 43*(1), 115–135. doi:10.100711747-014-0403-8

Hsiao, C.-H., Chang, J.-J., & Tang, K.-Y. (2016). Exploring the Influential Factors in Continuance Usage of Mobile Social Apps: Satisfaction, Habit and Customer Value Perspectives. *Telematics and Informatics, 33*(2), 342–355. doi:10.1016/j.tele.2015.08.014

Ivanova, O., Flores-Zamora, J., Khelladi, I., & Ivanaj, S. (2018). The generational cohort effect in the context of responsible consumption. *Management Decision, 57*(5), 1162–1183. doi:10.1108/MD-12-2016-0915

Kasasa. (2019, April 22). *Boomers, Gen X, Gen Y, and Gen Z Explained*. Retrieved from https://communityrising.kasasa.com/gen-x-gen-y-gen-z/

Koo, C., Wati, Y., & Chung, N. (2013). A Study of Mobile and Internet Banking Service: Applying for IS Success Model. *A Study of Mobile and Internet Banking Service*, 65-86.

Kourouthanassisa, P., Boletsisb, C., Bardakia, C., & Chasanidou, D. (2015). Tourists responses to mobile augmented reality travel guides: The role of emotions on adoption behavior. *Pervasive and Mobile Computing*, *18*, 71–87. doi:10.1016/j.pmcj.2014.08.009

Kuan, A. L., Ann, H. N., Mohd Badri, E. N., & Tang, K. Y. (2014, May). *UTAUT2 Influencing The Behavioural Intention To Adopt Mobile Applications.* Retrieved from utar: http://eprints.utar.edu.my/1277/

Lai, I. K. (2015). Traveler Acceptance of an App-Based Mobile Tour Guide. *Journal of Hospitality & Tourism Research (Washington, D.C.)*, *39*(3), 401–432. doi:10.1177/1096348013491596

Lee, S. (2020, June 17). *MP Moots Single Contact-Tracing App For Better Accountability*. Retrieved from CodeBlue: https://codeblue.galencentre.org/2020/06/17/mp-moots-single-contact-tracing-app-for-better-accountability/

Lissitsa, S., & Kol, O. (2016). Generation X vs. Generation Y – A decade of online shopping. *Journal of Retailing and Consumer Services*, *31*, 304–312. doi:10.1016/j.jretconser.2016.04.015

Litsa, T. (2018, June 4). *Analysis of Mary Meeker's Internet Trends Report 2018 - Part 2*. Retrieved from Clickz Web Site: https://www.clickz.com/analysis-of-mary-meekers-internet-trends-report-2018-part-2/214809/

MacHale, D. (2019). *Mobile Travel Trends 2019.* Retrieved 2019, from Travelport Digital: https://info-digital.travelport.com/trends2019

Manuel, N. (2018, September 28). *The Role of Augmented Reality in Mobile App Development*. Retrieved from Simple Programmer: https://simpleprogrammer.com/augmented-reality-mobile/

Mardiana, S., Tjakraatmadja, J. H., & Aprianingsih, A. (2015). DeLone-McLean Information System Success Model Revisited: The Separation of Intention to Use - Use and Integration of Technology Acceptance Models. *International Journal of Economics and Financial Issues*, 172-182.

Maria, S., Loureiro, C., Guerreiro, J., & Ali, F. (2019). 20 years of research on virtual reality and augmented reality in tourism context: A text-mining approach. *Tourism Management*, 1–21.

Mariani, M., Baggio, R., Fuhs, M., & Hoepken, W. (2018). Business Intelligence and Big Data in Hospitality and Tourism: A Systematic Literature Review. *International Journal of Contemporary Hospitality Management*, *30*(12), 3514–3554. doi:10.1108/IJCHM-07-2017-0461

Maurits, N. (2012). *From Neurology to Methodology and Back: An Introduction to Clinical Neuroengineering*. Springer. doi:10.1007/978-1-4614-1132-1

Merel, T. (2017, January 13). *After Mixed Year, Mobile AR to Drive $108 Billion VR/AR Market by 2021*. Retrieved from thewrap: https://www.thewrap.com/mixed-year-mobile-ar-drive-108-billion-vrar-market-2021-guest-column/

Olsson, T., Lagerstam, E., Kärkkäinen, T., & Väänänen-Vainio-Mattila, K. (2013). Expected user experience of mobile augmented reality services: A user study in the context of shopping centres. *Personal and Ubiquitous Computing*, *17*(2), 287–304. doi:10.100700779-011-0494-x

Parasuraman, A. (2000). Technology Readiness Index (Tri): A Multiple-Item Scale to Measure Readiness to Embrace New Technologies. *Journal of Service Research*, *2*(4), 307–320. doi:10.1177/109467050024001

Parasuraman, A., & Colby, C. (2002). Techno-Ready Marketing: How and Why Your Customers Adopt Technology. *Journal of Consumer Marketing*, *19*(4), 359–361. doi:10.1108/jcm.2002.19.4.359.1

Parasuraman, A., & Zeithaml, V. A. (1988). SERVQUAL: A multiple-item scale for measuring consumer perceptions of service quality. *Journal of Retailing*, 12–40.

Pascu, L. (2016, July 12). *IoT Could Generate up to $11.1 Trillion in Economic Value per Year by 2025*. Retrieved from Hot for Security: https://hotforsecurity.bitdefender.com/blog/iot-could-generate-up-to-11-1-trillion-in-economic-value-per-year-by-2025-15887.html

Peter, S., & McLean, E. R. (2009). A meta-analytic assessment of the DeLone and McLean IS success model: An examination of IS success at the individual level. *Information & Management*, *46*(3), 159–166. doi:10.1016/j.im.2008.12.006

Pradhan, M., Oh, J., & Lee, H. (2018). Understanding Travelers' Behavior for Sustainable Smart Tourism: A Technology Readiness Perspective. *Sustainability*, *10*(11), 4259. doi:10.3390u10114259

Prensky, M. (2001). Digital Natives, Digital Immigrants Part 1. *On the Horizon*, *9*(5), 1-6.

Quintal, V., Phau, I., Sims, D., & Cheah, I. (2016). Factors influencing generation Y's purchase intentions of prototypical versus me-too brands. *Journal of Retailing and Consumer Services*, *30*, 175–183. doi:10.1016/j.jretconser.2016.01.019

Ramon, P.-S., Santiago, F.-C., Javier, S.-G., & Emilio, R. (2019). User Acceptance of Mobile Apps for Restaurants: An Expanded and Extended UTAUT-2. *Sustainability*.

Rios, R. (2017, December 13). *Single-Purpose or Multi-Purpose App – Which Option Is Better?* Retrieved from DZone: https://dzone.com/articles/single-purpose-or-multi-purpose-app-which-option-i

Robinson, V. M., & Schanzel, H. A. (2019). A tourism inflex: Generation Z travel experiences. *Journal of Tourism Futures*, *5*(2), 127–141. doi:10.1108/JTF-01-2019-0014

Savage, P. (2020, May 19). *eCommerce Personalization: Why Demographic-Based Personalization is on Its Way Out (And What to Do Instead)*. Retrieved from Coredna: https://www.coredna.com/blogs/ecommerce-personalization#14

Sekaran, U., & Bougie, R. (2016). Research Methods For Business: A Skill Building Approach (7th ed.). Wiley.com.

Sharma, Y. (2018). *Online Travel Market to reach $1,091 Billion, Globally, by 2022*. Retrieved from Allied Market Research: https://www.alliedmarketresearch.com/press-release/online-travel.html

Singh, R. (2018, Jun 5). *Malaysia Airports embraces big data with launch of new mobile app*. Retrieved from malaymail: https://www.malaymail.com/news/life/2018/06/05/malaysia-airports-embraces-big-data-with-launch-of-new-mobile-app/1638636

Smit, C., Roberts-Lombard, M., & Mpinganjira, M. (2018). Technology readiness and mobile self-service technology adoption in the airline industry: An emerging market perspective. *Acta Commercii*, *18*(1), 1–12. doi:10.4102/ac.v18i1.580

Stfalcon.com. (2018, Dec 6). *Top 10 Travel Industry Trends in 2019*. Retrieved from medium.com: https://medium.com/swlh/top-10-travel-industry-trends-in-2019-d43d157de7b9

Sukhraj, R. (2017, November 13). *38 Mobile Marketing Statistics to Help You Plan for 2019*. Retrieved from Impact Web Site: https://www.impactbnd.com/blog/mobile-marketing-statistics

Travel Apps: The Revolution in Traveling. (2016, September 7). Retrieved from Shoutem: https://medium.com/shoutem/travel-apps-the-revolution-in-traveling-cde012e79dd4

Trice, A. (2014, Jun 23). *Mobile UX: Refining Perceived Performance*. Retrieved from Dzone / Performance Zone: https://dzone.com/articles/mobile-ux-refining-perceived

Urbach, N., & Ahlemann, F. (2010). Structural Equation Modeling in Information Systems Research Using Partial Least Squares. *Journal of Information Technology Theory and Application*, *11*(2), 5–40.

Urbach, N., & Mueller, B. (September 2011). The Updated DeLone and McLean Model of Information Systems Success. Information Systems Theory: Explaining and Predicting Our Digital Society, 1, 1-18.

Vakhnenko, H. (2019). *Top 5 Ideas in the Travel App Development*. Retrieved from agilie: https://agilie.com/en/blog/top-5-ideas-in-the-travel-app-development

Venkatesh, V., Morris, M. G., Davis, G. B., & Davis, F. D. (2003). User Acceptance of Information Technology: Toward a Unified View. *MIS Quarterly: Management Information Systems*, *27*(3), 425–478. doi:10.2307/30036540

Venkatesh, V., Thong, J. Y., & Xu, X. (2012). Consumer Acceptance and Use of Information Technology: Extending the Unified Theory of Acceptance and Use of Technology. *MIS Quarterly: Management Information Systems*, *36*(1), 157–178. doi:10.2307/41410412

Wixom, B. H., & Todd, P. A. (2005). A Theoretical Integration of User Satisfaction and Technology Acceptance. *Information Systems Research*, *16*(1), 85–102. doi:10.1287/isre.1050.0042

Yoo, C., Kwon, S., Na, H., & Chang, B. (2017). Factors Affecting the Adoption of Gamified Smart Tourism Applications: An Integrative Approach. *Sustainability*, *9*(12), 1–21. doi:10.3390u9122162

Zaied, A. N. (2012). An Integrated Success Model for Evaluating Information System in Public Sectors. *Journal of Emerging Trends in Computing and Information Sciences*, 814-825.

Conclusion

COVID-19 notably continues to impact our communities, people and businesses. Safety measures have been taken to minimise outbreaks from COVID-19 including shutting down restaurants, cinemas, hotels, sporting and entertainment venues, shops and shopping centres to name a few. All of which have led to disruption to the travel and tourism ecosystem worldwide. Further, the impact of COVID-19 and the associated measures, protocols and restrictions exercised either mandatory or voluntary by every individual have induced some behaviour and societal change.

New knowledge is therefore sought to provide insight to the industry on how to transform and enhance their operations from a technological perspective with a view to accommodating the emerging customers' needs and wants due to COVID-19 pandemic. Various studies, observations and assessments re impact of COVID-19 to hospitality can be found at: https://www.igi-global.com/book/handbook-research-impacts-implications-covid/268789.

For example, some recent developments in artificial intelligence (AI) and social service robot technologies have enabled the use of AI technologies in service delivery and the COVID-19 pandemic may amplify the popularity of such technology for public safety reasons.

Considering the significance of novel technologies, applications and solutions to change and contribute to the recovery and sustainability of the tourism and hospitality industry, this book presents advances in AI, mobile, blockchain, virtual and augmented reality, and other emerging technology and how they could be devised in the tourism and hospitality sector, which spans across the 9 chapters. This includes identification and discussion of both physical and psychological factors that can influence customers' and employees' acceptance of such technology for various tourism and hospitality services.

We hope that readers of this book will find the coverage and discussion of the recent advancements in various aspects of digital solutions in the tourism and hospitality sector useful in mitigating and emerging from the impact of COVID-19, as well as assisting hospitality scholars with their research focus to develop innovative solutions for the industry.

Compilation of References

ABS-CBN News. (2009). *Japan launches 'ramen robot chef'*. https://news.abs-cbn.com/lifestyle/08/05/09/japan-launches-ramen-robot-chef

Abubakar, A. I., Omeke, K. G., Ozturk, M., Hussain, S., & Imran, M. A. (2020). The Role of Artificial Intelligence Driven 5G Networks in COVID-19 Outbreak: Opportunities, Challenges, and Future Outlook. *Frontiers in Communications and Networks*, *1*, 575065. Advance online publication. doi:10.3389/frcmn.2020.575065

Adeola, O., & Evans, O. (2019). Digital tourism: Mobile phones, internet and tourism in Africa. *Tourism Recreation Research*, *44*(2), 190–202. doi:10.1080/02508281.2018.1562662

Agarwal, R., & Prasad, J. (1998). A Conceptual and Operational Definition of Personal Innovativeness in the Domain of Information Technology. *Information Systems Research*, *9*(2), 204–215. doi:10.1287/isre.9.2.204

Aksan Aldanmaz, E., & Sever, R. (2017). Gıdaların dizaynında 3 boyutlu yazıcı teknolojisi uygulamaları. AB2017, XIX. In Akademik Bilişim Konferansı, Aksaray.

Alcatel Lucent Enterprise. (2020, October). *The Internet of Things for the Hospitality Industry*. Solution Brief IoT in Hospitality. Retrieved from https://www.al-enterprise.com/-/media/assets/internet/documents/iot-for-hospitality-solutionbrief-en.pdf

Aliperti, G., & Cruz, A. M. (2020). Promoting built-for-disaster-purpose mobile applications: An interdisciplinary literature review to increase their penetration rate among tourists. *Journal of Hospitality and Tourism Management*, *44*, 193–210. doi:10.1016/j.jhtm.2020.06.006

Alnemari, A., Arodi, S., Sosa, V., Pandey, S., Romanowski, C., Raj, R., & Mishra, S. (2018). *Protecting Infrastructure Data via Enhanced Access Control, Blockchain and Differential Privacy*. Springer.

Alotaibi, R., Houghton, L., & Sandhu, K. (2017). Factors Influencing Users' Intentions to Use Mobile Government Applications in Saudi Arabia: TAM Applicability. *International Journal of Advanced Computer Science and Applications*, *8*(7), 200–211. doi:10.14569/IJACSA.2017.080727

Aluri, A. (2016). *The Future of IoT in Hospitality*. Tech Trends Connectivity. Retrieved from https://www.hftp.org/hitec/i/downloads/TechTalkIoT.pdf

Anderson, G. (Ed.). (2004). *Reinventing the museum: Historical and contemporary perspectives on the paradigm shift*. Editions Rowman Altamira.

Appel, L., Appel, E., Bogler, O., Wiseman, M., Cohen, L., Ein, N., Abrams, H. B., & Campos, J. L. (2020). Older adults with cognitive and/or physical impairments can benefit from immersive virtual reality experiences: A feasibility study. *Frontiers in medicine*, *6*, 329. doi:10.3389/fmed.2019.00329 PMID:32010701

Arıkan, I. (n.d.). *Gastronomide 3D gıda yazıcısı uygulamaları*. https://www.fiyatimbu.com/blog/gastronomide-3d-gida-yazicisi-uygulamalari_213

Arnaboldi, M., Melisa, L. & Lema, D. (2021). The participatory turn in museums: The online facet. *Poetics*. . doi:10.1016/j.poetic.2021.101536

Asghar, T., & Nauman, A. (2010). *Augmented Reality in Museum Environments: A Case Study at Lund University Historical Museum* [Unpublished Master Thesis]. Division of ergonomics and aerosol technology, Department of Design Sciences, Lund University, Sweden.

Asif, M., Sabeel, M., Rahman, M. U., & Khan, Z. (2015). Waiter robot–solution to restaurant automation. *MDSRC Proceedings*.

Attala, J. (2019, March 26). *10 Ways Smart Technology is Reshaping the Hotel Industry*. Hotel Management. Retrieved from: https://www.hotelmanagement.net/tech/10-ways-smart-technology-reshaping-hotel-industry

Avidar, R. (2018). Engagement, interactivity, and diffusion of innovations. *Handb. Commun. Engagem, 505*.

Azuma, R. T. (1997). A Survey of Augmented Reality. *Ed. Presence.*, *6*(4), 355–385. doi:10.1162/pres.1997.6.4.355

Azuma, R., Baillot, Y., Behringer, R., Feiner, S., Julier, S., & MacIntyre, B. (2001). *Recent advances in augmented reality*. Naval Research Lab. doi:10.1109/38.963459

Babich, N. (2018, February 12). A Comprehensive Guide To Mobile App Design. *Design & Development*.

Bach, C., & Scapin, D. L. (2004). Obstacles and Perspectives for Evaluating Mixed Reality Systems Usability. Acte du Dhir, W., A., & Al-kahtani, M. (2013). A case study on user experience (UX) evaluation of mobile augmented reality prototypes. *Journal of Universal Computer Science*, *19*, 1175–1196.

Barfield, W., Zeltzer, D., Sheridan, T., & Slater, M. (1995). Presence and performance within virtual environments. *Virtual environments and advanced interface design*, 473-513.

Barten, M. (2019). *How Augmented Reality is Revolutionising the Travel Industry*. Retrieved from Revfine Web Site: https://www.revfine.com/augmented-reality-travel-industry/

BBC. (2014). *'Smart' chopsticks unveiled in China.* https://www.bbc.com/news/technology-29062509

Beck, J. (2016). *Statistics about VR in travel and tourism.* Retrieved 2/20/2021, from http://www.virtual-reality-in-tourism.com/statistics-about-vr-in-travel/

Benbasat, I., & Barki, H. (2007). Quo vadis, TAM? *Journal of the Association for Information Systems, 8*(4), 211–218. doi:10.17705/1jais.00126

Bevan, N. (2009). What is the difference between the purpose of usability and user experience evaluation methods. *Proceedings of the Workshop UXEM '09.*

Biz4Intellia. (2022). *Intellia IoT (An End-To-End IoT Business Solution).* Intellia IoT. Retrieved from https://www.biz4intellia.com/

Blöcher, K., & Alt, R. (2020). AI and robotics in the European restaurant sector: Assessing potentials for process innovation in a high-contact service industry. *Electronic Markets.* Advance online publication. doi:10.100712525-020-00443-2

Booking global. (2018, April 17). *Where Sustainable Travel is Headed in 2018.* Retrieved from https://globalnews.booking.com/where-sustainable-travel-is-headed-in-2018/

Borràs, J., Moreno, A., & Valls, A. (2014). Intelligent tourism recommender systems: A survey. *Expert Systems with Applications, 41*(16), 7370–7389. doi:10.1016/j.eswa.2014.06.007

Bower, M., DeWitt, D., & Lai, J. W. (2020). Reasons associated with preservice teachers' intention to use immersive virtual reality in education. *British Journal of Educational Technology, 51*(6), 2214–2232. doi:10.1111/bjet.13009

Bozkurt Uzan, Ş., & Sevimli, Y. (2020). Gastronomideki robotik uygulamalar ve yapay zeka. *Tourism and Recreation, 2*(2), 46–58.

Braganza, A., Chen, W., Canhoto, A., & Sap, S. (2020). Productive employment and decent work: The impact of AI adoption on psychological contracts, job engagement and employee trust. *Journal of Business Research.* Advance online publication. doi:10.1016/j.jbusres.2020.08.018 PMID:32836565

Buhalis, D. (2003). eTourism: Information technology for strategic tourism management. Pearson Education.

Buhalis, D. (2019). Technology in tourism-from information communication technologies to eTourism and smart tourism towards ambient intelligence tourism: A perspective article. *Tourism Review, 75*(1), 267–272. doi:10.1108/TR-06-2019-0258

Buhalis, D., & Leung, R. (2017). Smart Hospitality—Interconnectivity and Interoperability Towards an Ecosystem. *International Journal of Hospitality Management,* 41–50.

Buhalis, D., & Sinarta, Y. (2019). Real-time co-creation and nowness service: Lessons from tourism and hospitality. *Journal of Travel & Tourism Marketing*, *36*(5), 563–582. doi:10.1080/10548408.2019.1592059

Bustamante, A., Sebastia, L., & Onaindia, E. (2020). BITOUR: A Business Intelligence Platform for Tourism Analysis. *ISPRS International Journal of Geo-Information*, *9*(11), 671. doi:10.3390/ijgi9110671

Calvignac, C., & Smolinski, J. (2020). How can the use of a mobile application change the course of a sightseeing tour? A question of pace, gaze and information processing. *Tourist Studies*, *20*(1), 49–74. doi:10.1177/1468797619885955

Calvo-Porral, C., & Pesqueira-Sanchez, R. (2019). Generational differences in technology behavior: Comparing millennials and Generation X. *Kybernetes*, *49*(11), 2755–2772. doi:10.1108/K-09-2019-0598

Camatti, N., Bertocchi, D., Carić, H., & van der Borg, J. (2020). A digital response system to mitigate overtourism. The case of Dubrovnik. *Journal of Travel & Tourism Marketing*, *37*(8–9), 887–901. doi:10.1080/10548408.2020.1828230

Carassa, A., Morganti, F., & Tirassa, M. (2004). *Movement, action, and situation: Presence in virtual environments*. Academic Press.

Carmigniani, J., Borko, F., Marco, A., Paolo, C., Ernesto, D., & Misa, I. (2011). Augmented Reality Technologies, Systems and Applications. *Multimedia Tools and Applications*, *51*(1), 341–377. doi:10.100711042-010-0660-6

Carvajal-Trujillo, E., Molinillo, S., & Liébana-Cabanillas, F. (2020). Determinants and risks of intentions to use mobile applications in museums: An application of fsQCA. *Current Issues in Tourism*, •••, 1–20. doi:10.1080/13683500.2020.1780200

Caudell, T. P., & Mizell, D. W. (1992). Augmented reality: An application of heads-up display technology to manual manufacturing processes. *System Sciences, 1992. Proceedings of the Twenty-Fifth Hawaii International Conference on,* 2(January), 659-669.

Çerkez, M., & Kızıldemir, Ö. (2020). Yiyecek–içecek işletmelerinde yapay zekâ kullanımı. *Türk Turizm Araştırmaları Dergisi*, *4*(2), 1264–1278. doi:10.26677/TR1010.2020.394

Cerović, M. P. (2019, September). The Importance of Innovation in Designing a Luxury Tourism Offering: The Perspective of Tourism Destination Stakeholders. *Tourism in Southern and Eastern Europe: Creating Innovative Tourism Experiences: The Way to Extend the Tourist Season.*

ChangeNOW. (2020, May 14). Blockchain Technology in Hospitality & Travel industry. *ChangeNOW*.

Chang, I.-C., Chou, P.-C., Yeh, R. K.-J., & Tseng, H.-T. (2016). Factors influencing Chinese tourists' intentions to use the Taiwan Medical Travel App. *Telematics and Informatics*, *33*(2), 401–409. doi:10.1016/j.tele.2015.09.007

Chang, K., Chang, C., Hou, H., Sung, Y., Chao, H., & Lee, C. (2014). Development and behavioral pattern analysis of a mobile guide system with augmented reality for painting appreciation instruction in an art museum. *Computers & Education*, *71*, 185–197. doi:10.1016/j.compedu.2013.09.022

Charfi, A. A. (2012). *L'expérience d'immersion en ligne dans les environnements marchands de réalité virtuelle*. Thèse de doctorat en gestion, Universite Paris-dauphine. France.

Charlotte, P.-D. (2019). *Best Practices for Researching Generation X, Y and Z*. Retrieved from FlexMR Web Site: https://blog.flexmr.net/researching-gen-x-y-and-z

Chauhan, D., & Saxena, P. (2021, December 19). *The Future of Hospitality - IOT will be the Next Big Thing*. SoftwareSuggest Blog. Retrieved from https://www.softwaresuggest.com/blog/iot-next-big-thing-future-hospitality/#

Chen, C.-C., & Tsai, J.-L. (2019). Determinants of behavioral intention to use the Personalized Location-based Mobile Tourism Application: An empirical study by integrating TAM with ISSM. *Future Generation Computer Systems*, *96*, 628–638. doi:10.1016/j.future.2017.02.028

Chen, M.-M., Murphy, H. C., & Knecht, S. (2016). An Importance Performance Analysis of smartphone applications for hotel chains. *Journal of Hospitality and Tourism Management*, *29*, 69–79. doi:10.1016/j.jhtm.2016.05.001

Cheong, R. (1995). The virtual threat to travel and tourism. *Tourism Management*, *16*(6), 417–422. doi:10.1016/0261-5177(95)00049-T

Choi, Y., Hickerson, B., & Lee, J. (2018). Investigation of the technology effects of online travel media on virtual travel experience and behavioral intention. *Journal of Travel & Tourism Marketing*, *35*(3), 320–335. doi:10.1080/10548408.2017.1340224

Chung, N., Hyunae, L., Jin-Young, K., & Chulmo, K. (2018). The Role of Augmented Reality for Experience-Influenced Environments: The Case of Cultural Heritage Tourism in Korea. *Journal of Travel Research*, *57*(5), 627–643. doi:10.1177/0047287517708255

Chylinski, M., Heller, J., Hilken, T., Keeling, D. I., Mahr, D., & De Ruyter, K. (2020, November). Augmented reality marketing: A technology-enabled approach to situated customer experience. *Australasian Marketing Journal*, *28*(4), 374–384. doi:10.1016/j.ausmj.2020.04.004

Cohen, J. (1988). *Statistical Power Analysis for the Behavioral Sciences* (2nd ed.). Routledge.

Cohen, J. (1992). A Power Primer. *Psychological Bulletin*, *112*(1), 155–159. doi:10.1037/0033-2909.112.1.155 PMID:19565683

Coppol, S. (2019, November 18). *Redefining the Hospitality Industry with Internet of Things*. Compare the Cloud. Retrieved from https://www.comparethecloud.net/articles/redefining-the-hospitality-industry-with-internet-of-things/

Craig, A. B. (2013). Understanding augmented reality: concepts and applications. Morgan Kaufmann.

Crosby, M. e. (2016). Blockchain technology: Beyond bitcoin. *Applied Innovation*, 6-10.

Current Population Estimates, Malaysia, 2020. (2020, July 15). Retrieved from Department of Statistics Malaysia: https://www.dosm.gov.my/v1/index.php?r=column/cthemeByCat&cat=155&bul_id=OVByWjg5YkQ3MWFZRTN5bDJiaEVhZz09&menu_id=L0pheU43NWJwRWVSZklWdzQ4TlhUUT09

Damala, A., Cubaud, P., Bationo, A., Houlier, P., & Marchal, I. (2008). Bridging the Gap between the Digital and the Physical: Design and Evaluation of a Mobile Augmented Reality Guide for the Museum Visit. In *Proceedings of the 3rd international conference on Digital Interactive Media in Entertainment and Arts - DIMEA '08*. Athens, Greece: ACM Press.

Davenport, T. H. (2018). *The AI Advantage How to Put the Artificial Intelligence Revolution to Work*. MIT Press. doi:10.7551/mitpress/11781.001.0001

Davenport, T. H., & Ronanki, R. (2018). Artificial intelligence for the real world. *Harvard Business Review*, (January-February), 2–10.

Davenport, T., Guha, A., Grewal, D., & Bressgott, T. (2020). How artificial intelligence will change the future of marketing. *Journal of the Academy of Marketing Science*, *48*(1), 24–42. doi:10.100711747-019-00696-0

Davis, E., & Bosma, T. (2020). *The Future of Travel & Hospitality*. Cognizant. Retrieved from https://www.cognizant.com/whitepapers/the-future-of-travel-and-hospitality-codex5236.pdf

Decortips. (2021). *Mutfakta yüksek teknoloji: Akıllı tava*. https://decortips.com/tr/mutfak/mutfakta-teknoloji-akilli-tava/

Del Vecchio, P., Secundo, G., Maruccia, Y., & Passiante, G. (2019). A system dynamic approach for the smart mobility of people: Implications in the age of big data. *Technological Forecasting and Social Change*, *149*, 119771. doi:10.1016/j.techfore.2019.119771

Deloitte. (2022). *Smart Sensors and Supply Chain Innovation: Advanced Platforms Create Digital Insights*. Perspectives. Retrieved from https://www2.deloitte.com/us/en/pages/operations/articles/smart-sensors-and-supply-chain.html

DeLone, W. H., & McLean, E. R. (1992). Information Systems Success: The Quest for the Dependent Variable. *Information Systems Research*, *3*(1), 60–95. doi:10.1287/isre.3.1.60

DeLone, W. H., & McLean, E. R. (2003). The DeLone and McLean Model of Information Systems Success: A Ten-Year Update. *Journal of Management Information Systems*, *19*(4), 9–30. doi:10.1080/07421222.2003.11045748

Demirci, A. E., & Ersoy, N. F. (2008). *Technology Readiness for Innovative High-Tech Products: How Consumers Perceive and Adopt New Technologies*. Academic Press.

Desouza, K. C., Dawson, G. S., & Chenok, D. (2020). Designing, developing, and deploying artificial intelligence systems: Lessons from and for the public sector. *Business Horizons*, *63*(2), 205–213. doi:10.1016/j.bushor.2019.11.004

DHL Trend Research. (2014). Retrieved from: https://www.dhl.com/content/dam/downloads/g0/about_us/logistics_insights/csi _augmented_reality_report_290414.pdf

Dieck, T. M. C., & Jung, T. (2015). A theoretical model of mobile augmented reality acceptance in urban heritage tourism. *Current Issues in Tourism*, *18*, 1–21.

Dieck, T. M. C., & Jung, T. (2016). Value of augmented reality to enhance the visitor experience: A case study of Manchester Jewish Museum. *Review of Tourism Research*, *7*, 1–5.

Dieck, T. M. C., & Jung, T. H. (2017). Value of Augmented Reality at Cultural Heritage Sites: A Stakeholder Approach. *Journal of Destination Marketing & Management*, *6*(2), 110–117. doi:10.1016/j.jdmm.2017.03.002

Dillet, R. (2014). *B4RM4N is a smart shaker that turns the art of cocktail mixing into a science*. https://techcrunch.com/2014/10/28/b4rm4n-is-a-smart-shaker-that-turns-the-art-of-cocktail-mixing-into-a-science/

Dogru, T. M. (n.d.). *Blockchain Technology & its Implications for the Hospitality Industry*. Boston Hospitality Review.

Dogtiev, A. (2019, February 16). *App Download and Usage Statistics (2018)*. Retrieved from BusinessofApps Web Site: https://www.businessofapps.com/data/app-statistics/

Do, H.-N., Shih, W., & Ha, Q.-A. (2020, August). Effects of mobile augmented reality apps on impulse buying behavior: An investigation in the tourism field. *Heliyon*, *6*(8), e04667. Advance online publication. doi:10.1016/j.heliyon.2020.e04667 PMID:32923709

Dong, X., & McIntyre, S. H. (2014, November). The Second Machine Age: Work, Progress, and Prosperity in a Time of Brilliant Technologies. *Quantitative Finance*, *14*(11), 1895–1896. doi:10.1080/14697688.2014.946440

Dorcic, J., Komsic, J., & Markovic, S. (2018). Mobile Technologies and Applications Towards Smart Tourism. *Tourism Review*.

Dorcic, J., Komsic, J., & Markovic, S. (2019). Mobile technologies and applications towards smart tourism–state of the art. *Tourism Review*, *74*(1), 82–103. doi:10.1108/TR-07-2017-0121

Dorie, A., & Loranger, D. (2020). The multi-generation Generational differences in channel activity. *International Journal of Retail & Distribution Management*, *48*(4), 395–416. doi:10.1108/IJRDM-06-2019-0196

Douglas, A., Lubbe, B., & van Rooyen, A. (2018). Business travellers' use of mobile travel applications: A generational analysis. *Information Technology & Tourism*, *18*(1–4), 113–132. doi:10.100740558-017-0103-6

Economic Impact Reports. (2019). https://wttc.org/Research/Economic-Impact#:~:text=Over%20the%20past%20five%20years,10.3%25%20of%20global%20GDP

Errichiello, L., Micera, R., Atzeni, M., & Del Chiappa, G. (2019). Exploring the implications of wearable virtual reality technology for museum visitors' experience: A cluster analysis. *International Journal of Tourism Research*, *21*(5), 590–605. doi:10.1002/jtr.2283

Escobar-Rodrígueza, T., & Carvajal-Trujillo, E. (2014). Online purchasing tickets for low cost carriers: An application of the unified theory of acceptance and use of technology (UTAUT) model. *Tourism Management*, *43*, 70–88. doi:10.1016/j.tourman.2014.01.017

Euronews. (2020). *No social distancing worries at 'UK's first' sci-fi robot-themed diner*. https://www.euronews.com/travel/2020/09/24/no-social-distancing-worries-at-uk-s-first-sci-fi-robot-themed-diner

Evli, A. (2020). *3D gıda baskısı tüketimde yeni bir çağ açıyor*. https://teknoloji.org/3d-gida-baskisi-tuketimde-yeni-bir-cag-aciyor/

Evrard, Y., Pras, B., Roux, E., Desmet, P., Dussaix, A. M., & Lilien, G. L. (2009). *Market-Fondements et méthodes des recherches en marketing* (4th ed.). Dunod.

Fadlullah, Z. M., Tang, F., Mao, B., Kato, N., Akashi, O., Inoue, T., & Mizutani, K. (2017). State-of-the-Art Deep Learning: Evolving Machine Intelligence Toward Tomorrow's Intelligent Network Traffic Control Systems. *IEEE Communications Surveys and Tutorials*, *19*(4), 2432–2455. doi:10.1109/COMST.2017.2707140

Falcao, R. P. Q., Ferreira, J. B., & Carrazedo Marques da Costa Filho, M. (2019). The influence of ubiquitous connectivity, trust, personality and generational effects on mobile tourism purchases. *Information Technology & Tourism*, *21*(4), 483–514. doi:10.100740558-019-00154-1

Fan, X., Chai, Z., Deng, N., & Dong, X. (2020). Adoption of augmented reality in online retailing and consumers' product attitude: A cognitive perspective. *Journal of Retailing and Consumer Services*, *53*(March), 101986. Advance online publication. doi:10.1016/j.jretconser.2019.101986

Fardous, J., Du, J. T., Hansen, P., Choo, K.-K. R., & Huang, S. (Sam). (. (2019). Group trip planning and information seeking behaviours by mobile social media users: A study of tourists in Australia, Bangladesh and China. *Journal of Information Science*. Advance online publication. doi:10.1177/0165551519890515

Fernández-Durán, J. (2016). Defining generational cohorts for marketing in Mexico. *Journal of Business Research*, *69*(2), 435–444. doi:10.1016/j.jbusres.2015.06.049

Flavián, C., Ibáñez-Sánchez, S., & Orús, C. (2019). The impact of virtual, augmented and mixed reality technologies on the customer experience. *Journal of Business Research*, *100*(July), 547–560. doi:10.1016/j.jbusres.2018.10.050

Fortino, G. G. (n.d.). *Advances in Body Sensor Networks: Sensors, Systems, and Applications*. MDPI.

Fourhooks. (2015, April 28). *The Generation Guide - Millennials, Gen X, Y, Z and Baby Boomers*. Retrieved from fourhooks: http://fourhooks.com/marketing/the-generation-guide-millennials-gen-x-y-z-and-baby-boomers-art5910718593/

Foxley, W. (2019, November 7). Digital Travel Firm Webjet Has Launched Its Booking Verification Blockchain. *CoinDesk.*

Frank, A. G., Mendes, G. H. S., Ayala, N. F., & Ghezzi, A. (2019). Servitization and Industry 4.0 convergence in the digital transformation of product firms: A business model innovation perspective. *Technological Forecasting and Social Change, 141*, 341–351. doi:10.1016/j.techfore.2019.01.014

Friedman, M., & Sen, S. (2017, June). *Hospitality in the Digital Era.* Retrieved from https://www.cognizant.com/whitepapers/hospitality-in-the-digital-era-codex2543.pdf

Fuchs, H., & Ackerman, J. (1999). *Displays for augmented reality: Historical remarks and future prospects. Mixed Reality Merging Real and Virtual Worlds*. Ohmsha Ltd.

Fuste-Forne, F. (2021). Robot chefs in gastronomy tourism: What's on the menu? *Tourism Management Perspectives, 37*, 37. doi:10.1016/j.tmp.2020.100774

Gafni, R. (2008). Framework for quality metrics in mobile-wireless information systems. *Interdisciplinary Journal of Information, Knowledge, and Management, 3*, 23–38. doi:10.28945/88

García, M. V., Blasco López, M. F., & Sastre Castillo, M. Á. (2019). Determinants of the acceptance of mobile learning as an element of human capital training in organisations. *Technological Forecasting and Social Change, 149*, 119783. Advance online publication. doi:10.1016/j.techfore.2019.119783

Ghaderi, Z., Hatamifar, P., & Ghahramani, L. (2019). How smartphones enhance local tourism experiences? *Asia Pacific Journal of Tourism Research, 24*(8), 778–788. doi:10.1080/10941665.2019.1630456

Gibbs, C., Gretzel, U., & Saltzman, J. (2016). An experience-based taxonomy of branded hotel mobile application features. *Information Technology & Tourism, 16*(2), 175–199. doi:10.100740558-016-0052-5

Gizli İlimler Kütüphanesi. (n.d.). *Robotik kültür ve turizm.* https://www.gizliilimler.org/Dijital-Sehirler-Cagi-ve-Robotlar-Sehri-Neom.htm

Glauser, W. (2013). Doctors among early adopters of Google Glass. *Canadian Medical Association Journal, 185*(16), 1385. doi:10.1503/cmaj.109-4607

Gonzalez, R., Gasco, J., & Llopis, J. (2020). Information and communication technologies and human resources in hospitality and tourism. *International Journal of Contemporary Hospitality Management, 32*(11), 3545–3579. doi:10.1108/IJCHM-04-2020-0272

Gordillo, A., Gallego, D., Barra, E., & Quemada, J. (2013). The City as a Learning Gamified Platform. *IEEE Frontiers in Education Conference (FIE).* 10.1109/FIE.2013.6684850

Gretzel, U., Reino, S., Kopera, S., & Koo, C. (2016). Smart Tourism Challenges. *Journal of Tourism*, 41-47.

Gretzel, U., Sigala, M., Xiang, Z., & Koo, C. (2015). Smart tourism: Foundations and developments. *Electronic Markets, 25*(3), 179–188. doi:10.100712525-015-0196-8

Gupta, A., Dogra, N., & George, B. (2017). What determines tourist adoption of smartphone apps?: An analysis based on the UTAUT-2 framework. *Journal of Hospitality and Tourism Technology*, 50–64.

Gupta, M. (2017). *Blockchain for dummies*. John Wiley & Sons, Inc.

Gursoy, D. (2018). Future of hospitality marketing and management research. *Tourism Management Perspectives, 25*, 185–188. doi:10.1016/j.tmp.2017.11.008

Guttentag, D. A. (2010). Virtual reality: Applications and implications for tourism. *Tourism Management, 31*(5), 637–651. doi:10.1016/j.tourman.2009.07.003

Haenlein, M., & Kaplan, A. (2019). A brief history of artificial intelligence: On the past, present, and future of artificial intelligence. *California Management Review, 61*(4), 5–14. doi:10.1177/0008125619864925

Hair, J., Hult, G. T., Ringle, C. M., & Sarstedt, M. (2017). *A Primer on Partial Least Squares Structural Equation Modeling (PLS-SEM)* (2nd ed.). SAGE Publications.

Hammady, R., Ma, M., & Powell, A. (2018). User Experience of Markerless Augmented Reality Applications in Cultural Heritage Museums: 'Museum Eye' as a Case Study. Academic Press.

Han, D.-I., Dieck, M. C. T., & Jung, T. (2018). User Experience Model for Augmented Reality Applications in Urban Heritage Tourism. *Journal of Heritage Tourism, 13*(1), 46–61. doi:10.1080/1743873X.2016.1251931

Hande, P. (2020, May 5). *Personalization in Travel – AI-powered Recommendations To Help Your Brand Take Off*. Retrieved from Smartech: https://netcoresmartech.com/blog/personalization-in-travel

Hao, F., Xiao, Q., & Chon, K. (2020). COVID-19 and China's Hotel Industry: Impacts, a Disaster Management Framework, and Post-Pandemic Agenda. *International Journal of Hospitality Management, 90*, 102636. doi:10.1016/j.ijhm.2020.102636 PMID:32834356

Harrison, C. M. (2008). *Exploring emotional web experience: More than just usability and good design* (Doctoral dissertation). University of York.

Hassan, A., & Jung, T. (2018). *Augmented reality as an emerging application in tourism marketing education. In Virtual and Augmented Reality: Concepts, Methodologies, Tools, and Applications*. IGI Global.

Hassenzahl, M. (2008). User experience (UX): towards an experiential perspective on product quality. *Proceedings of the 20th International Conference of the Association Francophone d'Interaction Homme-Machine on*. 10.1145/1512714.1512717

Hassenzahl, M. (2003). The thing and I: understanding the relationship between user and product. In M. A. Blythe, K. Overbeeke, A. F. Monk, & P. C. Wright (Eds.), *Funology* (pp. 31–42). Kluwer Academic Publishers. doi:10.1007/1-4020-2967-5_4

Hazarhun, E., & Yılmaz, Ö. (2020). Restoranlarda dijital dönüşüm: Touch restoran örneği. *Gastroia: Journal of Gastronomy and Travel Research*, *4*(3), 384–399.

Heidenreich, S., & Kraemer, T. (2015). Passive innovation resistance: The curse of innovation? Investigating consequences for innovative consumer behavior. *Journal of Economic Psychology*, *51*, 134–151. doi:10.1016/j.joep.2015.09.003

Heidenreich, S., & Spieth, P. (2013). Why innovations fail—The case of passive and active innovation resistance. *International Journal of Innovation Management*, *17*(05), 1350021. doi:10.1142/S1363919613500217

Henseler, J., Ringle, C. M., & Sarstedt, M. (2015). A new criterion for assessing discriminant validity in variance-based structural equation modeling. *Journal of the Academy of Marketing Science*, *43*(1), 115–135. doi:10.100711747-014-0403-8

Hirschman, E., & Holbrook, M. B. (1992). *The semiotics of consumption*. Editions Mouton de Gruyter.

Hjalager, A.-M. (1996). Tourism and the environment: The innovation connection. *Journal of Sustainable Tourism*, *4*(4), 201–218. doi:10.1080/09669589608667268

Hjalager, A.-M. (2002). Repairing innovation defectiveness in tourism. *Tourism Management*, *23*(5), 465–474. doi:10.1016/S0261-5177(02)00013-4

Högberg, J., Wästlund, E., Aas, T. H., Hjemdahl, K., & Nordgård, D. (2020). Herding the Hordes: Using Location-Based Services and Mobile Messaging to Affect Visitor Behavior. *Journal of Hospitality & Tourism Research (Washington, D.C.)*, *44*(5), 870–878. doi:10.1177/1096348020912449

Hou, Y., Zhang, K., & Li, G. (2021). Service robots or human staff: How social crowding shapes tourist preferences. *Tourism Management*, *83*, 104242. Advance online publication. doi:10.1016/j.tourman.2020.104242

Hoyer, W., Kroschke, D.M., Schmitt, B., Kraume, K., & Shankar, V. (2020). Transforming the customer experience through new technologies. *Journal of Interaction Marketing*. . doi:10.1016/j.intmar.2020.04.001

Hsiao, C.-H., Chang, J.-J., & Tang, K.-Y. (2016). Exploring the Influential Factors in Continuance Usage of Mobile Social Apps: Satisfaction, Habit and Customer Value Perspectives. *Telematics and Informatics*, *33*(2), 342–355. doi:10.1016/j.tele.2015.08.014

Huang, M. H., & Rust, R. T. (2018). Artificial Intelligence in Service. *Journal of Service Research*, *21*(2), 155–172. doi:10.1177/1094670517752459

Huang, Y. C., Chang, L. L., Yu, C. P., & Chen, J. (2019). Examining an extended technology acceptance model with experience construct on hotel consumers' adoption of mobile applications. *Journal of Hospitality Marketing & Management, 28*(8), 957–980. doi:10.1080/19368623.2019.1580172

Hudson, S., Matson-Barkat, S., Pallamin, N., & Jegou, G. (2019). With or without you? Interaction and immersion in a virtual reality experience. *Journal of Business Research, 100*, 459–468. doi:10.1016/j.jbusres.2018.10.062

Hunter, W. C., Chung, N., Gretzel, U., & Koo, C. (2015). Constructivist research in smart tourism. *Asia Pacific Journal of Information Systems, 25*(1), 105–120. doi:10.14329/apjis.2015.25.1.105

ICOM. (2007, August 24). *The World Museum Community - Museum Definition.* Retrieved from ICOM International Council of Museums: https://icom.museum/the-vision/museumdefinition/

Igor. (2020, February 18). *IOT Smart Hotels: Enabling Innovation in the Hospitality Industry.* Igor. Retrieved from https://www.igor-tech.com/news-and-insights/articles/iot-smart-hotels-enabling-innovation-in-the-hospitality-industry

IJsselsteijn, W. A., De Ridder, H., Freeman, J., & Avons, S. E. (2000). *Presence: concept, determinants, and measurement.* Paper presented at the Human vision and electronic imaging V.

Imbardelli, A. P. (2019, April 29). *How Smart Rooms can Transform Hotel Brands.* LinkedIn. Retrieved from https://www.linkedin.com/pulse/how-smart-rooms-can-transform-hotel-brands-a-patrick-imbardelli

Independent Türkçe. (2019). *Teknoloji hizmet sektöründe: Robot garsonlar işbaşı yaptı.* https://www.indyturk.com/node/89406/haber/teknoloji-hizmet-sekt%C3%B6r%C3%BCnde-robot-garsonlar-i%C5%9Fba%C5%9F%C4%B1-yapt%C4%B1

Infosys. (2018). *Role of AI in Travel and Hospitality Industry.* Infosys.

Institute of Museum and Library Services. (n.d.). *Status of technology and digitization in the nation's museums and libraries.* Retrieved from: https://www.imls.gov/resources/TechDig05/Technology% 2BDigitization.pdf

International Organization for Standardization. (2010a). *Ergonomics of human-system interaction —Part 11: Usability: Definitions and concepts* (ISO/DIS 9241-11.2, 3.1). Retrieved from: https://www.iso.org/obp/ui/#iso:std:iso:9241:-11:dis:ed-2:v2:en

International Organization for Standardization. (2010b). *Ergonomics of human-system interaction — Part 210: Human-centred design for interactive systems* (ISO 9241-210:2010, 2.15). Author.

Ivanov, S. W. (2017, May 17-19). Adoption of Robots and Service Automation by Tourism and Hospitality Companies. *INVTUR 2017*, 1-19.

Ivanov, S., Gretzel, U., Berezina, K., Sigala, M., & Webster, C. (2019). Progress on robotics in hospitality and tourism: a review of the literature. *Journal of Hospitality and Tourism Technology, 10*(4). doi:10.1108/JHTT-08-2018-0087

Ivanova, O., Flores-Zamora, J., Khelladi, I., & Ivanaj, S. (2018). The generational cohort effect in the context of responsible consumption. *Management Decision, 57*(5), 1162–1183. doi:10.1108/MD-12-2016-0915

Izmaylov, M. A. (n.d.). *A Practical Application of Blockchain for the Travel Industry*. WindingTree.

Jiang, Y., & Kim, Y. (2015). Developing multi-dimensional green value: Extending social exchange theory to explore customers' purchase intention in green hotels–evidence from Korea. *International Journal of Contemporary Hospitality Management, 27*(2), 308–334. doi:10.1108/IJCHM-08-2013-0383

Jordan, M. I., & Mitchell, T. M. (2015). Machine learning: Trends, perspectives, and prospects. *Science, 349*(6245), 255–260. doi:10.1126cience.aaa8415 PMID:26185243

Jung, T., Chung, N., & Leue, M. C. (2015). The determinants of recommendations to use augmented reality technologies: The case of a Korean theme park. *Tourism Management, 49*, 75–86. doi:10.1016/j.tourman.2015.02.013

Jung, T., Dieck, M. C. T., Hyunae, L., & Chung, N. (2016). *Effects of Virtual Reality and Augmented Reality on Visitor Experiences in Museum. In Information and Communication Technologies in Tourism 2016*. Springer International Publishing.

Jung, T., tom Dieck, M. C., Lee, H., & Chung, N. (2016). *Effects of virtual reality and augmented reality on visitor experiences in museum. In Information and communication technologies in tourism 2016*. Springer.

Kaimal, G., Carroll-Haskins, K., Berberian, M., Dougherty, A., Carlton, N., & Ramakrishnan, A. (2020). Virtual Reality in Art Therapy: A Pilot Qualitative Study of the Novel Medium and Implications for Practice. *Art Therapy: Journal of the American Art Therapy Association, 37*(1), 16–24. doi:10.1080/07421656.2019.1659662

Kakatkar, C., Bilgram, V., & Füller, J. (2020). Innovation analytics: Leveraging artificial intelligence in the innovation process. *Business Horizons, 63*(2), 171–181. doi:10.1016/j.bushor.2019.10.006

Kansakar, P. M. (2018, February). Technology in Hospitality Industry: Prospects and Challenges. *IEEE Consumer Electronics Magazine*.

Kansakar, P. M., Munir, A., & Shabani, N. (2019). Technology in the Hospitality Industry: Prospects and Challenges. *IEEE Consumer Electronics Magazine, 8*(3), 60–65. doi:10.1109/MCE.2019.2892245

Kapiki, S. (2010, January). Energy Management in Hospitality: A Study of the Thessaloniki Hotels. *Economics and Organization of Future Enterprise*, 78-97.

Kasasa. (2019, April 22). *Boomers, Gen X, Gen Y, and Gen Z Explained*. Retrieved from https://communityrising.kasasa.com/gen-x-gen-y-gen-z/

Kesim, M., & Ozarslan, Y. (2012). Augmented Reality in Education: Current Technologies and the Potential for Education. *Procedia: Social and Behavioral Sciences*, *47*, 297–302. doi:10.1016/j.sbspro.2012.06.654

Khanna, A. S. A. (2020). *Blockchain Technology for Hospitality Industry*. Springer. doi:10.1007/978-3-030-63396-7_7

Kim, J. Y., Chung, N., & Ahn, K. M. (2019). The impact of mobile tour information services on destination travel intention. *Information Development*, *35*(1), 107–120. doi:10.1177/0266666917730437

Kim, J., Wei, S., & Ruys, H. (2003). Segmenting the market of West Australian senior tourists using an artificial neural network. *Tourism Management*, *24*(1), 25–34. doi:10.1016/S0261-5177(02)00050-X

Kim, M., Park, M., Kim, J., Kwon, Y., Sohn, D., Yoon, H., & Seo, K. (2021). On the common and different expectations on robot service in restaurant between customers and employees. *ACM/IEEE International Conference on Human-Robot Interaction*, 262–265. 10.1145/3434074.3447172

Kiryakova, G., Angelova, N., & Yordanova, L. (2017). The potential of augmented reality to change the business. *Trakia Journal of Sciences*, *15*(1, Suppl.1), 394–401. doi:10.15547/tjs.2017.s.01.066

Kishore, N., Kiang, M. V., Engø-Monsen, K., Vembar, N., Schroeder, A., Balsari, S., & Buckee, C. O. (2020). Measuring mobility to monitor travel and physical distancing interventions: A common framework for mobile phone data analysis. *The Lancet. Digital Health*, *2*(11), e622–e628. doi:10.1016/S2589-7500(20)30193-X PMID:32905027

Kiyokawa, K. (2012). Trends and Vision of Head Mounted Display in Augmented Reality. In *International Symposium on Ubiquitous Virtual Reality. IEEE.* 10.1109/ISUVR.2012.11

Kock, F., Josiassen, A., & Assaf, A. G. (2016). Advancing destination image: The destination content model. *Annals of Tourism Research*, *61*, 28–44. doi:10.1016/j.annals.2016.07.003

Kontogianni, A., & Alepis, E. (2019). Moments of interest: A novel cloud-based crowdsourcing application enhancing smart tourism recommendations. In *11th Computer Science and Electronic Engineering Conference (CEEC)* (Vol. 11, pp. 144-149). Colchester, UK: University of Essex. 10.1109/CEEC47804.2019.8974337

Koo, C., Wati, Y., & Chung, N. (2013). A Study of Mobile and Internet Banking Service: Applying for IS Success Model. *A Study of Mobile and Internet Banking Service*, 65-86.

Kotler, N., & Kotler, P. (2000). Can museums be all things to all people? Missions, goals, and marketing's role. *Museum Management and Curatorship*, *18*(3), 271–287. doi:10.1080/09647770000301803

Kourouthanassisa, P., Boletsisb, C., Bardakia, C., & Chasanidou, D. (2015). Tourists responses to mobile augmented reality travel guides: The role of emotions on adoption behavior. *Pervasive and Mobile Computing*, *18*, 71–87. doi:10.1016/j.pmcj.2014.08.009

Kowalewski, D. M. (2017, March 14). Blockchain Will Transform Customer Loyalty Programs. *Harvard Business Review*.

Kuan, A. L., Ann, H. N., Mohd Badri, E. N., & Tang, K. Y. (2014, May). *UTAUT2 Influencing The Behavioural Intention To Adopt Mobile Applications*. Retrieved from utar: http://eprints.utar.edu.my/1277/

Kumar, V. K. (2016, March). Sanative Chatbot for Health Seekers. *International Journal of Engineering and Computer Science, 5*(3), 16022 - 16025.

Kumar, R. L., Li, A., & Wang, W. (2018). Learning and Optimizing through Dynamic Pricing. *Journal of Revenue and Pricing Management, Springer, 17*(2), 63–67. doi:10.105741272-017-0120-2

Kung, W.-S. (2020). Reconceptualizing the digital humanities in Asia: New representations of art, history and culture (2nd ed.). Springer. doi:10.1007/978-981-15-4642-6

Kuniavsky, M. (2007). *User experience and HCI. In The Human-Computer Interaction Handbook: Fundamentals, Evolving Technologies, and Emerging Applications* (2nd ed.). Lawrence Erlbaum Associates Inc.

Küstür, S. (2020). *Hollanda'da bir restoran sosyal mesafe için robot garsonlar kullanıyor*. https://www.teknoblog.com/sosyal-mesafe-icin-robot-garson-cozumu/

Kwak, M. K., Lee, J., & Cha, S. S. (2021). Senior consumer motivations and perceived value of robot service restaurants in Korea. *Sustainability, 13*(5), 13. doi:10.3390u13052755

Kwon, J. M., & Bae, J. (2013). Mobile applications in the hospitality industry. *Journal of Hospitality and Tourism Technology, 4*(1), 81–92. doi:10.1108/17579881311302365

LaGrandeur, K. a. (2017). *Surviving the Machine Age - Intelligent Technology and the Transformation of Human Work*. Springer.

Lai, I. K. (2015). Traveler Acceptance of an App-Based Mobile Tour Guide. *Journal of Hospitality & Tourism Research (Washington, D.C.), 39*(3), 401–432. doi:10.1177/1096348013491596

Landman, P. (2021). *Top Hospitality & Hotel Trends of 2021 and Beyond*. Retrieved from Xotels.com: https://www.xotels.com/en/hotel-management/hotel-trends

Langford, G. W. (2019). *US Travel and Hospitality Outlook*. Deloitte.

Laurell, C., Sandström, C., Berthold, A., & Larsson, D. (2019). Exploring barriers to adoption of Virtual Reality through Social Media Analytics and Machine Learning–An assessment of technology, network, price and trialability. *Journal of Business Research, 100*, 469–474. doi:10.1016/j.jbusres.2019.01.017

LaValle, S. M., Yershova, A., Katsev, M., & Antonov, M. (2014). *Head tracking for the Oculus Rift*. Paper presented at the 2014 IEEE International Conference on Robotics and Automation (ICRA). 10.1109/ICRA.2014.6906608

Law, R., Buhalis, D., & Cobanoglu, C. (2014). Progress on information and communication technologies in hospitality and tourism. *International Journal of Contemporary Hospitality Management, 26*(5), 727–750. doi:10.1108/IJCHM-08-2013-0367

Lee, S. (2020, June 17). *MP Moots Single Contact-Tracing App For Better Accountability.* Retrieved from CodeBlue: https://codeblue.galencentre.org/2020/06/17/mp-moots-single-contact-tracing-app-for-better-accountability/

Lee, H., Jung, T. H., Dieck, M. C. T., & Chung, N. (2020). Experiencing immersive virtual reality in museums. *Information & Management, 57*(5), 103229. Advance online publication. doi:10.1016/j.im.2019.103229

Lee, I., & Shin, Y. J. (2020). Machine learning for enterprises: Applications, algorithm selection, and challenges. *Business Horizons, 63*(2), 157–170. doi:10.1016/j.bushor.2019.10.005

Lei, S. I., Ye, S., Wang, D., & Law, R. (2020). Engaging Customers in Value Co-Creation Through Mobile Instant Messaging in the Tourism and Hospitality Industry. *Journal of Hospitality & Tourism Research (Washington, D.C.), 44*(2), 229–251. doi:10.1177/1096348019893066

Lessiter, J., Freeman, J., Keogh, E., & Davidoff, J. (2001). A cross-media presence questionnaire: The ITC-Sense of Presence Inventory. *Presence (Cambridge, Mass.), 10*(3), 282–297. doi:10.1162/105474601300343612

Li, C., Guo, S., Wang, C., & Zhang, J. (2019). Veni, vidi, vici: The impact of social media on virtual acculturation in tourism context. *Technological Forecasting and Social Change, 145*(August), 513–522. Advance online publication. doi:10.1016/j.techfore.2019.01.013

Liebana-Cabanillas, F., Carvajal-Trujillo, E., Villarejo-Ramos, Á. F., & Higueras-Castillo, E. (2020). Antecedents of the intention to use NFC mobile applications in tourism. *Journal of Hospitality and Tourism Technology, 11*(2), 369–383. doi:10.1108/JHTT-03-2019-0048

Li, J., Bonn, M. A., & Ye, B. H. (2019). Hotel employee's artificial intelligence and robotics awareness and its impact on turnover intention: The moderating roles of perceived organizational support and competitive psychological climate. *Tourism Management, 73*, 172–181. doi:10.1016/j.tourman.2019.02.006

Li, J., Xu, L., Tang, L., Wang, S., & Li, L. (2018). Big data in tourism research: A literature review. *Tourism Management, 68*, 301–323. doi:10.1016/j.tourman.2018.03.009

Li, L., Yu, F., Shi, D., Shi, J., Tian, Z., Yang, J., ... Jiang, Q. (2017). Application of virtual reality technology in clinical medicine. *American Journal of Translational Research, 9*(9), 3867. PMID:28979666

Li, S., Tian, W., Lundberg, C., Gkritzali, A., & Sundström, M. (2020). Two Tales of One City: Fantasy Proneness, Authenticity, and Loyalty of On-Screen Tourism Destinations. *Journal of Travel Research.*

Lissitsa, S., & Kol, O. (2016). Generation X vs. Generation Y – A decade of online shopping. *Journal of Retailing and Consumer Services, 31*, 304–312. doi:10.1016/j.jretconser.2016.04.015

Litsa, T. (2018, June 4). *Analysis of Mary Meeker's Internet Trends Report 2018 - Part 2*. Retrieved from Clickz Web Site: https://www.clickz.com/analysis-of-mary-meekers-internet-trends-report-2018-part-2/214809/

Li, X., Law, R., Xie, G., & Wang, S. (2021). Review of tourism forecasting research with internet data. *Tourism Management, 83*, 104245. Advance online publication. doi:10.1016/j.tourman.2020.104245

Lomas, N. (2014). *Foodini ιs a 3D printer that lets you print dishes with fresh ingredients*. Academic Press.

Loureiro, S. M. C., Guerreiro, J., Eloy, S., Langaro, D., & Panchapakesan, P. (2019). Understanding the use of Virtual Reality in Marketing: A text mining-based review. *Journal of Business Research, 100*(July), 514–530. doi:10.1016/j.jbusres.2018.10.055

Ludlow, B. L. (2015). Virtual reality: Emerging applications and future directions. *Rural Special Education Quarterly, 34*(3), 3–10. doi:10.1177/875687051503400302

Lu, Y., Yuan, F., Lin, J., & Yuan, K. (2020). TouristGo: A location-based mobile game to improve tourist experience by visiting path optimisation. *Personal and Ubiquitous Computing, 24*(3), 405–418. doi:10.100700779-019-01327-9

MacHale, D. (2019). *Mobile Travel Trends 2019*. Retrieved 2019, from Travelport Digital: https://info-digital.travelport.com/trends2019

Magazine, X. (2017, September). Benefits of using Blockchain in the Hospitality Travel Industry. *XcelLab Magazine*.

Makar, M. a. (2014). *Automatic Message Selection with a Chatbot*. Academic Press.

Manuel, N. (2018, September 28). *The Role of Augmented Reality in Mobile App Development*. Retrieved from Simple Programmer: https://simpleprogrammer.com/augmented-reality-mobile/

Marbouh, D. A. (2020). *Blockchain for COVID-19: Review, Opportunities, and a Trusted Tracking System*. National Center for Biotechnology Information, U.S. National Library of Medicine.

Marchiori, E., Niforatos, E., & Preto, L. (2018). Analysis of users' heart rate data and self-reported perceptions to understand effective virtual reality characteristics. *Information Technology & Tourism, 18*(1), 133–155. doi:10.100740558-018-0104-0

Mardiana, S., Tjakraatmadja, J. H., & Aprianingsih, A. (2015). DeLone-McLean Information System Success Model Revisited: The Separation of Intention to Use - Use and Integration of Technology Acceptance Models. *International Journal of Economics and Financial Issues*, 172-182.

Mariani, M., Baggio, R., Fuhs, M., & Hoepken, W. (2018). Business Intelligence and Big Data in Hospitality and Tourism: A Systematic Literature Review. *International Journal of Contemporary Hospitality Management, 30*(12), 3514–3554. doi:10.1108/IJCHM-07-2017-0461

Maria, S., Loureiro, C., Guerreiro, J., & Ali, F. (2019). 20 years of research on virtual reality and augmented reality in tourism context: A text-mining approach. *Tourism Management*, 1–21.

Martin, B. A. S., Jin, H. S., Wang, D., Nguyen, H., Zhan, K., & Wang, Y. X. (2020). The influence of consumer anthropomorphism on attitudes towards artificial intelligence trip advisors. *Journal of Hospitality and Tourism Management*, *44*, 108–111. doi:10.1016/j.jhtm.2020.06.004

Marty, P. F. (2008). Museum websites and museum visitors: Digital museum resources and their use. *Museum Management and Curatorship*, *23*(1), 81–99. doi:10.1080/09647770701865410

Maurits, N. (2012). *From Neurology to Methodology and Back: An Introduction to Clinical Neuroengineering*. Springer. doi:10.1007/978-1-4614-1132-1

Mazars. (2021). *Using Blockchain to Promote Trust in the Hospitality Industry*. Academic Press.

McMahan, A. (2003). Immersion, engagement and presence. *The Video Game Theory Reader*, *67*, 86.

Mei, Q. (2004). A knowledge processing oriented life cycle study from a digital museum system. *Proceedings of the 42nd annual Southeast regional conference*, 116-121. 10.1145/986537.986565

Meisrow, R., & Blumenthal, J. (2019, April 18). *Why IOT is the "Smart" Solution for the Hospitality Industry*. PhocusWire. Retrieved from https://www.phocuswire.com/PwC-opinion-IoT-for-hospitality

Merel, T. (2017, January 13). *After Mixed Year, Mobile AR to Drive $108 Billion VR/AR Market by 2021*. Retrieved from thewrap: https://www.thewrap.com/mixed-year-mobile-ar-drive-108-billion-vrar-market-2021-guest-column/

Mihelj, M., Novak, D., & Beguš, S. (2014). *Virtual reality technology and applications*. Academic Press.

Milgram, P., & Kishino, F. (1994). A taxonomy of mixed reality visual displays. *IEICE Transactions on Information and Systems*, *77*, 1321–1329.

Miller, E. (2019). *5 Innovations that are Reshaping the Hospitality Industry*. Retrieved from: https://innovationmanagement.se/2019/01/24/5-innovations-that-are-reshaping-the-hospitality-industry/

Ministry for Europe and Foreign Affairs. (2019, May). *The Ministry's Actions to Promote Tourism*. Ministry for Europe and Foreign Affairs.

Miso Robotics. (n.d.). *About us*. https://misorobotics.com/about-us/

Mogali, S. (2015). Internet of things and its role in smart kitchen. *4th National Conference on Scientometrics*, 1-11.

Morosan, C., & DeFranco, A. (2016a). Co-creating value in hotels using mobile devices: A conceptual model with empirical validation. *International Journal of Hospitality Management*, *52*, 131–142. doi:10.1016/j.ijhm.2015.10.004

Morosan, C., & DeFranco, A. (2016b). Modeling guests' intentions to use mobile apps in hotels: The roles of personalization, privacy, and involvement. *International Journal of Contemporary Hospitality Management, 28*(9), 1968–1991. doi:10.1108/IJCHM-07-2015-0349

Moseley, R. (2016). *Immersive brain entrainment in virtual worlds: actualizing meditative states. In Emerging Trends and Advanced Technologies for Computational Intelligence*. Springer.

Murphy, J., Gretzel, U., & Pesonen, J. (2019). Marketing robot services in hospitality and tourism: The role of anthropomorphism. *Journal of Travel & Tourism Marketing, 36*(7), 784–795. doi:10.1080/10548408.2019.1571983

Murray, J. H., & Murray, J. H. (2017). *Hamlet on the holodeck: The future of narrative in cyberspace*. MIT Press.

Nash, E. B., Edwards, G. W., Thompson, J. A., & Barfield, W. (2000). A review of presence and performance in virtual environments. *International Journal of Human-Computer Interaction, 12*(1), 1–41. doi:10.1207/S15327590IJHC1201_1

NationMaster. (2019). *France - International Tourism Expenditure*. Retrieved from: https://www.nationmaster.com/nmx/timeseries/france-international-tourism-expenditures

Natural Machines. (n.d.). *How long does it take to print something? What is the print speed?* https://www.naturalmachines.com/faq

Navío-Marco, J., Ruiz-Gómez, L. M., & Sevilla-Sevilla, C. (2018). Progress in information technology and tourism management: 30 years on and 20 years after the internet-Revisiting Buhalis & Law's landmark study about eTourism. *Tourism Management, 69*, 460–470. doi:10.1016/j.tourman.2018.06.002

Nayik, G., Muzaffar, K., & Gull, A. (2015). Robotics and food technology: A mini review. *Journal of Nutrition & Food Sciences, 5*(4), 1–5.

Neuburger, L., & Egger, R. (2018). Augmented reality: Providing a different dimension for museum visitors. In *Augmented Reality and Virtual Reality* (pp. 65–77). Springer. doi:10.1007/978-3-319-64027-3_5

Neuburger, L., & Roman, E. (2017). *An Afternoon at the Museum: Through the Lens of Augmented Reality. In Information and Communication Technologies in Tourism 2017*. Springer International Publishing.

Neuhofer, B., Buhalis, D., & Ladkin, A. (2014). A Typology of Technology-Enhanced Tourism Experiences. *International Journal of Tourism Research, 16*(4), 340–350. doi:10.1002/jtr.1958

New Software Solutions. (2020). *Akıllı turizm otelcilik ve yapay zeka uygulamaları*. https://www.new.com.tr/blog/akilli-turizm-otelcilik-ve-yapay-zeka-uygulamalari/

Ng, Y.-L., Ma, F., Ho, F. K., Ip, P., & Fu, K. (2019). Effectiveness of virtual and augmented reality-enhanced exercise on physical activity, psychological outcomes, and physical performance: A systematic review and meta-analysis of randomized controlled trials. *Computers in Human Behavior, 99*, 278–291. doi:10.1016/j.chb.2019.05.026

Nicolau, J. L., & Santa-María, M. J. (2013). The effect of innovation on hotel market value. *International Journal of Hospitality Management, 32*, 71–79. doi:10.1016/j.ijhm.2012.04.005

No, E., & Kim, J. K. (2014). Determinants of the adoption for travel information on smartphone. *International Journal of Tourism Research, 16*(6), 534–545. doi:10.1002/jtr.1945

Nweke, H. F., Teh, Y. W., Al-garadi, M. A., & Alo, U. R. (2018). Deep learning algorithms for human activity recognition using mobile and wearable sensor networks: State of the art and research challenges. *Expert Systems with Applications, 105*, 233–261. doi:10.1016/j.eswa.2018.03.056

Olsen, T. L., & Tomlin, B. (2020). Industry 4.0: Opportunities and challenges for operations management. *Manufacturing & Service Operations Management, 22*(1), 113–122. doi:10.1287/msom.2019.0796

Olsson, T., Ihamäki, P., Lagerstam, E., Ventä-Olkkonen, L., & Väänänen-Vainio-Mattila, K. (2009). User expectations for mobile mixed reality services: an initial user study. *Proceedings from the European Conference on Cognitive Ergonomics: Designing beyond the Product*, 177–184.

Olsson, T., Lagerstam, E., Kärkkäinen, T., & Väänänen-Vainio-Mattila, K. (2013). Expected user experience of mobile augmented reality services: A user study in the context of shopping centres. *Personal and Ubiquitous Computing, 17*(2), 287–304. doi:10.100700779-011-0494-x

Oracle Hospitality. (2016). *Hotel 2025*. Retrieved from https://www.oracle.com/webfolder/s/delivery_production/docs/FY16h1/doc31/Hotels-2025-v5a.pdf

Otoo, F. E., Kim, S., & Choi, Y. (2020). Understanding senior tourists' preferences and characteristics based on their overseas travel motivation clusters. *Journal of Travel & Tourism Marketing, 37*(2), 246–257. doi:10.1080/10548408.2020.1740136

Özer, B. (2019). *Gıda işlemede yeni bir teknolojik yaklaşım: 3-boyutlu gıda yazıcıları*. https://sutdunyasi.com/gida-islemede-yeni-bir-teknolojik-yaklasim-3-boyutlu-gida-yazicilari/

Özgüneş, R., Bozok, D., & Küçükaltan, D. (2020). Yiyecek içecek sektöründe ileri teknoloji ve pandemik düzene doğru: Yakın gelecekte bir robota 'eline sağlık!' diyebilir miyiz? *Afyon Kocatepe Üniversitesi Sosyal Bilimler Dergisi, 22*(4), 1124–1139. doi:10.32709/akusosbil.797343

Pallud, J., & Monod, E. (2010). User experience of museum technologies: The phenomenological scales. *European Journal of Information Systems, 19*(5), 562–580. doi:10.1057/ejis.2010.37

Panarello, A., Tapas, N., Merlino, G., Longo, F., & Puliafito, A. (2018). Blockchain and IoT Integration: A Systematic Survey. *MDPI*.

Parasuraman, A. (2000). Technology Readiness Index (Tri): A Multiple-Item Scale to Measure Readiness to Embrace New Technologies. *Journal of Service Research*, *2*(4), 307–320. doi:10.1177/109467050024001

Parasuraman, A., & Colby, C. (2002). Techno-Ready Marketing: How and Why Your Customers Adopt Technology. *Journal of Consumer Marketing*, *19*(4), 359–361. doi:10.1108/jcm.2002.19.4.359.1

Parasuraman, A., & Zeithaml, V. A. (1988). SERVQUAL: A multiple-item scale for measuring consumer perceptions of service quality. *Journal of Retailing*, 12–40.

Pardon. (n.d.). *Dijital QR Menü*. https://pardon-app.com/dijital-qr-menu/

Park, S., & Huang, Y. (2017). Motivators and inhibitors in booking a hotel via smartphones. *International Journal of Contemporary Hospitality Management*, *29*(1), 161–178. doi:10.1108/IJCHM-03-2015-0103

Park, S., Xu, Y., Jiang, L., Chen, Z., & Huang, S. (2020). Spatial structures of tourism destinations: A trajectory data mining approach leveraging mobile big data. *Annals of Tourism Research*, *84*, 102973. Advance online publication. doi:10.1016/j.annals.2020.102973

Parvez, M. O. (2020). Use of machine learning technology for tourist and organizational services: high-tech innovation in the hospitality industry. *Journal of Tourism Futures*. doi:10.1108/JTF-09-2019-0083

Pascu, L. (2016, July 12). *IoT Could Generate up to $11.1 Trillion in Economic Value per Year by 2025*. Retrieved from Hot for Security: https://hotforsecurity.bitdefender.com/blog/iot-could-generate-up-to-11-1-trillion-in-economic-value-per-year-by-2025-15887.html

Patrício, L., Fisk, R. F., & Cunha, J. F. (2003). Improving satisfaction with bank service offerings: Measuring the contribution of each delivery channel. *Managing Service Quality*, *13*(6), 471–482. doi:10.1108/09604520310506531

Pauget, B., & Dammak, A (2019). The implementation of the Internet of Things: what impact on organizations? *Technol Forecast Soc Change*, *140*, 140-146. . doi:10.1016/j.techfore.2018.03.012

Peacock, D., & Brownbill, J. (2007). *Audiences, visitors, users: Reconceptualising users of museum on-line content and services*. Museums and the Web.

Perry Hobson, J., & Williams, A. P. (1995). Virtual reality: A new horizon for the tourism industry. *Journal of Vacation Marketing*, *1*(2), 124–135. doi:10.1177/135676679500100202

Pestek, A., & Sarvan, M. (2020). Virtual reality and modern tourism. *Journal of Tourism Futures*.

Peter, S., & McLean, E. R. (2009). A meta-analytic assessment of the DeLone and McLean IS success model: An examination of IS success at the individual level. *Information & Management*, *46*(3), 159–166. doi:10.1016/j.im.2008.12.006

Peters, G. W. (2016). Understanding modern banking ledgers through blockchain technologies: Future of transaction processing and smart contracts on the internet of money. In *Banking Beyond Banks and Money* (pp. 239–278). Springer. doi:10.1007/978-3-319-42448-4_13

Pierdicca, R., Paolanti, M., & Frontoni, E. (2019). eTourism: ICT and its role for tourism management. *Journal of Hospitality and Tourism Technology*, *10*(1), 90–106. doi:10.1108/JHTT-07-2017-0043

Pill Digital. (n.d.). *Restoran ve cafeler için QR kod ile menu.* https://www.pill.com.tr/icerik/100/restoran-ve-cafeler-icin-qr-kod-ile-menu

Pine, B. J. II, & Gilmore, J. H. (1999). *The Experience Economy: Work Is Theater and Every Business a Stage.* Harvard Business School Press.

Pires, I. M., Garcia, N. M., Pombo, N., Flórez-Revuelta, F., Spinsante, S., & Teixeira, M. C. (2018). Identification of activities of daily living through data fusion on motion and magnetic sensors embedded on mobile devices. *Pervasive and Mobile Computing*, *47*, 78–93. doi:10.1016/j.pmcj.2018.05.005

Pop, C., Cioara, T., Antal, M., Anghel, I., Salomie, I., & Bertoncini, M. (2018). Blockchain Based Decentralized Management of Demand Response Programs in Smart Energy Grids. *MDPI.*

Portalés, C., José, L. L., & Pérez C. (2009). Photogrammetry and Augmented Reality for Cultural Heritage Applications. *The Photogrammetric Record*, *24*(128), 316-331. doi:10.1111/j.1477-9730.2009.00549

Potter, K. (2019, March 3). *The Pros and Cons of IOT in the Hotel Industry.* Hotel Technology News. Retrieved from https://hoteltechnologynews.com/2019/03/pros-cons-iot-hospitality-industry/

Pr Outlet Plus. (2016). *IBM watson ve hilton iş birliği.* https://www.proutletplus.com/2016/09/03/ibm-watson-ve-hilton-is-birligi/

Pradhan, M., Oh, J., & Lee, H. (2018). Understanding Travelers' Behavior for Sustainable Smart Tourism: A Technology Readiness Perspective. *Sustainability*, *10*(11), 4259. doi:10.3390u10114259

Prensky, M. (2001). Digital Natives, Digital Immigrants Part 1. *On the Horizon, 9*(5), 1-6.

Proente Otomasyon. (n.d.). *Endüstri alanında yaşanan gelişmelerin sınıflandırılması.* https://proente.com/endustri-alaninda-yasanan-gelismelerin-siniflandirilmasi/

Protel. (n.d.). *QR menu order ile temassız sipariş & ödeme.* https://www.protel.com.tr/restoran-otomasyonu/qr-menu-order-temassiz-siparis-odeme-sistemi/

Pulli, P., Zheng, X., Antoniac, P., Hickey, S., Manninen, T., Martikainen, O., & Kuroda, T. (2007). *Design and development of mobile services platform for senior citizens.* Paper presented at the 13th International Conference on Concurrent Enterprising, Nottingham, UK.

PWC. (2018). *Blockchain.* Digital Services.

PWC. (2018). *Fourth Industrial Revolution for the Earth Series: Building block(chain)s for a better planet.* Author.

Qin, H., Peak, D. A., & Prybutok, V. (2021). A virtual market in your pocket: How does mobile augmented reality (MAR) influence consumer decision making? *Journal of Retailing and Consumer Services*, *58*(January), 102337. Advance online publication. doi:10.1016/j.jretconser.2020.102337

Qomariyah, N. N., Sari, S. A., & Fajar, A. N. (2020). Sonia: An integrated indonesia online tourism system in new normal era. *International Journal of Innovative Computing, Information, & Control*, *16*(6), 1829–1843. doi:10.24507/ijicic.16.06.1829

Quintal, V., Phau, I., Sims, D., & Cheah, I. (2016). Factors influencing generation Y's purchase intentions of prototypical versus me-too brands. *Journal of Retailing and Consumer Services*, *30*, 175–183. doi:10.1016/j.jretconser.2016.01.019

Radianti, J., Majchrzak, T. A., Fromm, J., & Wohlgenannt, I. (2020). A systematic review of immersive virtual reality applications for higher education: Design elements, lessons learned, and research agenda. *Computers & Education*, *147*, 103778. doi:10.1016/j.compedu.2019.103778

Ramon, P.-S., Santiago, F.-C., Javier, S.-G., & Emilio, R. (2019). User Acceptance of Mobile Apps for Restaurants: An Expanded and Extended UTAUT-2. *Sustainability*.

Rangnekar, P. (2020, April 16). *How the Hospitality Industry is Enhancing with IOT Technology?* Biz4solutions. Retrieved from https://www.biz4solutions.com/blog/how-the-hospitality-industry-is-enhancing-with-iot-technology/

Rather, R. A. (2020). Customer experience and engagement in tourism destinations: The experiential marketing perspective. *Journal of Travel & Tourism Marketing*, *37*(1), 15–32. doi:10.1080/10548408.2019.1686101

Raun, J., Ahas, R., & Tiru, M. (2016). Measuring tourism destinations using mobile tracking data. *Tourism Management*, *57*, 202–212. doi:10.1016/j.tourman.2016.06.006

Reif, J., & Schmücker, D. (2020). Exploring new ways of visitor tracking using big data sources: Opportunities and limits of passive mobile data for tourism. *Journal of Destination Marketing & Management*, *18*, 100481. Advance online publication. doi:10.1016/j.jdmm.2020.100481

Reis, J., Melão, N., Salvadorinho, J., Soares, B., & Rosete, A. (2020). Service robots in the hospitality industry: The case of Henn-na hotel, Japan. *Technology in Society*, *63*, 101423. Advance online publication. doi:10.1016/j.techsoc.2020.101423

Revfine.com. (2021). *Hospitality Industry: The No. 1 Hospitality Information Guide*. Retrieved from: https://www.revfine.com/hospitality-industry/

Rios, R. (2017, December 13). *Single-Purpose or Multi-Purpose App – Which Option Is Better?* Retrieved from DZone: https://dzone.com/articles/single-purpose-or-multi-purpose-app-which-option-i

Rivera, M., Croes, R., & Zhong, Y. Y. (2016). Developing mobile services: A look at first-time and repeat visitors in a small island destination. *International Journal of Contemporary Hospitality Management, 28*(12), 2721–2747. doi:10.1108/IJCHM-02-2015-0052

Robinson, V. M., & Schanzel, H. A. (2019). A tourism inflex: Generation Z travel experiences. *Journal of Tourism Futures, 5*(2), 127–141. doi:10.1108/JTF-01-2019-0014

Rodrigues, J. M. F., Lessa, J., Gregório, M., Ramos, C., & Cardoso, P. J. S. (2016). An Initial Framework for a Museum Application for Senior Citizens. In *The 7th International Conference on Software Development and Technologies for Enhancing Accessibility and Fighting Info-exclusion - DSAI 2016.* ACM Press. 10.1145/3019943.3020002

Rodríguez Sánchez, I., Williams, A. M., & García Andreu, H. (2020). Customer resistance to tourism innovations: Entrepreneurs' understanding and management strategies. *Journal of Travel Research, 59*(3), 450–464. doi:10.1177/0047287519843188

Romelaer, P. (2005). L'entretien de recherche. Management des ressources humaines: méthodes de recherche en sciences humaines et sociales. *De Boeck*, 101-137.

Rusu, C., Rusu, V., Roncagliolo, S., Apablaza, J., & Rusu, V. Z. (2015). User Experience Evaluations: Challenges for Newcomers. In. Lecture Notes in Computer Science: Vol. 9186. *Design, User Experience, and Usability: Design Discourse*. Springer., doi:10.1007/978-3-319-20886-2_23.

Saluveer, E., Raun, J., Tiru, M., Altin, L., Kroon, J., Snitsarenko, T., Aasa, A., & Silm, S. (2020). Methodological framework for producing national tourism statistics from mobile positioning data. *Annals of Tourism Research, 81*, 102895. Advance online publication. doi:10.1016/j.annals.2020.102895

Samala, N., Katkam, B. S., Bellamkonda, R. S., & Rodriguez, R. V. (2020). Impact of AI and robotics in the tourism sector: a critical insight. *Journal of Tourism Futures*. doi:10.1108/JTF-07-2019-0065

Samara, D., Magnisalis, I., & Peristeras, V. (2020). Artificial intelligence and big data in tourism: A systematic literature review. *Journal of Hospitality and Tourism Technology, 11*(2), 343–367. doi:10.1108/JHTT-12-2018-0118

Saravanan, S., & Sadhu Ramakrishnan, B. (2016). Preserving privacy in the context of location based services through location hider in mobile-tourism. *Information Technology & Tourism, 16*(2), 229–248. doi:10.100740558-016-0056-1

Savage, P. (2020, May 19). *eCommerce Personalization: Why Demographic-Based Personalization is on Its Way Out (And What to Do Instead)*. Retrieved from Coredna: https://www.coredna.com/blogs/ecommerce-personalization#14

Sekaran, U., & Bougie, R. (2016). Research Methods For Business: A Skill Building Approach (7th ed.). Wiley.com.

Selvam, A., Yap, T. T.-V., Ng, H., Tong, H.-L., & Ho, C. C. (2016). Augmented Reality for Information Retrieval Aimed at Museum Exhibitions using Smartphones. *Journal of Engineering and Applied Sciences (Asian Research Publishing Network), 100*(3), 635–639.

Seo, K. H., & Lee, J. H. (2021). The emergence of service robots at restaurants: Integrating trust, perceived risk and satisfaction. *Sustainability, 13*(8), 4431. doi:10.3390u13084431

Service Robots. (2020). *Robot technology in the restaurant industry*. https://www.servicerobots.com/blog/robot-technology-in-the-restaurant-industry/

Seyitoğlu, F., & Ivanov, S. (2020). Service robots as a tool for physical distancing in tourism. *Current Issues in Tourism, 1–4*. Advance online publication. doi:10.1080/13683500.2020.1774518

Sharma, Y. (2018). *Online Travel Market to reach $1,091 Billion, Globally, by 2022*. Retrieved from Allied Market Research: https://www.alliedmarketresearch.com/press-release/online-travel.html

Shaw, K. (2018). *Zume pizza names new president, adds new robot to pizza-making staff*. https://www.roboticsbusinessreview.com/news/zume-pizza-adds-robot-pizza-making-staff/

Shin, H., & Kang, J. (2020). Reducing perceived health risk to attract hotel customers in the COVID-19 pandemic era: Focused on technology innovation for social distancing and cleanliness. *International Journal of Hospitality Management, 91*, 102664. doi:10.1016/j.ijhm.2020.102664 PMID:32921871

Short, T. (2015). *Guest Preferences for Technology Use in Hotels*. Software Advice. Retrieved from https://www.softwareadvice.com/hotel-management/industryview/technology-use-report-2015/

Simulation in Hospitality: Sparking Travelers' Imagination. (2021). *ediiie*. Retrieved from https://www.ediiie.com/hospitality.html

Singh, R. (2018, Jun 5). *Malaysia Airports embraces big data with launch of new mobile app*. Retrieved from malaymail: https://www.malaymail.com/news/life/2018/06/05/malaysia-airports-embraces-big-data-with-launch-of-new-mobile-app/1638636

Slater, M. (2009). Place illusion and plausibility can lead to realistic behaviour in immersive virtual environments. *Philosophical Transactions of the Royal Society of London. Series B, Biological Sciences, 364*(1535), 3549–3557. doi:10.1098/rstb.2009.0138 PMID:19884149

Smartypans. (n.d.). https://smartypans.io/

Smit, C., Roberts-Lombard, M., & Mpinganjira, M. (2018). Technology readiness and mobile self-service technology adoption in the airline industry: An emerging market perspective. *Acta Commercii, 18*(1), 1–12. doi:10.4102/ac.v18i1.580

Song, L. S. (2009). A Staged Service Innovation Model. *Innovation Technology Entrepreneurship & Marketing, 40*(3), 571–599.

Srivastava, A. (n.d.). *How blockchain is transforming the hospitality industry*. Today's Hotelier.

Steinicke, F. (2016). *Being really virtual*. Springer. doi:10.1007/978-3-319-43078-2

Stephens, M. (2014). *Energy management for hospitality industry to reach $970M by 2020.* Retrieved from: https://insights.ehotelier.com/news/2014/03/13/energy-management-for-hospitality-industry-to-reach-970m-by-2020/

Stfalcon.com. (2018, Dec 6). *Top 10 Travel Industry Trends in 2019.* Retrieved from medium.com: https://medium.com/swlh/top-10-travel-industry-trends-in-2019-d43d157de7b9

Stienmetz, J. L., Levy, S. E., & Boo, S. (2013). Factors Influencing the Usability of Mobile Destination Management Organization Websites. *Journal of Travel Research, 52*(4), 453–464. doi:10.1177/0047287512467698

Stylidis, D., Belhassen, Y., & Shani, A. (2015). Three tales of a city: Stakeholders' images of Eilat as a tourist destination. *Journal of Travel Research, 54*(6), 702–716. doi:10.1177/0047287514532373

Stylus. (2014). *Smart chopsticks.* https://www.stylus.com/wvmyhn

Suh, A., & Prophet, J. (2018). The state of immersive technology research: A literature analysis. *Computers in Human Behavior, 86*, 77–90. doi:10.1016/j.chb.2018.04.019

Sukhraj, R. (2017, November 13). *38 Mobile Marketing Statistics to Help You Plan for 2019.* Retrieved from Impact Web Site: https://www.impactbnd.com/blog/mobile-marketing-statistics

Sundarapandiyan, M., Karthik, S., & Daniel, J. A. (2019). IOT based smart kitchen. *International Journal of Computer Science Trends and Technology, 7*(2), 13–16.

Sun, J., Peng, Z., Zhou, W., Fuh, J., Hong, G., & Chiu, A. (2015b). A review on 3D printing for customized food fabrication. *Procedia Manufacturing, 1*, 308–319. doi:10.1016/j.promfg.2015.09.057

Sun, J., Zhou, W., Huang, D., Fuh, J. Y. H., & Hong, G. S. (2015a). An overview of 3D printing technologies for food fabrication. *Food and Bioprocess Technology, 8*(8), 1605–1615. doi:10.100711947-015-1528-6

Sutherland, I. E. (1968). A head-mounted three dimensional display. *Proceedings of the December 9-11, 1968, fall joint computer conference, part I.*

Su, X., Sperli, G., Moscato, V., Picariello, A., Esposito, C., & Choi, C. (2019). An Edge Intelligence Empowered Recommender System Enabling Cultural Heritage Applications. *IEEE Transactions on Industrial Informatics, 15*(7), 4266–4275. doi:10.1109/TII.2019.2908056

Taş, D., & Olum, E. (2020). Yiyecek-içecek sektöründe sürdürülebilirlik ve yenilikçi yaklaşımlar. *Türk Turizm Araştırmaları Dergisi, 4*(3), 3082–3098. doi:10.26677/TR1010.2020.527

Team Trilyo. (2018, April 27). *What is the Role of Internet of Things (IOT) in Hotels.* Trilyo Blog. Retrieved from https://www.trilyo.com/blog/what-is-the-role-of-internet-of-things-iot-in-hospitality-industry/

Tess. (2016). *Cooking revolution: Are food 3D printers changing the face of high end cuisine?* http://www.3ders.org/articles/20160301-cooking-revolution-are-food-3d-printers-changing-the-face-of-high-end-cuisine.html

Travel Apps: The Revolution in Traveling. (2016, September 7). Retrieved from Shoutem: https://medium.com/shoutem/travel-apps-the-revolution-in-traveling-cde012e79dd4

Trice, A. (2014, Jun 23). *Mobile UX: Refining Perceived Performance.* Retrieved from Dzone / Performance Zone: https://dzone.com/articles/mobile-ux-refining-perceived

Trivago. (2019, April 18). *Sustainable Hospitality: Eco-Friendly Industry Trends and Tips.* Trivago Business Blog. Retrieved from https://businessblog.trivago.com/sustainable-hospitality-trends-eco-friendly-hotel-tips/

Tscheu, F., & Buhalis, D. (2016). *Augmented Reality at Cultural Heritage Sites. Information and Communication Technologies in Tourism.* Cham Springer International Publishing.

Türkiye İş Bankası. (2020). *Nesnelerin interneti nedir ve hayatımızı nasıl değiştirecek?* https://www.isbank.com.tr/blog/nesnelerin-interneti-nedir

Tussyadiah, I. (2020). A review of research into automation in tourism: Launching the Annals of Tourism Research Curated Collection on Artificial Intelligence and Robotics in Tourism. *Annals of Tourism Research, 81*, 102883. doi:10.1016/j.annals.2020.102883

Tussyadiah, I. P., Wang, D., Jung, T. H., & tom Dieck, M. C. (2018). Virtual reality, presence, and attitude change: Empirical evidence from tourism. *Tourism Management, 66*, 140–154. doi:10.1016/j.tourman.2017.12.003

Tussyadiah, I., & Miller, G. (2019). Nudged by a robot: Responses to agency and feedback. *Annals of Tourism Research, 78*, 102752. Advance online publication. doi:10.1016/j.annals.2019.102752

Unilever Food Solutions. (n.d.). *Normalleşme döneminde dijital menüler.* https://www.unileverfoodsolutions.com.tr/konsept-uygulamalarimiz/normallesme-doneminde-restoranlar-icin-10-altin-kural/normallesme-doneminde-dijital-menuler.html

Universal Robots. (n.d.). *Universal robots' cobots prepare delicious pizzas.* https://www.universal-robots.com/about-universal-robots/news-centre/universal-robots-cobots-prepare-delicious-pizzas/

Urbach, N., & Mueller, B. (September 2011). The Updated DeLone and McLean Model of Information Systems Success. Information Systems Theory: Explaining and Predicting Our Digital Society, 1, 1-18.

Urbach, N., & Ahlemann, F. (2010). Structural Equation Modeling in Information Systems Research Using Partial Least Squares. *Journal of Information Technology Theory and Application, 11*(2), 5–40.

Vailshery, L. S. (2021, January 22). *Number of digital voice assistants in use worldwide from 2019 to 2024 (in billions).* Statista.

Vakhnenko, H. (2019). *Top 5 Ideas in the Travel App Development*. Retrieved from agilie: https://agilie.com/en/blog/top-5-ideas-in-the-travel-app-development

van der Meulen, R. (2015, November 10). *Gartner says 6.4 Billion Connected "Things" Will be in Use in 2016, Up 30 Percent from 2015*. Newsroom Press Releases. Retrieved from https://www.gartner.com/en/newsroom/press-releases/2015-11-10-gartner-says-6-billion-connected-things-will-be-in-use-in-2016-up-30-percent-from-2015

van Esch, P., & Black, J. S. (2019). Factors that influence new generation candidates to engage with and complete digital, AI-enabled recruiting. *Business Horizons*, *62*(6), 729–739. doi:10.1016/j.bushor.2019.07.004

Van Krevelen, D., & Poelman, R. (2010). A survey of augmented reality technologies, applications and limitations. *The International Journal of Virtual Reality: a Multimedia Publication for Professionals*, *9*(2), 1–20. doi:10.20870/IJVR.2010.9.2.2767

Vatan, A., & Dogan, S. (2021). What do hotel employees think about service robots? A qualitative study in Turkey. *Tourism Management Perspectives*, *37*, 100775. Advance online publication. doi:10.1016/j.tmp.2020.100775

Venkatesh, V., Morris, M. G., Davis, G. B., & Davis, F. D. (2003). User Acceptance of Information Technology: Toward a Unified View. *MIS Quarterly: Management Information Systems*, *27*(3), 425–478. doi:10.2307/30036540

Venkatesh, V., Thong, J. Y., & Xu, X. (2012). Consumer Acceptance and Use of Information Technology: Extending the Unified Theory of Acceptance and Use of Technology. *MIS Quarterly: Management Information Systems*, *36*(1), 157–178. doi:10.2307/41410412

Vu, H. Q., Li, G., Law, R., & Zhang, Y. (2018). Tourist Activity Analysis by Leveraging Mobile Social Media Data. *Journal of Travel Research*, *57*(7), 883–898. doi:10.1177/0047287517722232

Vuuren, M. V. (2009). *Eating robot ramen in nagoya*. https://www.reformatt.com/blog/nagoya

Wang, D., Xiang, Z., & Fesenmaier, D. R. (2014). Adapting to the mobile world: A model of smartphone use. *Annals of Tourism Research*, *48*, 11–26. doi:10.1016/j.annals.2014.04.008

Wang, D., Xiang, Z., & Fesenmaier, D. R. (2016). Smartphone Use in Everyday Life and Travel. *Journal of Travel Research*, *55*(1), 52–63. doi:10.1177/0047287514535847

Wang, D., Xiang, Z., Law, R., & Ki, T. P. (2016). Assessing Hotel-Related Smartphone Apps Using Online Reviews. *Journal of Hospitality Marketing & Management*, *25*(3), 291–313. doi:10.1080/19368623.2015.1012282

Wang, R., Luo, J., & Huang, S. (2020). Developing an artificial intelligence framework for online destination image photos identification. *Journal of Destination Marketing & Management*, *18*, 100512. Advance online publication. doi:10.1016/j.jdmm.2020.100512

Wang, W., Kumar, N., Chen, J., Gong, Z., Kong, X., Wei, W., & Gao, H. (2020). Realizing the Potential of Internet of Things for Smart Tourism with 5G and AI. *IEEE Network*, *34*(6), 295–301. doi:10.1109/MNET.011.2000250

Wang, Y. S., Li, H. T., Li, C. R., & Zhang, D. Z. (2016). Factors affecting hotels' adoption of mobile reservation systems: A technology-organization-environment framework. *Tourism Management*, *53*, 163–172. doi:10.1016/j.tourman.2015.09.021

Wei, W., Qi, R., & Zhang, L. (2019). Effects of virtual reality on theme park visitors' experience and behaviors: A presence perspective. *Tourism Management*, *71*, 282–293. doi:10.1016/j.tourman.2018.10.024

Wikipedia. (n.d.). *Nesnelerin interneti*. https://tr.wikipedia.org/wiki/Nesnelerin_interneti

Wilson, P. F., Stott, J., Warnett, J. M., Attridge, A., Smith, M. P., & Williams, M. A. (2018). Museum visitor preference for the physical properties of 3D printed replicas. *Journal of Cultural Heritage*, *32*(July–August), 176–185. doi:10.1016/j.culher.2018.02.002

Wixom, B. H., & Todd, P. A. (2005). A Theoretical Integration of User Satisfaction and Technology Acceptance. *Information Systems Research*, *16*(1), 85–102. doi:10.1287/isre.1050.0042

Xu, F., Nash, N., & Whitmarsh, L. (2020). Big data or small data? A methodological review of sustainable tourism. *Journal of Sustainable Tourism*, *28*(2), 144–163. doi:10.1080/09669582.2019.1631318

Xu, F., Tian, F., Buhalis, D., Weber, J., & Zhang, H. (2016). Tourists as Mobile Gamers: Gamification for Tourism Marketing. *Journal of Travel & Tourism Marketing*, *33*(8), 1124–1142. doi:10.1080/10548408.2015.1093999

Xu, Y., Li, J., Belyi, A., & Park, S. (2021). Characterizing destination networks through mobility traces of international tourists— A case study using a nationwide mobile positioning dataset. *Tourism Management*, *82*, 104195. Advance online publication. doi:10.1016/j.tourman.2020.104195

Yaga, D. (2018). *Blockchain Technology Overview*. U.S. Department of Commerce.

Yaprak Aydın, H., Kılıç, A., & Tekin, A. R. (2019). Geleneksel Türk gıdalarının 3B yazıcı ile yazdırılması. *International Journal of 3D Printing Technologies and Digital Industry*, *3*(1), 1-10.

Yıldız, E., & Davutoğlu, N. (2020). Turizm 4.0'dan turizm gastronomi 4.0'a giden yolda: Geleceğin restoranları ve yönetimi. *Akademik Sosyal Araştırmalar Dergisi*, *8*(109), 301–318.

Yim, M. Y.-C., & Park, S.-Y. (2019). "I am not satisfied with my body, so I like augmented reality (AR)": Consumer responses to AR-based product presentations. *Journal of Business Research*, *100*, 581–589. doi:10.1016/j.jbusres.2018.10.041

Yoo, C., Kwon, S., Na, H., & Chang, B. (2017). Factors Affecting the Adoption of Gamified Smart Tourism Applications: An Integrative Approach. *Sustainability*, *9*(12), 1–21. doi:10.3390u9122162

Yuce, A., Arasli, H., Ozturen, A., & Daskin, M. (2020). Feeling the Service Product Closer: Triggering Visit Intention via Virtual Reality. *Sustainability, 12*(16), 6632. doi:10.3390u12166632

Yusoff, N. H. M., Isvaramurty, A., & Razalli, H. (2019). Mobile application for tourist's personal travelling management in Kuala Lumpur. In *IEEE 9th International Conference on System Engineering and Technology (ICSET)* (Vol. 9, pp. 354-359). Shah Alam, Malaysia: Institut Teknologi Bandung. 10.1109/ICSEngT.2019.8906443

Zaied, A. N. (2012). An Integrated Success Model for Evaluating Information System in Public Sectors. *Journal of Emerging Trends in Computing and Information Sciences*, 814-825.

Zheng, W., Liao, Z., & Lin, Z. (2020). Navigating through the complex transport system: A heuristic approach for city tourism recommendation. *Tourism Management, 81*, 104162. Advance online publication. doi:10.1016/j.tourman.2020.104162

Zheng, X., & Pulli, P. (2005). Extending quality function deployment to enterprise mobile services design and development. *Journal of Control Engineering and Applied Informatics, 7*(2), 42–49.

Zoric, M. a. (n.d.). *Achieving Growth in Europe's Evolving HVAC and Filters Industries.* Business Sweden.

Related References

To continue our tradition of advancing information science and technology research, we have compiled a list of recommended IGI Global readings. These references will provide additional information and guidance to further enrich your knowledge and assist you with your own research and future publications.

Abdul Razak, R., & Mansor, N. A. (2021). Instagram Influencers in Social Media-Induced Tourism: Rethinking Tourist Trust Towards Tourism Destination. In M. Dinis, L. Bonixe, S. Lamy, & Z. Breda (Eds.), *Impact of New Media in Tourism* (pp. 135-144). IGI Global. https://doi.org/10.4018/978-1-7998-7095-1.ch009

Abir, T., & Khan, M. Y. (2022). Importance of ICT Advancement and Culture of Adaptation in the Tourism and Hospitality Industry for Developing Countries. In C. Ramos, S. Quinteiro, & A. Gonçalves (Eds.), *ICT as Innovator Between Tourism and Culture* (pp. 30–41). IGI Global. https://doi.org/10.4018/978-1-7998-8165-0.ch003

Abir, T., & Khan, M. Y. (2022). Importance of ICT Advancement and Culture of Adaptation in the Tourism and Hospitality Industry for Developing Countries. In C. Ramos, S. Quinteiro, & A. Gonçalves (Eds.), *ICT as Innovator Between Tourism and Culture* (pp. 30–41). IGI Global. https://doi.org/10.4018/978-1-7998-8165-0.ch003

Abtahi, M. S., Behboudi, L., & Hasanabad, H. M. (2017). Factors Affecting Internet Advertising Adoption in Ad Agencies. *International Journal of Innovation in the Digital Economy*, *8*(4), 18–29. doi:10.4018/IJIDE.2017100102

Afenyo-Agbe, E., & Mensah, I. (2022). Principles, Benefits, and Barriers to Community-Based Tourism: Implications for Management. In I. Mensah & E. Afenyo-Agbe (Eds.), *Prospects and Challenges of Community-Based Tourism and Changing Demographics* (pp. 1–29). IGI Global. doi:10.4018/978-1-7998-7335-8.ch001

Agbo, V. M. (2022). Distributive Justice Issues in Community-Based Tourism. In I. Mensah & E. Afenyo-Agbe (Eds.), *Prospects and Challenges of Community-Based Tourism and Changing Demographics* (pp. 107–129). IGI Global. https://doi.org/10.4018/978-1-7998-7335-8.ch005

Agrawal, S. (2017). The Impact of Emerging Technologies and Social Media on Different Business(es): Marketing and Management. In O. Rishi & A. Sharma (Eds.), *Maximizing Business Performance and Efficiency Through Intelligent Systems* (pp. 37–49). Hershey, PA: IGI Global. doi:10.4018/978-1-5225-2234-8.ch002

Ahmad, A., & Johari, S. (2022). Georgetown as a Gastronomy Tourism Destination: Visitor Awareness Towards Revisit Intention of Nasi Kandar Restaurant. In M. Valeri (Ed.), *New Governance and Management in Touristic Destinations* (pp. 71–83). IGI Global. https://doi.org/10.4018/978-1-6684-3889-3.ch005

Alkhatib, G., & Bayouq, S. T. (2021). A TAM-Based Model of Technological Factors Affecting Use of E-Tourism. *International Journal of Tourism and Hospitality Management in the Digital Age*, *5*(2), 50–67. https://doi.org/10.4018/IJTHMDA.20210701.oa1

Altinay Ozdemir, M. (2021). Virtual Reality (VR) and Augmented Reality (AR) Technologies for Accessibility and Marketing in the Tourism Industry. In C. Eusébio, L. Teixeira, & M. Carneiro (Eds.), *ICT Tools and Applications for Accessible Tourism* (pp. 277-301). IGI Global. https://doi.org/10.4018/978-1-7998-6428-8.ch013

Anantharaman, R. N., Rajeswari, K. S., Angusamy, A., & Kuppusamy, J. (2017). Role of Self-Efficacy and Collective Efficacy as Moderators of Occupational Stress Among Software Development Professionals. *International Journal of Human Capital and Information Technology Professionals*, *8*(2), 45–58. doi:10.4018/IJHCITP.2017040103

Aninze, F., El-Gohary, H., & Hussain, J. (2018). The Role of Microfinance to Empower Women: The Case of Developing Countries. *International Journal of Customer Relationship Marketing and Management*, *9*(1), 54–78. doi:10.4018/IJCRMM.2018010104

Antosova, G., Sabogal-Salamanca, M., & Krizova, E. (2021). Human Capital in Tourism: A Practical Model of Endogenous and Exogenous Territorial Tourism Planning in Bahía Solano, Colombia. In V. Costa, A. Moura, & M. Mira (Eds.), *Handbook of Research on Human Capital and People Management in the Tourism Industry* (pp. 282–302). IGI Global. https://doi.org/10.4018/978-1-7998-4318-4.ch014

Arsenijević, O. M., Orčić, D., & Kastratović, E. (2017). Development of an Optimization Tool for Intangibles in SMEs: A Case Study from Serbia with a Pilot Research in the Prestige by Milka Company. In M. Vemić (Ed.), *Optimal Management Strategies in Small and Medium Enterprises* (pp. 320–347). Hershey, PA: IGI Global. doi:10.4018/978-1-5225-1949-2.ch015

Aryanto, V. D., Wismantoro, Y., & Widyatmoko, K. (2018). Implementing Eco-Innovation by Utilizing the Internet to Enhance Firm's Marketing Performance: Study of Green Batik Small and Medium Enterprises in Indonesia. *International Journal of E-Business Research, 14*(1), 21–36. doi:10.4018/IJEBR.2018010102

Asero, V., & Billi, S. (2022). New Perspective of Networking in the DMO Model. In M. Valeri (Ed.), *New Governance and Management in Touristic Destinations* (pp. 105–118). IGI Global. https://doi.org/10.4018/978-1-6684-3889-3.ch007

Atiku, S. O., & Fields, Z. (2017). Multicultural Orientations for 21st Century Global Leadership. In N. Baporikar (Ed.), *Management Education for Global Leadership* (pp. 28–51). Hershey, PA: IGI Global. doi:10.4018/978-1-5225-1013-0.ch002

Atiku, S. O., & Fields, Z. (2018). Organisational Learning Dimensions and Talent Retention Strategies for the Service Industries. In N. Baporikar (Ed.), *Global Practices in Knowledge Management for Societal and Organizational Development* (pp. 358–381). Hershey, PA: IGI Global. doi:10.4018/978-1-5225-3009-1.ch017

Atsa'am, D. D., & Kuset Bodur, E. (2021). Pattern Mining on How Organizational Tenure Affects the Psychological Capital of Employees Within the Hospitality and Tourism Industry: Linking Employees' Organizational Tenure With PsyCap. *International Journal of Tourism and Hospitality Management in the Digital Age, 5*(2), 17–28. https://doi.org/10.4018/IJTHMDA.2021070102

Ávila, L., & Teixeira, L. (2018). The Main Concepts Behind the Dematerialization of Business Processes. In M. Khosrow-Pour, D.B.A. (Ed.), Encyclopedia of Information Science and Technology, Fourth Edition (pp. 888-898). Hershey, PA: IGI Global. https://doi.org/ doi:10.4018/978-1-5225-2255-3.ch076

Ayorekire, J., Mugizi, F., Obua, J., & Ampaire, G. (2022). Community-Based Tourism and Local People's Perceptions Towards Conservation: The Case of Queen Elizabeth Conservation Area, Uganda. In I. Mensah & E. Afenyo-Agbe (Eds.), *Prospects and Challenges of Community-Based Tourism and Changing Demographics* (pp. 56–82). IGI Global. https://doi.org/10.4018/978-1-7998-7335-8.ch003

Baleiro, R. (2022). Tourist Literature and the Architecture of Travel in Olga Tokarczuk and Patti Smith. In R. Baleiro & R. Pereira (Eds.), *Global Perspectives on Literary Tourism and Film-Induced Tourism* (pp. 202-216). IGI Global. https://doi.org/10.4018/978-1-7998-8262-6.ch011

Barat, S. (2021). Looking at the Future of Medical Tourism in Asia. *International Journal of Tourism and Hospitality Management in the Digital Age*, *5*(1), 19–33. https://doi.org/10.4018/IJTHMDA.2021010102

Barbosa, C. A., Magalhães, M., & Nunes, M. R. (2021). Travel Instagramability: A Way of Choosing a Destination? In M. Dinis, L. Bonixe, S. Lamy, & Z. Breda (Eds.), *Impact of New Media in Tourism* (pp. 173-190). IGI Global. https://doi.org/10.4018/978-1-7998-7095-1.ch011

Bari, M. W., & Khan, Q. (2021). Pakistan as a Destination of Religious Tourism. In E. Alaverdov & M. Bari (Eds.), *Global Development of Religious Tourism* (pp. 1-10). IGI Global. https://doi.org/10.4018/978-1-7998-5792-1.ch001

Bartens, Y., Chunpir, H. I., Schulte, F., & Voß, S. (2017). Business/IT Alignment in Two-Sided Markets: A COBIT 5 Analysis for Media Streaming Business Models. In S. De Haes & W. Van Grembergen (Eds.), *Strategic IT Governance and Alignment in Business Settings* (pp. 82–111). Hershey, PA: IGI Global. doi:10.4018/978-1-5225-0861-8.ch004

Bashayreh, A. M. (2018). Organizational Culture and Organizational Performance. In W. Lee & F. Sabetzadeh (Eds.), *Contemporary Knowledge and Systems Science* (pp. 50–69). Hershey, PA: IGI Global. doi:10.4018/978-1-5225-5655-8.ch003

Bechthold, L., Lude, M., & Prügl, R. (2021). Crisis Favors the Prepared Firm: How Organizational Ambidexterity Relates to Perceptions of Organizational Resilience. In A. Zehrer, G. Glowka, K. Schwaiger, & V. Ranacher-Lackner (Eds.), *Resiliency Models and Addressing Future Risks for Family Firms in the Tourism Industry* (pp. 178–205). IGI Global. https://doi.org/10.4018/978-1-7998-7352-5.ch008

Bedford, D. A. (2018). Sustainable Knowledge Management Strategies: Aligning Business Capabilities and Knowledge Management Goals. In N. Baporikar (Ed.), *Global Practices in Knowledge Management for Societal and Organizational Development* (pp. 46–73). Hershey, PA: IGI Global. doi:10.4018/978-1-5225-3009-1.ch003

Bekjanov, D., & Matyusupov, B. (2021). Influence of Innovative Processes in the Competitiveness of Tourist Destination. In J. Soares (Ed.), *Innovation and Entrepreneurial Opportunities in Community Tourism* (pp. 243–263). IGI Global. https://doi.org/10.4018/978-1-7998-4855-4.ch014

Bharwani, S., & Musunuri, D. (2018). Reflection as a Process From Theory to Practice. In M. Khosrow-Pour, D.B.A. (Ed.), Encyclopedia of Information Science and Technology, Fourth Edition (pp. 1529-1539). Hershey, PA: IGI Global. doi:10.4018/978-1-5225-2255-3.ch132

Bhatt, G. D., Wang, Z., & Rodger, J. A. (2017). Information Systems Capabilities and Their Effects on Competitive Advantages: A Study of Chinese Companies. *Information Resources Management Journal*, *30*(3), 41–57. doi:10.4018/IRMJ.2017070103

Bhushan, M., & Yadav, A. (2017). Concept of Cloud Computing in ESB. In R. Bhadoria, N. Chaudhari, G. Tomar, & S. Singh (Eds.), *Exploring Enterprise Service Bus in the Service-Oriented Architecture Paradigm* (pp. 116–127). Hershey, PA: IGI Global. doi:10.4018/978-1-5225-2157-0.ch008

Bhushan, S. (2017). System Dynamics Base-Model of Humanitarian Supply Chain (HSCM) in Disaster Prone Eco-Communities of India: A Discussion on Simulation and Scenario Results. *International Journal of System Dynamics Applications*, *6*(3), 20–37. doi:10.4018/IJSDA.2017070102

Binder, D., & Miller, J. W. (2021). A Generations' Perspective on Employer Branding in Tourism. In V. Costa, A. Moura, & M. Mira (Eds.), *Handbook of Research on Human Capital and People Management in the Tourism Industry* (pp. 152–174). IGI Global. https://doi.org/10.4018/978-1-7998-4318-4.ch008

Birch Freeman, A. A., Mensah, I., & Antwi, K. B. (2022). Smiling vs. Frowning Faces: Community Participation for Sustainable Tourism in Ghanaian Communities. In I. Mensah & E. Afenyo-Agbe (Eds.), *Prospects and Challenges of Community-Based Tourism and Changing Demographics* (pp. 83–106). IGI Global. https://doi.org/10.4018/978-1-7998-7335-8.ch004

Biswas, A., & De, A. K. (2017). On Development of a Fuzzy Stochastic Programming Model with Its Application to Business Management. In S. Trivedi, S. Dey, A. Kumar, & T. Panda (Eds.), *Handbook of Research on Advanced Data Mining Techniques and Applications for Business Intelligence* (pp. 353–378). Hershey, PA: IGI Global. doi:10.4018/978-1-5225-2031-3.ch021

Boragnio, A., & Faracce Macia, C. (2021). "Taking Care of Yourself at Home": Use of E-Commerce About Food and Care During the COVID-19 Pandemic in the City of Buenos Aires. In M. Korstanje (Ed.), *Socio-Economic Effects and Recovery Efforts for the Rental Industry: Post-COVID-19 Strategies* (pp. 45–71). IGI Global. https://doi.org/10.4018/978-1-7998-7287-0.ch003

Borges, V. D. (2021). Happiness: The Basis for Public Policy in Tourism. In A. Perinotto, V. Mayer, & J. Soares (Eds.), *Rebuilding and Restructuring the Tourism Industry: Infusion of Happiness and Quality of Life* (pp. 1–25). IGI Global. https://doi.org/10.4018/978-1-7998-7239-9.ch001

Bücker, J., & Ernste, K. (2018). Use of Brand Heroes in Strategic Reputation Management: The Case of Bacardi, Adidas, and Daimler. In A. Erdemir (Ed.), *Reputation Management Techniques in Public Relations* (pp. 126–150). Hershey, PA: IGI Global. doi:10.4018/978-1-5225-3619-2.ch007

Buluk Eşitti, B. (2021). COVID-19 and Alternative Tourism: New Destinations and New Tourism Products. In M. Demir, A. Dalgıç, & F. Ergen (Eds.), *Handbook of Research on the Impacts and Implications of COVID-19 on the Tourism Industry* (pp. 786–805). IGI Global. https://doi.org/10.4018/978-1-7998-8231-2.ch038

Bureš, V. (2018). Industry 4.0 From the Systems Engineering Perspective: Alternative Holistic Framework Development. In R. Brunet-Thornton & F. Martinez (Eds.), *Analyzing the Impacts of Industry 4.0 in Modern Business Environments* (pp. 199–223). Hershey, PA: IGI Global. doi:10.4018/978-1-5225-3468-6.ch011

Buzady, Z. (2017). Resolving the Magic Cube of Effective Case Teaching: Benchmarking Case Teaching Practices in Emerging Markets – Insights from the Central European University Business School, Hungary. In D. Latusek (Ed.), *Case Studies as a Teaching Tool in Management Education* (pp. 79–103). Hershey, PA: IGI Global. doi:10.4018/978-1-5225-0770-3.ch005

Camillo, A. (2021). *Legal Matters, Risk Management, and Risk Prevention: From Forming a Business to Legal Representation.* IGI Global. doi:10.4018/978-1-7998-4342-9.ch004

Căpusneanu, S., & Topor, D. I. (2018). Business Ethics and Cost Management in SMEs: Theories of Business Ethics and Cost Management Ethos. In I. Oncioiu (Ed.), *Ethics and Decision-Making for Sustainable Business Practices* (pp. 109–127). Hershey, PA: IGI Global. doi:10.4018/978-1-5225-3773-1.ch007

Chan, R. L., Mo, P. L., & Moon, K. K. (2018). Strategic and Tactical Measures in Managing Enterprise Risks: A Study of the Textile and Apparel Industry. In K. Strang, M. Korstanje, & N. Vajjhala (Eds.), *Research, Practices, and Innovations in Global Risk and Contingency Management* (pp. 1–19). Hershey, PA: IGI Global. doi:10.4018/978-1-5225-4754-9.ch001

Charlier, S. D., Burke-Smalley, L. A., & Fisher, S. L. (2018). Undergraduate Programs in the U.S: A Contextual and Content-Based Analysis. In J. Mendy (Ed.), *Teaching Human Resources and Organizational Behavior at the College Level* (pp. 26–57). Hershey, PA: IGI Global. doi:10.4018/978-1-5225-2820-3.ch002

Chumillas, J., Güell, M., & Quer, P. (2022). The Use of ICT in Tourist and Educational Literary Routes: The Role of the Guide. In C. Ramos, S. Quinteiro, & A. Gonçalves (Eds.), *ICT as Innovator Between Tourism and Culture* (pp. 15–29). IGI Global. https://doi.org/10.4018/978-1-7998-8165-0.ch002

Dahlberg, T., Kivijärvi, H., & Saarinen, T. (2017). IT Investment Consistency and Other Factors Influencing the Success of IT Performance. In S. De Haes & W. Van Grembergen (Eds.), *Strategic IT Governance and Alignment in Business Settings* (pp. 176–208). Hershey, PA: IGI Global. doi:10.4018/978-1-5225-0861-8.ch007

Damnjanović, A. M. (2017). Knowledge Management Optimization through IT and E-Business Utilization: A Qualitative Study on Serbian SMEs. In M. Vemić (Ed.), *Optimal Management Strategies in Small and Medium Enterprises* (pp. 249–267). Hershey, PA: IGI Global. doi:10.4018/978-1-5225-1949-2.ch012

Daneshpour, H. (2017). Integrating Sustainable Development into Project Portfolio Management through Application of Open Innovation. In M. Vemić (Ed.), *Optimal Management Strategies in Small and Medium Enterprises* (pp. 370–387). Hershey, PA: IGI Global. doi:10.4018/978-1-5225-1949-2.ch017

Daniel, A. D., & Reis de Castro, V. (2018). Entrepreneurship Education: How to Measure the Impact on Nascent Entrepreneurs. In A. Carrizo Moreira, J. Guilherme Leitão Dantas, & F. Manuel Valente (Eds.), *Nascent Entrepreneurship and Successful New Venture Creation* (pp. 85–110). Hershey, PA: IGI Global. doi:10.4018/978-1-5225-2936-1.ch004

David, R., Swami, B. N., & Tangirala, S. (2018). Ethics Impact on Knowledge Management in Organizational Development: A Case Study. In N. Baporikar (Ed.), *Global Practices in Knowledge Management for Societal and Organizational Development* (pp. 19–45). Hershey, PA: IGI Global. doi:10.4018/978-1-5225-3009-1.ch002

De Uña-Álvarez, E., & Villarino-Pérez, M. (2022). Fostering Ecocultural Resources, Identity, and Tourism in Inland Territories (Galicia, NW Spain). In G. Fernandes (Ed.), *Challenges and New Opportunities for Tourism in Inland Territories: Ecocultural Resources and Sustainable Initiatives* (pp. 1-16). IGI Global. https://doi.org/10.4018/978-1-7998-7339-6.ch001

Delias, P., & Lakiotaki, K. (2018). Discovering Process Horizontal Boundaries to Facilitate Process Comprehension. *International Journal of Operations Research and Information Systems*, *9*(2), 1–31. doi:10.4018/IJORIS.2018040101

Denholm, J., & Lee-Davies, L. (2018). Success Factors for Games in Business and Project Management. In *Enhancing Education and Training Initiatives Through Serious Games* (pp. 34–68). Hershey, PA: IGI Global. doi:10.4018/978-1-5225-3689-5.ch002

Deshpande, M. (2017). Best Practices in Management Institutions for Global Leadership: Policy Aspects. In N. Baporikar (Ed.), *Management Education for Global Leadership* (pp. 1–27). Hershey, PA: IGI Global. doi:10.4018/978-1-5225-1013-0.ch001

Deshpande, M. (2018). Policy Perspectives for SMEs Knowledge Management. In N. Baporikar (Ed.), *Knowledge Integration Strategies for Entrepreneurship and Sustainability* (pp. 23–46). Hershey, PA: IGI Global. doi:10.4018/978-1-5225-5115-7.ch002

Dezdar, S. (2017). ERP Implementation Projects in Asian Countries: A Comparative Study on Iran and China. *International Journal of Information Technology Project Management*, *8*(3), 52–68. doi:10.4018/IJITPM.2017070104

Domingos, D., Respício, A., & Martinho, R. (2017). Reliability of IoT-Aware BPMN Healthcare Processes. In C. Reis & M. Maximiano (Eds.), *Internet of Things and Advanced Application in Healthcare* (pp. 214–248). Hershey, PA: IGI Global. doi:10.4018/978-1-5225-1820-4.ch008

Dosumu, O., Hussain, J., & El-Gohary, H. (2017). An Exploratory Study of the Impact of Government Policies on the Development of Small and Medium Enterprises in Developing Countries: The Case of Nigeria. *International Journal of Customer Relationship Marketing and Management*, *8*(4), 51–62. doi:10.4018/IJCRMM.2017100104

Durst, S., Bruns, G., & Edvardsson, I. R. (2017). Retaining Knowledge in Smaller Building and Construction Firms. *International Journal of Knowledge and Systems Science*, *8*(3), 1–12. doi:10.4018/IJKSS.2017070101

Edvardsson, I. R., & Durst, S. (2017). Outsourcing, Knowledge, and Learning: A Critical Review. *International Journal of Knowledge-Based Organizations*, *7*(2), 13–26. doi:10.4018/IJKBO.2017040102

Edwards, J. S. (2018). Integrating Knowledge Management and Business Processes. In M. Khosrow-Pour, D.B.A. (Ed.), Encyclopedia of Information Science and Technology, Fourth Edition (pp. 5046-5055). Hershey, PA: IGI Global. doi:10.4018/978-1-5225-2255-3.ch437

Eichelberger, S., & Peters, M. (2021). Family Firm Management in Turbulent Times: Opportunities for Responsible Tourism. In A. Zehrer, G. Glowka, K. Schwaiger, & V. Ranacher-Lackner (Eds.), *Resiliency Models and Addressing Future Risks for Family Firms in the Tourism Industry* (pp. 103–124). IGI Global. https://doi.org/10.4018/978-1-7998-7352-5.ch005

Eide, D., Hjalager, A., & Hansen, M. (2022). Innovative Certifications in Adventure Tourism: Attributes and Diffusion. In R. Augusto Costa, F. Brandão, Z. Breda, & C. Costa (Eds.), *Planning and Managing the Experience Economy in Tourism* (pp. 161-175). IGI Global. https://doi.org/10.4018/978-1-7998-8775-1.ch009

Ejiogu, A. O. (2018). Economics of Farm Management. In *Agricultural Finance and Opportunities for Investment and Expansion* (pp. 56–72). Hershey, PA: IGI Global. doi:10.4018/978-1-5225-3059-6.ch003

Ekanem, I., & Abiade, G. E. (2018). Factors Influencing the Use of E-Commerce by Small Enterprises in Nigeria. *International Journal of ICT Research in Africa and the Middle East*, *7*(1), 37–53. doi:10.4018/IJICTRAME.2018010103

Ekanem, I., & Alrossais, L. A. (2017). Succession Challenges Facing Family Businesses in Saudi Arabia. In P. Zgheib (Ed.), *Entrepreneurship and Business Innovation in the Middle East* (pp. 122–146). Hershey, PA: IGI Global. doi:10.4018/978-1-5225-2066-5.ch007

El Faquih, L., & Fredj, M. (2017). Ontology-Based Framework for Quality in Configurable Process Models. *Journal of Electronic Commerce in Organizations*, *15*(2), 48–60. doi:10.4018/JECO.2017040104

Faisal, M. N., & Talib, F. (2017). Building Ambidextrous Supply Chains in SMEs: How to Tackle the Barriers? *International Journal of Information Systems and Supply Chain Management*, *10*(4), 80–100. doi:10.4018/IJISSCM.2017100105

Fernandes, T. M., Gomes, J., & Romão, M. (2017). Investments in E-Government: A Benefit Management Case Study. *International Journal of Electronic Government Research*, *13*(3), 1–17. doi:10.4018/IJEGR.2017070101

Figueira, L. M., Honrado, G. R., & Dionísio, M. S. (2021). Human Capital Management in the Tourism Industry in Portugal. In V. Costa, A. Moura, & M. Mira (Eds.), *Handbook of Research on Human Capital and People Management in the Tourism Industry* (pp. 1–19). IGI Global. doi:10.4018/978-1-7998-4318-4.ch001

Gao, S. S., Oreal, S., & Zhang, J. (2018). Contemporary Financial Risk Management Perceptions and Practices of Small-Sized Chinese Businesses. In I. Management Association (Ed.), Global Business Expansion: Concepts, Methodologies, Tools, and Applications (pp. 917-931). Hershey, PA: IGI Global. doi:10.4018/978-1-5225-5481-3.ch041

Garg, R., & Berning, S. C. (2017). Indigenous Chinese Management Philosophies: Key Concepts and Relevance for Modern Chinese Firms. In B. Christiansen & G. Koc (Eds.), *Transcontinental Strategies for Industrial Development and Economic Growth* (pp. 43–57). Hershey, PA: IGI Global. doi:10.4018/978-1-5225-2160-0.ch003

Gencer, Y. G. (2017). Supply Chain Management in Retailing Business. In U. Akkucuk (Ed.), *Ethics and Sustainability in Global Supply Chain Management* (pp. 197–210). Hershey, PA: IGI Global. doi:10.4018/978-1-5225-2036-8.ch011

Gera, R., Arora, S., & Malik, S. (2021). Emotional Labor in the Tourism Industry: Strategies, Antecedents, and Outcomes. In V. Costa, A. Moura, & M. Mira (Eds.), *Handbook of Research on Human Capital and People Management in the Tourism Industry* (pp. 73–91). IGI Global. https://doi.org/10.4018/978-1-7998-4318-4.ch004

Giacosa, E. (2018). The Increasing of the Regional Development Thanks to the Luxury Business Innovation. In L. Carvalho (Ed.), *Handbook of Research on Entrepreneurial Ecosystems and Social Dynamics in a Globalized World* (pp. 260–273). Hershey, PA: IGI Global. doi:10.4018/978-1-5225-3525-6.ch011

Glowka, G., Tusch, M., & Zehrer, A. (2021). The Risk Perception of Family Business Owner-Manager in the Tourism Industry: A Qualitative Comparison of the Intra-Firm Senior and Junior Generation. In A. Zehrer, G. Glowka, K. Schwaiger, & V. Ranacher-Lackner (Eds.), *Resiliency Models and Addressing Future Risks for Family Firms in the Tourism Industry* (pp. 126–153). IGI Global. https://doi.org/10.4018/978-1-7998-7352-5.ch006

Glykas, M., & George, J. (2017). Quality and Process Management Systems in the UAE Maritime Industry. *International Journal of Productivity Management and Assessment Technologies*, *5*(1), 20–39. doi:10.4018/IJPMAT.2017010102

Glykas, M., Valiris, G., Kokkinaki, A., & Koutsoukou, Z. (2018). Banking Business Process Management Implementation. *International Journal of Productivity Management and Assessment Technologies, 6*(1), 50–69. doi:10.4018/IJPMAT.2018010104

Gomes, J., & Romão, M. (2017). The Balanced Scorecard: Keeping Updated and Aligned with Today's Business Trends. *International Journal of Productivity Management and Assessment Technologies, 5*(2), 1–15. doi:10.4018/IJPMAT.2017070101

Gomes, J., & Romão, M. (2017). Aligning Information Systems and Technology with Benefit Management and Balanced Scorecard. In S. De Haes & W. Van Grembergen (Eds.), *Strategic IT Governance and Alignment in Business Settings* (pp. 112–131). Hershey, PA: IGI Global. doi:10.4018/978-1-5225-0861-8.ch005

Goyal, A. (2021). Communicating and Building Destination Brands With New Media. In M. Dinis, L. Bonixe, S. Lamy, & Z. Breda (Eds.), *Impact of New Media in Tourism* (pp. 1-20). IGI Global. https://doi.org/10.4018/978-1-7998-7095-1.ch001

Grefen, P., & Turetken, O. (2017). Advanced Business Process Management in Networked E-Business Scenarios. *International Journal of E-Business Research, 13*(4), 70–104. doi:10.4018/IJEBR.2017100105

Guasca, M., Van Broeck, A. M., & Vanneste, D. (2021). Tourism and the Social Reintegration of Colombian Ex-Combatants. In J. da Silva, Z. Breda, & F. Carbone (Eds.), *Role and Impact of Tourism in Peacebuilding and Conflict Transformation* (pp. 66-86). IGI Global. https://doi.org/10.4018/978-1-7998-5053-3.ch005

Haider, A., & Saetang, S. (2017). Strategic IT Alignment in Service Sector. In S. Rozenes & Y. Cohen (Eds.), *Handbook of Research on Strategic Alliances and Value Co-Creation in the Service Industry* (pp. 231–258). Hershey, PA: IGI Global. doi:10.4018/978-1-5225-2084-9.ch012

Hajilari, A. B., Ghadaksaz, M., & Fasghandis, G. S. (2017). Assessing Organizational Readiness for Implementing ERP System Using Fuzzy Expert System Approach. *International Journal of Enterprise Information Systems, 13*(1), 67–85. doi:10.4018/IJEIS.2017010105

Haldorai, A., Ramu, A., & Murugan, S. (2018). Social Aware Cognitive Radio Networks: Effectiveness of Social Networks as a Strategic Tool for Organizational Business Management. In H. Bansal, G. Shrivastava, G. Nguyen, & L. Stanciu (Eds.), *Social Network Analytics for Contemporary Business Organizations* (pp. 188–202). Hershey, PA: IGI Global. doi:10.4018/978-1-5225-5097-6.ch010

Hall, O. P. Jr. (2017). Social Media Driven Management Education. *International Journal of Knowledge-Based Organizations, 7*(2), 43–59. doi:10.4018/IJKBO.2017040104

Hanifah, H., Halim, H. A., Ahmad, N. H., & Vafaei-Zadeh, A. (2017). Innovation Culture as a Mediator Between Specific Human Capital and Innovation Performance Among Bumiputera SMEs in Malaysia. In N. Ahmad, T. Ramayah, H. Halim, & S. Rahman (Eds.), *Handbook of Research on Small and Medium Enterprises in Developing Countries* (pp. 261–279). Hershey, PA: IGI Global. doi:10.4018/978-1-5225-2165-5.ch012

Hartlieb, S., & Silvius, G. (2017). Handling Uncertainty in Project Management and Business Development: Similarities and Differences. In Y. Raydugin (Ed.), *Handbook of Research on Leveraging Risk and Uncertainties for Effective Project Management* (pp. 337–362). Hershey, PA: IGI Global. doi:10.4018/978-1-5225-1790-0.ch016

Hass, K. B. (2017). Living on the Edge: Managing Project Complexity. In Y. Raydugin (Ed.), *Handbook of Research on Leveraging Risk and Uncertainties for Effective Project Management* (pp. 177–201). Hershey, PA: IGI Global. doi:10.4018/978-1-5225-1790-0.ch009

Hawking, P., & Carmine Sellitto, C. (2017). Developing an Effective Strategy for Organizational Business Intelligence. In M. Tavana (Ed.), *Enterprise Information Systems and the Digitalization of Business Functions* (pp. 222–237). Hershey, PA: IGI Global. doi:10.4018/978-1-5225-2382-6.ch010

Hawking, P., & Sellitto, C. (2017). A Fast-Moving Consumer Goods Company and Business Intelligence Strategy Development. *International Journal of Enterprise Information Systems, 13*(2), 22–33. doi:10.4018/IJEIS.2017040102

Hawking, P., & Sellitto, C. (2017). Business Intelligence Strategy: Two Case Studies. *International Journal of Business Intelligence Research, 8*(2), 17–30. doi:10.4018/IJBIR.2017070102

Hee, W. J., Jalleh, G., Lai, H., & Lin, C. (2017). E-Commerce and IT Projects: Evaluation and Management Issues in Australian and Taiwanese Hospitals. *International Journal of Public Health Management and Ethics, 2*(1), 69–90. doi:10.4018/IJPHME.2017010104

Hernandez, A. A. (2018). Exploring the Factors to Green IT Adoption of SMEs in the Philippines. *Journal of Cases on Information Technology, 20*(2), 49–66. doi:10.4018/JCIT.2018040104

Hollman, A., Bickford, S., & Hollman, T. (2017). Cyber InSecurity: A Post-Mortem Attempt to Assess Cyber Problems from IT and Business Management Perspectives. *Journal of Cases on Information Technology*, *19*(3), 42–70. doi:10.4018/JCIT.2017070104

Ibrahim, F., & Zainin, N. M. (2021). Exploring the Technological Impacts: The Case of Museums in Brunei Darussalam. *International Journal of Tourism and Hospitality Management in the Digital Age*, *5*(1), 1–18. https://doi.org/10.4018/IJTHMDA.2021010101

Igbinakhase, I. (2017). Responsible and Sustainable Management Practices in Developing and Developed Business Environments. In Z. Fields (Ed.), *Collective Creativity for Responsible and Sustainable Business Practice* (pp. 180–207). Hershey, PA: IGI Global. doi:10.4018/978-1-5225-1823-5.ch010

Iwata, J. J., & Hoskins, R. G. (2017). Managing Indigenous Knowledge in Tanzania: A Business Perspective. In P. Jain & N. Mnjama (Eds.), *Managing Knowledge Resources and Records in Modern Organizations* (pp. 198–214). Hershey, PA: IGI Global. doi:10.4018/978-1-5225-1965-2.ch012

Jain, P. (2017). Ethical and Legal Issues in Knowledge Management Life-Cycle in Business. In P. Jain & N. Mnjama (Eds.), *Managing Knowledge Resources and Records in Modern Organizations* (pp. 82–101). Hershey, PA: IGI Global. doi:10.4018/978-1-5225-1965-2.ch006

James, S., & Hauli, E. (2017). Holistic Management Education at Tanzanian Rural Development Planning Institute. In N. Baporikar (Ed.), *Management Education for Global Leadership* (pp. 112–136). Hershey, PA: IGI Global. doi:10.4018/978-1-5225-1013-0.ch006

Janošková, M., Csikósová, A., & Čulková, K. (2018). Measurement of Company Performance as Part of Its Strategic Management. In R. Leon (Ed.), *Managerial Strategies for Business Sustainability During Turbulent Times* (pp. 309–335). Hershey, PA: IGI Global. doi:10.4018/978-1-5225-2716-9.ch017

Jean-Vasile, A., & Alecu, A. (2017). Theoretical and Practical Approaches in Understanding the Influences of Cost-Productivity-Profit Trinomial in Contemporary Enterprises. In A. Jean Vasile & D. Nicolò (Eds.), *Sustainable Entrepreneurship and Investments in the Green Economy* (pp. 28–62). Hershey, PA: IGI Global. doi:10.4018/978-1-5225-2075-7.ch002

Joia, L. A., & Correia, J. C. (2018). CIO Competencies From the IT Professional Perspective: Insights From Brazil. *Journal of Global Information Management*, *26*(2), 74–103. doi:10.4018/JGIM.2018040104

Juma, A., & Mzera, N. (2017). Knowledge Management and Records Management and Competitive Advantage in Business. In P. Jain & N. Mnjama (Eds.), *Managing Knowledge Resources and Records in Modern Organizations* (pp. 15–28). Hershey, PA: IGI Global. doi:10.4018/978-1-5225-1965-2.ch002

K., I., & A, V. (2018). Monitoring and Auditing in the Cloud. In K. Munir (Ed.), *Cloud Computing Technologies for Green Enterprises* (pp. 318-350). Hershey, PA: IGI Global. https://doi.org/ doi:10.4018/978-1-5225-3038-1.ch013

Kabra, G., Ghosh, V., & Ramesh, A. (2018). Enterprise Integrated Business Process Management and Business Intelligence Framework for Business Process Sustainability. In A. Paul, D. Bhattacharyya, & S. Anand (Eds.), *Green Initiatives for Business Sustainability and Value Creation* (pp. 228–238). Hershey, PA: IGI Global. doi:10.4018/978-1-5225-2662-9.ch010

Kaoud, M. (2017). Investigation of Customer Knowledge Management: A Case Study Research. *International Journal of Service Science, Management, Engineering, and Technology*, *8*(2), 12–22. doi:10.4018/IJSSMET.2017040102

Katuu, S. (2018). A Comparative Assessment of Enterprise Content Management Maturity Models. In N. Gwangwava & M. Mutingi (Eds.), *E-Manufacturing and E-Service Strategies in Contemporary Organizations* (pp. 93–118). Hershey, PA: IGI Global. doi:10.4018/978-1-5225-3628-4.ch005

Khan, M. Y., & Abir, T. (2022). The Role of Social Media Marketing in the Tourism and Hospitality Industry: A Conceptual Study on Bangladesh. In C. Ramos, S. Quinteiro, & A. Gonçalves (Eds.), *ICT as Innovator Between Tourism and Culture* (pp. 213–229). IGI Global. https://doi.org/10.4018/978-1-7998-8165-0.ch013

Kinnunen, S., Ylä-Kujala, A., Marttonen-Arola, S., Kärri, T., & Baglee, D. (2018). Internet of Things in Asset Management: Insights from Industrial Professionals and Academia. *International Journal of Service Science, Management, Engineering, and Technology*, *9*(2), 104–119. doi:10.4018/IJSSMET.2018040105

Klein, A. Z., Sabino de Freitas, A., Machado, L., Freitas, J. C. Jr, Graziola, P. G. Jr, & Schlemmer, E. (2017). Virtual Worlds Applications for Management Education. In L. Tomei (Ed.), *Exploring the New Era of Technology-Infused Education* (pp. 279–299). Hershey, PA: IGI Global. doi:10.4018/978-1-5225-1709-2.ch017

Kővári, E., Saleh, M., & Steinbachné Hajmásy, G. (2022). The Impact of Corporate Digital Responsibility (CDR) on Internal Stakeholders' Satisfaction in Hungarian Upscale Hotels. In M. Valeri (Ed.), *New Governance and Management in Touristic Destinations* (pp. 35–51). IGI Global. https://doi.org/10.4018/978-1-6684-3889-3.ch003

Kożuch, B., & Jabłoński, A. (2017). Adopting the Concept of Business Models in Public Management. In M. Lewandowski & B. Kożuch (Eds.), *Public Sector Entrepreneurship and the Integration of Innovative Business Models* (pp. 10–46). Hershey, PA: IGI Global. doi:10.4018/978-1-5225-2215-7.ch002

Kumar, J., Adhikary, A., & Jha, A. (2017). Small Active Investors' Perceptions and Preferences Towards Tax Saving Mutual Fund Schemes in Eastern India: An Empirical Note. *International Journal of Asian Business and Information Management*, *8*(2), 35–45. doi:10.4018/IJABIM.2017040103

Latusi, S., & Fissore, M. (2021). Pilgrimage Routes to Happiness: Comparing the Camino de Santiago and Via Francigena. In A. Perinotto, V. Mayer, & J. Soares (Eds.), *Rebuilding and Restructuring the Tourism Industry: Infusion of Happiness and Quality of Life* (pp. 157–182). IGI Global. https://doi.org/10.4018/978-1-7998-7239-9.ch008

Lavassani, K. M., & Movahedi, B. (2017). Applications Driven Information Systems: Beyond Networks toward Business Ecosystems. *International Journal of Innovation in the Digital Economy*, *8*(1), 61–75. doi:10.4018/IJIDE.2017010104

Lazzareschi, V. H., & Brito, M. S. (2017). Strategic Information Management: Proposal of Business Project Model. In G. Jamil, A. Soares, & C. Pessoa (Eds.), *Handbook of Research on Information Management for Effective Logistics and Supply Chains* (pp. 59–88). Hershey, PA: IGI Global. doi:10.4018/978-1-5225-0973-8.ch004

Lechuga Sancho, M. P., & Martín Navarro, A. (2022). Evolution of the Literature on Social Responsibility in the Tourism Sector: A Systematic Literature Review. In G. Fernandes (Ed.), *Challenges and New Opportunities for Tourism in Inland Territories: Ecocultural Resources and Sustainable Initiatives* (pp. 169–186). IGI Global. https://doi.org/10.4018/978-1-7998-7339-6.ch010

Lederer, M., Kurz, M., & Lazarov, P. (2017). Usage and Suitability of Methods for Strategic Business Process Initiatives: A Multi Case Study Research. *International Journal of Productivity Management and Assessment Technologies*, *5*(1), 40–51. doi:10.4018/IJPMAT.2017010103

Lee, I. (2017). A Social Enterprise Business Model and a Case Study of Pacific Community Ventures (PCV). In V. Potocan, M. Üngan, & Z. Nedelko (Eds.), *Handbook of Research on Managerial Solutions in Non-Profit Organizations* (pp. 182–204). Hershey, PA: IGI Global. doi:10.4018/978-1-5225-0731-4.ch009

Leon, L. A., Seal, K. C., Przasnyski, Z. H., & Wiedenman, I. (2017). Skills and Competencies Required for Jobs in Business Analytics: A Content Analysis of Job Advertisements Using Text Mining. *International Journal of Business Intelligence Research, 8*(1), 1–25. doi:10.4018/IJBIR.2017010101

Levy, C. L., & Elias, N. I. (2017). SOHO Users' Perceptions of Reliability and Continuity of Cloud-Based Services. In M. Moore (Ed.), *Cybersecurity Breaches and Issues Surrounding Online Threat Protection* (pp. 248–287). Hershey, PA: IGI Global. doi:10.4018/978-1-5225-1941-6.ch011

Levy, M. (2018). Change Management Serving Knowledge Management and Organizational Development: Reflections and Review. In N. Baporikar (Ed.), *Global Practices in Knowledge Management for Societal and Organizational Development* (pp. 256–270). Hershey, PA: IGI Global. doi:10.4018/978-1-5225-3009-1.ch012

Lewandowski, M. (2017). Public Organizations and Business Model Innovation: The Role of Public Service Design. In M. Lewandowski & B. Kożuch (Eds.), *Public Sector Entrepreneurship and the Integration of Innovative Business Models* (pp. 47–72). Hershey, PA: IGI Global. doi:10.4018/978-1-5225-2215-7.ch003

Lhannaoui, H., Kabbaj, M. I., & Bakkoury, Z. (2017). A Survey of Risk-Aware Business Process Modelling. *International Journal of Risk and Contingency Management, 6*(3), 14–26. doi:10.4018/IJRCM.2017070102

Li, J., Sun, W., Jiang, W., Yang, H., & Zhang, L. (2017). How the Nature of Exogenous Shocks and Crises Impact Company Performance?: The Effects of Industry Characteristics. *International Journal of Risk and Contingency Management, 6*(4), 40–55. doi:10.4018/IJRCM.2017100103

Lopez-Fernandez, M., Perez-Perez, M., Serrano-Bedia, A., & Cobo-Gonzalez, A. (2021). Small and Medium Tourism Enterprise Survival in Times of Crisis: "El Capricho de Gaudí. In D. Toubes & N. Araújo-Vila (Eds.), *Risk, Crisis, and Disaster Management in Small and Medium-Sized Tourism Enterprises* (pp. 103–129). IGI Global. doi:10.4018/978-1-7998-6996-2.ch005

Mahajan, A., Maidullah, S., & Hossain, M. R. (2022). Experience Toward Smart Tour Guide Apps in Travelling: An Analysis of Users' Reviews on Audio Odigos and Trip My Way. In R. Augusto Costa, F. Brandão, Z. Breda, & C. Costa (Eds.), *Planning and Managing the Experience Economy in Tourism* (pp. 255-273). IGI Global. https://doi.org/10.4018/978-1-7998-8775-1.ch014

Malega, P. (2017). Small and Medium Enterprises in the Slovak Republic: Status and Competitiveness of SMEs in the Global Markets and Possibilities of Optimization. In M. Vemić (Ed.), *Optimal Management Strategies in Small and Medium Enterprises* (pp. 102–124). Hershey, PA: IGI Global. doi:10.4018/978-1-5225-1949-2.ch006

Malewska, K. M. (2017). Intuition in Decision-Making on the Example of a Non-Profit Organization. In V. Potocan, M. Üngan, & Z. Nedelko (Eds.), *Handbook of Research on Managerial Solutions in Non-Profit Organizations* (pp. 378–399). Hershey, PA: IGI Global. doi:10.4018/978-1-5225-0731-4.ch018

Maroofi, F. (2017). Entrepreneurial Orientation and Organizational Learning Ability Analysis for Innovation and Firm Performance. In N. Baporikar (Ed.), *Innovation and Shifting Perspectives in Management Education* (pp. 144–165). Hershey, PA: IGI Global. doi:10.4018/978-1-5225-1019-2.ch007

Marques, M., Moleiro, D., Brito, T. M., & Marques, T. (2021). Customer Relationship Management as an Important Relationship Marketing Tool: The Case of the Hospitality Industry in Estoril Coast. In M. Dinis, L. Bonixe, S. Lamy, & Z. Breda (Eds.), Impact of New Media in Tourism (pp. 39-56). IGI Global. https://doi.org/doi:10.4018/978-1-7998-7095-1.ch003

Martins, P. V., & Zacarias, M. (2017). A Web-based Tool for Business Process Improvement. *International Journal of Web Portals*, *9*(2), 68–84. doi:10.4018/IJWP.2017070104

Matthies, B., & Coners, A. (2017). Exploring the Conceptual Nature of e-Business Projects. *Journal of Electronic Commerce in Organizations*, *15*(3), 33–63. doi:10.4018/JECO.2017070103

Mayer, V. F., Fraga, C. C., & Silva, L. C. (2021). Contributions of Neurosciences to Studies of Well-Being in Tourism. In A. Perinotto, V. Mayer, & J. Soares (Eds.), *Rebuilding and Restructuring the Tourism Industry: Infusion of Happiness and Quality of Life* (pp. 108–128). IGI Global. https://doi.org/10.4018/978-1-7998-7239-9.ch006

McKee, J. (2018). Architecture as a Tool to Solve Business Planning Problems. In M. Khosrow-Pour, D.B.A. (Ed.), Encyclopedia of Information Science and Technology, Fourth Edition (pp. 573-586). Hershey, PA: IGI Global. doi:10.4018/978-1-5225-2255-3.ch050

McMurray, A. J., Cross, J., & Caponecchia, C. (2018). The Risk Management Profession in Australia: Business Continuity Plan Practices. In N. Bajgoric (Ed.), *Always-On Enterprise Information Systems for Modern Organizations* (pp. 112–129). Hershey, PA: IGI Global. doi:10.4018/978-1-5225-3704-5.ch006

Meddah, I. H., & Belkadi, K. (2018). Mining Patterns Using Business Process Management. In R. Hamou (Ed.), *Handbook of Research on Biomimicry in Information Retrieval and Knowledge Management* (pp. 78–89). Hershey, PA: IGI Global. doi:10.4018/978-1-5225-3004-6.ch005

Melian, A. G., & Camprubí, R. (2021). The Accessibility of Museum Websites: The Case of Barcelona. In C. Eusébio, L. Teixeira, & M. Carneiro (Eds.), *ICT Tools and Applications for Accessible Tourism* (pp. 234–255). IGI Global. https://doi.org/10.4018/978-1-7998-6428-8.ch011

Mendes, L. (2017). TQM and Knowledge Management: An Integrated Approach Towards Tacit Knowledge Management. In D. Jaziri-Bouagina & G. Jamil (Eds.), *Handbook of Research on Tacit Knowledge Management for Organizational Success* (pp. 236–263). Hershey, PA: IGI Global. doi:10.4018/978-1-5225-2394-9.ch009

Menezes, V. D., & Cavagnaro, E. (2021). Communicating Sustainable Initiatives in the Hotel Industry: The Case of the Hotel Jakarta Amsterdam. In F. Brandão, Z. Breda, R. Costa, & C. Costa (Eds.), *Handbook of Research on the Role of Tourism in Achieving Sustainable Development Goals* (pp. 224-234). IGI Global. https://doi.org/10.4018/978-1-7998-5691-7.ch013

Menezes, V. D., & Cavagnaro, E. (2021). Communicating Sustainable Initiatives in the Hotel Industry: The Case of the Hotel Jakarta Amsterdam. In F. Brandão, Z. Breda, R. Costa, & C. Costa (Eds.), *Handbook of Research on the Role of Tourism in Achieving Sustainable Development Goals* (pp. 224-234). IGI Global. https://doi.org/10.4018/978-1-7998-5691-7.ch013

Mitas, O., Bastiaansen, M., & Boode, W. (2022). If You're Happy, I'm Happy: Emotion Contagion at a Tourist Information Center. In R. Augusto Costa, F. Brandão, Z. Breda, & C. Costa (Eds.), *Planning and Managing the Experience Economy in Tourism* (pp. 122-140). IGI Global. https://doi.org/10.4018/978-1-7998-8775-1.ch007

Mnjama, N. M. (2017). Preservation of Recorded Information in Public and Private Sector Organizations. In P. Jain & N. Mnjama (Eds.), *Managing Knowledge Resources and Records in Modern Organizations* (pp. 149–167). Hershey, PA: IGI Global. doi:10.4018/978-1-5225-1965-2.ch009

Mokoqama, M., & Fields, Z. (2017). Principles of Responsible Management Education (PRME): Call for Responsible Management Education. In Z. Fields (Ed.), *Collective Creativity for Responsible and Sustainable Business Practice* (pp. 229–241). Hershey, PA: IGI Global. doi:10.4018/978-1-5225-1823-5.ch012

Monteiro, A., Lopes, S., & Carbone, F. (2021). Academic Mobility: Bridging Tourism and Peace Education. In J. da Silva, Z. Breda, & F. Carbone (Eds.), *Role and Impact of Tourism in Peacebuilding and Conflict Transformation* (pp. 275-301). IGI Global. https://doi.org/10.4018/978-1-7998-5053-3.ch016

Muniapan, B. (2017). Philosophy and Management: The Relevance of Vedanta in Management. In P. Ordóñez de Pablos (Ed.), *Managerial Strategies and Solutions for Business Success in Asia* (pp. 124–139). Hershey, PA: IGI Global. doi:10.4018/978-1-5225-1886-0.ch007

Murad, S. E., & Dowaji, S. (2017). Using Value-Based Approach for Managing Cloud-Based Services. In A. Turuk, B. Sahoo, & S. Addya (Eds.), *Resource Management and Efficiency in Cloud Computing Environments* (pp. 33–60). Hershey, PA: IGI Global. doi:10.4018/978-1-5225-1721-4.ch002

Mutahar, A. M., Daud, N. M., Thurasamy, R., Isaac, O., & Abdulsalam, R. (2018). The Mediating of Perceived Usefulness and Perceived Ease of Use: The Case of Mobile Banking in Yemen. *International Journal of Technology Diffusion, 9*(2), 21–40. doi:10.4018/IJTD.2018040102

Naidoo, V. (2017). E-Learning and Management Education at African Universities. In N. Baporikar (Ed.), *Management Education for Global Leadership* (pp. 181–201). Hershey, PA: IGI Global. doi:10.4018/978-1-5225-1013-0.ch009

Naidoo, V., & Igbinakhase, I. (2018). Opportunities and Challenges of Knowledge Retention in SMEs. In N. Baporikar (Ed.), *Knowledge Integration Strategies for Entrepreneurship and Sustainability* (pp. 70–94). Hershey, PA: IGI Global. doi:10.4018/978-1-5225-5115-7.ch004

Naumov, N., & Costandachi, G. (2021). Creativity and Entrepreneurship: Gastronomic Tourism in Mexico. In J. Soares (Ed.), *Innovation and Entrepreneurial Opportunities in Community Tourism* (pp. 90–108). IGI Global. https://doi.org/10.4018/978-1-7998-4855-4.ch006

Nayak, S., & Prabhu, N. (2017). Paradigm Shift in Management Education: Need for a Cross Functional Perspective. In N. Baporikar (Ed.), *Management Education for Global Leadership* (pp. 241–255). Hershey, PA: IGI Global. doi:10.4018/978-1-5225-1013-0.ch012

Nedelko, Z., & Potocan, V. (2017). Management Solutions in Non-Profit Organizations: Case of Slovenia. In V. Potocan, M. Üngan, & Z. Nedelko (Eds.), *Handbook of Research on Managerial Solutions in Non-Profit Organizations* (pp. 1–22). Hershey, PA: IGI Global. doi:10.4018/978-1-5225-0731-4.ch001

Nedelko, Z., & Potocan, V. (2017). Priority of Management Tools Utilization among Managers: International Comparison. In V. Wang (Ed.), *Encyclopedia of Strategic Leadership and Management* (pp. 1083–1094). Hershey, PA: IGI Global. doi:10.4018/978-1-5225-1049-9.ch075

Nedelko, Z., Raudeliūnienė, J., & Črešnar, R. (2018). Knowledge Dynamics in Supply Chain Management. In N. Baporikar (Ed.), *Knowledge Integration Strategies for Entrepreneurship and Sustainability* (pp. 150–166). Hershey, PA: IGI Global. doi:10.4018/978-1-5225-5115-7.ch008

Nguyen, H. T., & Hipsher, S. A. (2018). Innovation and Creativity Used by Private Sector Firms in a Resources-Constrained Environment. In S. Hipsher (Ed.), *Examining the Private Sector's Role in Wealth Creation and Poverty Reduction* (pp. 219–238). Hershey, PA: IGI Global. doi:10.4018/978-1-5225-3117-3.ch010

Obicci, P. A. (2017). Risk Sharing in a Partnership. In *Risk Management Strategies in Public-Private Partnerships* (pp. 115–152). Hershey, PA: IGI Global. doi:10.4018/978-1-5225-2503-5.ch004

Obidallah, W. J., & Raahemi, B. (2017). Managing Changes in Service Oriented Virtual Organizations: A Structural and Procedural Framework to Facilitate the Process of Change. *Journal of Electronic Commerce in Organizations*, *15*(1), 59–83. doi:10.4018/JECO.2017010104

Ojo, O. (2017). Impact of Innovation on the Entrepreneurial Success in Selected Business Enterprises in South-West Nigeria. *International Journal of Innovation in the Digital Economy*, *8*(2), 29–38. doi:10.4018/IJIDE.2017040103

Okdinawati, L., Simatupang, T. M., & Sunitiyoso, Y. (2017). Multi-Agent Reinforcement Learning for Value Co-Creation of Collaborative Transportation Management (CTM). *International Journal of Information Systems and Supply Chain Management*, *10*(3), 84–95. doi:10.4018/IJISSCM.2017070105

Olivera, V. A., & Carrillo, I. M. (2021). Organizational Culture: A Key Element for the Development of Mexican Micro and Small Tourist Companies. In J. Soares (Ed.), *Innovation and Entrepreneurial Opportunities in Community Tourism* (pp. 227–242). IGI Global. doi:10.4018/978-1-7998-4855-4.ch013

Ossorio, M. (2022). Corporate Museum Experiences in Enogastronomic Tourism. In R. Augusto Costa, F. Brandão, Z. Breda, & C. Costa (Eds.), Planning and Managing the Experience Economy in Tourism (pp. 107-121). IGI Global. https://doi.org/doi:10.4018/978-1-7998-8775-1.ch006

Ossorio, M. (2022). Enogastronomic Tourism in Times of Pandemic. In G. Fernandes (Ed.), *Challenges and New Opportunities for Tourism in Inland Territories: Ecocultural Resources and Sustainable Initiatives* (pp. 241–255). IGI Global. https://doi.org/10.4018/978-1-7998-7339-6.ch014

Özekici, Y. K. (2022). ICT as an Acculturative Agent and Its Role in the Tourism Context: Introduction, Acculturation Theory, Progress of the Acculturation Theory in Extant Literature. In C. Ramos, S. Quinteiro, & A. Gonçalves (Eds.), *ICT as Innovator Between Tourism and Culture* (pp. 42–66). IGI Global. https://doi.org/10.4018/978-1-7998-8165-0.ch004

Pal, K. (2018). Building High Quality Big Data-Based Applications in Supply Chains. In A. Kumar & S. Saurav (Eds.), *Supply Chain Management Strategies and Risk Assessment in Retail Environments* (pp. 1–24). Hershey, PA: IGI Global. doi:10.4018/978-1-5225-3056-5.ch001

Palos-Sanchez, P. R., & Correia, M. B. (2018). Perspectives of the Adoption of Cloud Computing in the Tourism Sector. In J. Rodrigues, C. Ramos, P. Cardoso, & C. Henriques (Eds.), *Handbook of Research on Technological Developments for Cultural Heritage and eTourism Applications* (pp. 377–400). Hershey, PA: IGI Global. doi:10.4018/978-1-5225-2927-9.ch018

Papadopoulou, G. (2021). Promoting Gender Equality and Women Empowerment in the Tourism Sector. In F. Brandão, Z. Breda, R. Costa, & C. Costa (Eds.), Handbook of Research on the Role of Tourism in Achieving Sustainable Development Goals (pp. 152-174). IGI Global. https://doi.org/ doi:10.4018/978-1-7998-5691-7.ch009

Papp-Váry, Á. F., & Tóth, T. Z. (2022). Analysis of Budapest as a Film Tourism Destination. In R. Baleiro & R. Pereira (Eds.), *Global Perspectives on Literary Tourism and Film-Induced Tourism* (pp. 257-279). IGI Global. https://doi.org/10.4018/978-1-7998-8262-6.ch014

Patiño, B. E. (2017). New Generation Management by Convergence and Individual Identity: A Systemic and Human-Oriented Approach. In N. Baporikar (Ed.), *Innovation and Shifting Perspectives in Management Education* (pp. 119–143). Hershey, PA: IGI Global. doi:10.4018/978-1-5225-1019-2.ch006

Patro, C. S. (2021). Digital Tourism: Influence of E-Marketing Technology. In M. Dinis, L. Bonixe, S. Lamy, & Z. Breda (Eds.), *Impact of New Media in Tourism* (pp. 234-254). IGI Global. https://doi.org/10.4018/978-1-7998-7095-1.ch014

Pawliczek, A., & Rössler, M. (2017). Knowledge of Management Tools and Systems in SMEs: Knowledge Transfer in Management. In A. Bencsik (Ed.), *Knowledge Management Initiatives and Strategies in Small and Medium Enterprises* (pp. 180–203). Hershey, PA: IGI Global. doi:10.4018/978-1-5225-1642-2.ch009

Pejic-Bach, M., Omazic, M. A., Aleksic, A., & Zoroja, J. (2018). Knowledge-Based Decision Making: A Multi-Case Analysis. In R. Leon (Ed.), *Managerial Strategies for Business Sustainability During Turbulent Times* (pp. 160–184). Hershey, PA: IGI Global. doi:10.4018/978-1-5225-2716-9.ch009

Perano, M., Hysa, X., & Calabrese, M. (2018). Strategic Planning, Cultural Context, and Business Continuity Management: Business Cases in the City of Shkoder. In A. Presenza & L. Sheehan (Eds.), *Geopolitics and Strategic Management in the Global Economy* (pp. 57–77). Hershey, PA: IGI Global. doi:10.4018/978-1-5225-2673-5.ch004

Pereira, R., Mira da Silva, M., & Lapão, L. V. (2017). IT Governance Maturity Patterns in Portuguese Healthcare. In S. De Haes & W. Van Grembergen (Eds.), *Strategic IT Governance and Alignment in Business Settings* (pp. 24–52). Hershey, PA: IGI Global. doi:10.4018/978-1-5225-0861-8.ch002

Pérez-Uribe, R. I., Torres, D. A., Jurado, S. P., & Prada, D. M. (2018). Cloud Tools for the Development of Project Management in SMEs. In R. Perez-Uribe, C. Salcedo-Perez, & D. Ocampo-Guzman (Eds.), *Handbook of Research on Intrapreneurship and Organizational Sustainability in SMEs* (pp. 95–120). Hershey, PA: IGI Global. doi:10.4018/978-1-5225-3543-0.ch005

Petrisor, I., & Cozmiuc, D. (2017). Global Supply Chain Management Organization at Siemens in the Advent of Industry 4.0. In L. Saglietto & C. Cezanne (Eds.), *Global Intermediation and Logistics Service Providers* (pp. 123–142). Hershey, PA: IGI Global. doi:10.4018/978-1-5225-2133-4.ch007

Pierce, J. M., Velliaris, D. M., & Edwards, J. (2017). A Living Case Study: A Journey Not a Destination. In N. Silton (Ed.), *Exploring the Benefits of Creativity in Education, Media, and the Arts* (pp. 158–178). Hershey, PA: IGI Global. doi:10.4018/978-1-5225-0504-4.ch008

Pipia, S., & Pipia, S. (2021). Challenges of Religious Tourism in the Conflict Region: An Example of Jerusalem. In E. Alaverdov & M. Bari (Eds.), *Global Development of Religious Tourism* (pp. 135-148). IGI Global. https://doi.org/10.4018/978-1-7998-5792-1.ch009

Poulaki, P., Kritikos, A., Vasilakis, N., & Valeri, M. (2022). The Contribution of Female Creativity to the Development of Gastronomic Tourism in Greece: The Case of the Island of Naxos in the South Aegean Region. In M. Valeri (Ed.), *New Governance and Management in Touristic Destinations* (pp. 246–258). IGI Global. https://doi.org/10.4018/978-1-6684-3889-3.ch015

Radosavljevic, M., & Andjelkovic, A. (2017). Multi-Criteria Decision Making Approach for Choosing Business Process for the Improvement: Upgrading of the Six Sigma Methodology. In J. Stanković, P. Delias, S. Marinković, & S. Rochhia (Eds.), *Tools and Techniques for Economic Decision Analysis* (pp. 225–247). Hershey, PA: IGI Global. doi:10.4018/978-1-5225-0959-2.ch011

Radovic, V. M. (2017). Corporate Sustainability and Responsibility and Disaster Risk Reduction: A Serbian Overview. In M. Camilleri (Ed.), *CSR 2.0 and the New Era of Corporate Citizenship* (pp. 147–164). Hershey, PA: IGI Global. doi:10.4018/978-1-5225-1842-6.ch008

Raghunath, K. M., Devi, S. L., & Patro, C. S. (2018). Impact of Risk Assessment Models on Risk Factors: A Holistic Outlook. In K. Strang, M. Korstanje, & N. Vajjhala (Eds.), *Research, Practices, and Innovations in Global Risk and Contingency Management* (pp. 134–153). Hershey, PA: IGI Global. doi:10.4018/978-1-5225-4754-9.ch008

Raman, A., & Goyal, D. P. (2017). Extending IMPLEMENT Framework for Enterprise Information Systems Implementation to Information System Innovation. In M. Tavana (Ed.), *Enterprise Information Systems and the Digitalization of Business Functions* (pp. 137–177). Hershey, PA: IGI Global. doi:10.4018/978-1-5225-2382-6.ch007

Rao, Y., & Zhang, Y. (2017). The Construction and Development of Academic Library Digital Special Subject Databases. In L. Ruan, Q. Zhu, & Y. Ye (Eds.), *Academic Library Development and Administration in China* (pp. 163–183). Hershey, PA: IGI Global. doi:10.4018/978-1-5225-0550-1.ch010

Ravasan, A. Z., Mohammadi, M. M., & Hamidi, H. (2018). An Investigation Into the Critical Success Factors of Implementing Information Technology Service Management Frameworks. In K. Jakobs (Ed.), *Corporate and Global Standardization Initiatives in Contemporary Society* (pp. 200–218). Hershey, PA: IGI Global. doi:10.4018/978-1-5225-5320-5.ch009

Rezaie, S., Mirabedini, S. J., & Abtahi, A. (2018). Designing a Model for Implementation of Business Intelligence in the Banking Industry. *International Journal of Enterprise Information Systems, 14*(1), 77–103. doi:10.4018/IJEIS.2018010105

Richards, V., Matthews, N., Williams, O. J., & Khan, Z. (2021). The Challenges of Accessible Tourism Information Systems for Tourists With Vision Impairment: Sensory Communications Beyond the Screen. In C. Eusébio, L. Teixeira, & M. Carneiro (Eds.), *ICT Tools and Applications for Accessible Tourism* (pp. 26–54). IGI Global. https://doi.org/10.4018/978-1-7998-6428-8.ch002

Rodrigues de Souza Neto, V., & Marques, O. (2021). Rural Tourism Fostering Welfare Through Sustainable Development: A Conceptual Approach. In A. Perinotto, V. Mayer, & J. Soares (Eds.), *Rebuilding and Restructuring the Tourism Industry: Infusion of Happiness and Quality of Life* (pp. 38–57). IGI Global. https://doi.org/10.4018/978-1-7998-7239-9.ch003

Romano, L., Grimaldi, R., & Colasuonno, F. S. (2017). Demand Management as a Success Factor in Project Portfolio Management. In L. Romano (Ed.), *Project Portfolio Management Strategies for Effective Organizational Operations* (pp. 202–219). Hershey, PA: IGI Global. doi:10.4018/978-1-5225-2151-8.ch008

Rubio-Escuderos, L., & García-Andreu, H. (2021). Competitiveness Factors of Accessible Tourism E-Travel Agencies. In C. Eusébio, L. Teixeira, & M. Carneiro (Eds.), *ICT Tools and Applications for Accessible Tourism* (pp. 196–217). IGI Global. https://doi.org/10.4018/978-1-7998-6428-8.ch009

Rucci, A. C., Porto, N., Darcy, S., & Becka, L. (2021). Smart and Accessible Cities?: Not Always – The Case for Accessible Tourism Initiatives in Buenos Aries and Sydney. In C. Eusébio, L. Teixeira, & M. Carneiro (Eds.), *ICT Tools and Applications for Accessible Tourism* (pp. 115–145). IGI Global. https://doi.org/10.4018/978-1-7998-6428-8.ch006

Ruhi, U. (2018). Towards an Interdisciplinary Socio-Technical Definition of Virtual Communities. In M. Khosrow-Pour, D.B.A. (Ed.), Encyclopedia of Information Science and Technology, Fourth Edition (pp. 4278-4295). Hershey, PA: IGI Global. doi:10.4018/978-1-5225-2255-3.ch371

Ryan, L., Catena, M., Ros, P., & Stephens, S. (2021). Designing Entrepreneurial Ecosystems to Support Resource Management in the Tourism Industry. In V. Costa, A. Moura, & M. Mira (Eds.), *Handbook of Research on Human Capital and People Management in the Tourism Industry* (pp. 265–281). IGI Global. https://doi.org/10.4018/978-1-7998-4318-4.ch013

Sabuncu, I. (2021). Understanding Tourist Perceptions and Expectations During Pandemic Through Social Media Big Data. In M. Demir, A. Dalgıç, & F. Ergen (Eds.), *Handbook of Research on the Impacts and Implications of COVID-19 on the Tourism Industry* (pp. 330–350). IGI Global. https://doi.org/10.4018/978-1-7998-8231-2.ch016

Safari, M. R., & Jiang, Q. (2018). The Theory and Practice of IT Governance Maturity and Strategies Alignment: Evidence From Banking Industry. *Journal of Global Information Management*, *26*(2), 127–146. doi:10.4018/JGIM.2018040106

Sahoo, J., Pati, B., & Mohanty, B. (2017). Knowledge Management as an Academic Discipline: An Assessment. In B. Gunjal (Ed.), *Managing Knowledge and Scholarly Assets in Academic Libraries* (pp. 99–126). Hershey, PA: IGI Global. doi:10.4018/978-1-5225-1741-2.ch005

Saini, D. (2017). Relevance of Teaching Values and Ethics in Management Education. In N. Baporikar (Ed.), *Management Education for Global Leadership* (pp. 90–111). Hershey, PA: IGI Global. doi:10.4018/978-1-5225-1013-0.ch005

Sambhanthan, A. (2017). Assessing and Benchmarking Sustainability in Organisations: An Integrated Conceptual Model. *International Journal of Systems and Service-Oriented Engineering*, *7*(4), 22–43. doi:10.4018/IJSSOE.2017100102

Sambhanthan, A., & Potdar, V. (2017). A Study of the Parameters Impacting Sustainability in Information Technology Organizations. *International Journal of Knowledge-Based Organizations*, *7*(3), 27–39. doi:10.4018/IJKBO.2017070103

Sánchez-Fernández, M. D., & Manríquez, M. R. (2018). The Entrepreneurial Spirit Based on Social Values: The Digital Generation. In P. Isaias & L. Carvalho (Eds.), *User Innovation and the Entrepreneurship Phenomenon in the Digital Economy* (pp. 173–193). Hershey, PA: IGI Global. doi:10.4018/978-1-5225-2826-5.ch009

Sanchez-Ruiz, L., & Blanco, B. (2017). Process Management for SMEs: Barriers, Enablers, and Benefits. In M. Vemić (Ed.), *Optimal Management Strategies in Small and Medium Enterprises* (pp. 293–319). Hershey, PA: IGI Global. doi:10.4018/978-1-5225-1949-2.ch014

Sanz, L. F., Gómez-Pérez, J., & Castillo-Martinez, A. (2018). Analysis of the European ICT Competence Frameworks. In V. Ahuja & S. Rathore (Eds.), *Multidisciplinary Perspectives on Human Capital and Information Technology Professionals* (pp. 225–245). Hershey, PA: IGI Global. doi:10.4018/978-1-5225-5297-0.ch012

Sarvepalli, A., & Godin, J. (2017). Business Process Management in the Classroom. *Journal of Cases on Information Technology, 19*(2), 17–28. doi:10.4018/JCIT.2017040102

Saxena, G. G., & Saxena, A. (2021). Host Community Role in Medical Tourism Development. In M. Singh & S. Kumaran (Eds.), *Growth of the Medical Tourism Industry and Its Impact on Society: Emerging Research and Opportunities* (pp. 105–127). IGI Global. https://doi.org/10.4018/978-1-7998-3427-4.ch006

Saygili, E. E., Ozturkoglu, Y., & Kocakulah, M. C. (2017). End Users' Perceptions of Critical Success Factors in ERP Applications. *International Journal of Enterprise Information Systems, 13*(4), 58–75. doi:10.4018/IJEIS.2017100104

Saygili, E. E., & Saygili, A. T. (2017). Contemporary Issues in Enterprise Information Systems: A Critical Review of CSFs in ERP Implementations. In M. Tavana (Ed.), *Enterprise Information Systems and the Digitalization of Business Functions* (pp. 120–136). Hershey, PA: IGI Global. doi:10.4018/978-1-5225-2382-6.ch006

Schwaiger, K. M., & Zehrer, A. (2021). The COVID-19 Pandemic and Organizational Resilience in Hospitality Family Firms: A Qualitative Approach. In A. Zehrer, G. Glowka, K. Schwaiger, & V. Ranacher-Lackner (Eds.), *Resiliency Models and Addressing Future Risks for Family Firms in the Tourism Industry* (pp. 32–49). IGI Global. https://doi.org/10.4018/978-1-7998-7352-5.ch002

Scott, N., & Campos, A. C. (2022). Cognitive Science of Tourism Experiences. In R. Augusto Costa, F. Brandão, Z. Breda, & C. Costa (Eds.), Planning and Managing the Experience Economy in Tourism (pp. 1-21). IGI Global. https://doi.org/ doi:10.4018/978-1-7998-8775-1.ch001

Seidenstricker, S., & Antonino, A. (2018). Business Model Innovation-Oriented Technology Management for Emergent Technologies. In M. Khosrow-Pour, D.B.A. (Ed.), Encyclopedia of Information Science and Technology, Fourth Edition (pp. 4560-4569). Hershey, PA: IGI Global. doi:10.4018/978-1-5225-2255-3.ch396

Selvi, M. S. (2021). Changes in Tourism Sales and Marketing Post COVID-19. In M. Demir, A. Dalgıç, & F. Ergen (Eds.), *Handbook of Research on the Impacts and Implications of COVID-19 on the Tourism Industry* (pp. 437–460). IGI Global. doi:10.4018/978-1-7998-8231-2.ch021

Senaratne, S., & Gunarathne, A. D. (2017). Excellence Perspective for Management Education from a Global Accountants' Hub in Asia. In N. Baporikar (Ed.), *Management Education for Global Leadership* (pp. 158–180). Hershey, PA: IGI Global. doi:10.4018/978-1-5225-1013-0.ch008

Sensuse, D. I., & Cahyaningsih, E. (2018). Knowledge Management Models: A Summative Review. *International Journal of Information Systems in the Service Sector, 10*(1), 71–100. doi:10.4018/IJISSS.2018010105

Seth, M., Goyal, D., & Kiran, R. (2017). Diminution of Impediments in Implementation of Supply Chain Management Information System for Enhancing its Effectiveness in Indian Automobile Industry. *Journal of Global Information Management, 25*(3), 1–20. doi:10.4018/JGIM.2017070101

Seyal, A. H., & Rahman, M. N. (2017). Investigating Impact of Inter-Organizational Factors in Measuring ERP Systems Success: Bruneian Perspectives. In M. Tavana (Ed.), *Enterprise Information Systems and the Digitalization of Business Functions* (pp. 178–204). Hershey, PA: IGI Global. doi:10.4018/978-1-5225-2382-6.ch008

Shaqrah, A. A. (2018). Analyzing Business Intelligence Systems Based on 7s Model of McKinsey. *International Journal of Business Intelligence Research, 9*(1), 53–63. doi:10.4018/IJBIR.2018010104

Sharma, A. J. (2017). Enhancing Sustainability through Experiential Learning in Management Education. In N. Baporikar (Ed.), *Management Education for Global Leadership* (pp. 256–274). Hershey, PA: IGI Global. doi:10.4018/978-1-5225-1013-0.ch013

Shetty, K. P. (2017). Responsible Global Leadership: Ethical Challenges in Management Education. In N. Baporikar (Ed.), *Innovation and Shifting Perspectives in Management Education* (pp. 194–223). Hershey, PA: IGI Global. doi:10.4018/978-1-5225-1019-2.ch009

Sinthupundaja, J., & Kohda, Y. (2017). Effects of Corporate Social Responsibility and Creating Shared Value on Sustainability. *International Journal of Sustainable Entrepreneurship and Corporate Social Responsibility, 2*(1), 27–38. doi:10.4018/IJSECSR.2017010103

Škarica, I., & Hrgović, A. V. (2018). Implementation of Total Quality Management Principles in Public Health Institutes in the Republic of Croatia. *International Journal of Productivity Management and Assessment Technologies, 6*(1), 1–16. doi:10.4018/IJPMAT.2018010101

Skokic, V. (2021). How Small Hotel Owners Practice Resilience: Longitudinal Study Among Small Family Hotels in Croatia. In A. Zehrer, G. Glowka, K. Schwaiger, & V. Ranacher-Lackner (Eds.), *Resiliency Models and Addressing Future Risks for Family Firms in the Tourism Industry* (pp. 50–73). IGI Global. doi:10.4018/978-1-7998-7352-5.ch003

Smuts, H., Kotzé, P., Van der Merwe, A., & Loock, M. (2017). Framework for Managing Shared Knowledge in an Information Systems Outsourcing Context. *International Journal of Knowledge Management, 13*(4), 1–30. doi:10.4018/IJKM.2017100101

Sousa, M. J., Cruz, R., Dias, I., & Caracol, C. (2017). Information Management Systems in the Supply Chain. In G. Jamil, A. Soares, & C. Pessoa (Eds.), *Handbook of Research on Information Management for Effective Logistics and Supply Chains* (pp. 469–485). Hershey, PA: IGI Global. doi:10.4018/978-1-5225-0973-8.ch025

Spremic, M., Turulja, L., & Bajgoric, N. (2018). Two Approaches in Assessing Business Continuity Management Attitudes in the Organizational Context. In N. Bajgoric (Ed.), *Always-On Enterprise Information Systems for Modern Organizations* (pp. 159–183). Hershey, PA: IGI Global. doi:10.4018/978-1-5225-3704-5.ch008

Steenkamp, A. L. (2018). Some Insights in Computer Science and Information Technology. In *Examining the Changing Role of Supervision in Doctoral Research Projects: Emerging Research and Opportunities* (pp. 113–133). Hershey, PA: IGI Global. doi:10.4018/978-1-5225-2610-0.ch005

Stipanović, C., Rudan, E., & Zubović, V. (2022). Reaching the New Tourist Through Creativity: Sustainable Development Challenges in Croatian Coastal Towns. In M. Valeri (Ed.), *New Governance and Management in Touristic Destinations* (pp. 231–245). IGI Global. https://doi.org/10.4018/978-1-6684-3889-3.ch014

Tabach, A., & Croteau, A. (2017). Configurations of Information Technology Governance Practices and Business Unit Performance. *International Journal of IT/Business Alignment and Governance, 8*(2), 1–27. doi:10.4018/IJITBAG.2017070101

Talaue, G. M., & Iqbal, T. (2017). Assessment of e-Business Mode of Selected Private Universities in the Philippines and Pakistan. *International Journal of Online Marketing, 7*(4), 63–77. doi:10.4018/IJOM.2017100105

Tam, G. C. (2017). Project Manager Sustainability Competence. In *Managerial Strategies and Green Solutions for Project Sustainability* (pp. 178–207). Hershey, PA: IGI Global. doi:10.4018/978-1-5225-2371-0.ch008

Tambo, T. (2018). Fashion Retail Innovation: About Context, Antecedents, and Outcome in Technological Change Projects. In I. Management Association (Ed.), Fashion and Textiles: Breakthroughs in Research and Practice (pp. 233-260). Hershey, PA: IGI Global. https://doi.org/ doi:10.4018/978-1-5225-3432-7.ch010

Tantau, A. D., & Frăţilă, L. C. (2018). Information and Management System for Renewable Energy Business. In *Entrepreneurship and Business Development in the Renewable Energy Sector* (pp. 200–244). Hershey, PA: IGI Global. doi:10.4018/978-1-5225-3625-3.ch006

Teixeira, N., Pardal, P. N., & Rafael, B. G. (2018). Internationalization, Financial Performance, and Organizational Challenges: A Success Case in Portugal. In L. Carvalho (Ed.), *Handbook of Research on Entrepreneurial Ecosystems and Social Dynamics in a Globalized World* (pp. 379–423). Hershey, PA: IGI Global. doi:10.4018/978-1-5225-3525-6.ch017

Teixeira, P., Teixeira, L., Eusébio, C., Silva, S., & Teixeira, A. (2021). The Impact of ICTs on Accessible Tourism: Evidence Based on a Systematic Literature Review. In C. Eusébio, L. Teixeira, & M. Carneiro (Eds.), *ICT Tools and Applications for Accessible Tourism* (pp. 1–25). IGI Global. doi:10.4018/978-1-7998-6428-8.ch001

Trad, A., & Kalpić, D. (2018). The Business Transformation Framework, Agile Project and Change Management. In M. Khosrow-Pour, D.B.A. (Ed.), Encyclopedia of Information Science and Technology, Fourth Edition (pp. 620-635). Hershey, PA: IGI Global. https://doi.org/ doi:10.4018/978-1-5225-2255-3.ch054

Trad, A., & Kalpić, D. (2018). The Business Transformation and Enterprise Architecture Framework: The Financial Engineering E-Risk Management and E-Law Integration. In B. Sergi, F. Fidanoski, M. Ziolo, & V. Naumovski (Eds.), *Regaining Global Stability After the Financial Crisis* (pp. 46–65). Hershey, PA: IGI Global. doi:10.4018/978-1-5225-4026-7.ch003

Trengereid, V. (2022). Conditions of Network Engagement: The Quest for a Common Good. In R. Augusto Costa, F. Brandão, Z. Breda, & C. Costa (Eds.), *Planning and Managing the Experience Economy in Tourism* (pp. 69-84). IGI Global. https://doi.org/10.4018/978-1-7998-8775-1.ch004

Turulja, L., & Bajgoric, N. (2018). Business Continuity and Information Systems: A Systematic Literature Review. In N. Bajgoric (Ed.), *Always-On Enterprise Information Systems for Modern Organizations* (pp. 60–87). Hershey, PA: IGI Global. doi:10.4018/978-1-5225-3704-5.ch004

Vargas-Hernández, J. G. (2017). Professional Integrity in Business Management Education. In N. Baporikar (Ed.), *Management Education for Global Leadership* (pp. 70–89). Hershey, PA: IGI Global. doi:10.4018/978-1-5225-1013-0.ch004

Varnacı Uzun, F. (2021). The Destination Preferences of Foreign Tourists During the COVID-19 Pandemic and Attitudes Towards: Marmaris, Turkey. In M. Demir, A. Dalgıç, & F. Ergen (Eds.), *Handbook of Research on the Impacts and Implications of COVID-19 on the Tourism Industry* (pp. 285–306). IGI Global. https://doi.org/10.4018/978-1-7998-8231-2.ch014

Vasista, T. G., & AlAbdullatif, A. M. (2017). Role of Electronic Customer Relationship Management in Demand Chain Management: A Predictive Analytic Approach. *International Journal of Information Systems and Supply Chain Management, 10*(1), 53–67. doi:10.4018/IJISSCM.2017010104

Vieru, D., & Bourdeau, S. (2017). Survival in the Digital Era: A Digital Competence-Based Multi-Case Study in the Canadian SME Clothing Industry. *International Journal of Social and Organizational Dynamics in IT, 6*(1), 17–34. doi:10.4018/IJSODIT.2017010102

Vijayan, G., & Kamarulzaman, N. H. (2017). An Introduction to Sustainable Supply Chain Management and Business Implications. In M. Khan, M. Hussain, & M. Ajmal (Eds.), *Green Supply Chain Management for Sustainable Business Practice* (pp. 27–50). Hershey, PA: IGI Global. doi:10.4018/978-1-5225-0635-5.ch002

Vlachvei, A., & Notta, O. (2017). Firm Competitiveness: Theories, Evidence, and Measurement. In A. Vlachvei, O. Notta, K. Karantininis, & N. Tsounis (Eds.), *Factors Affecting Firm Competitiveness and Performance in the Modern Business World* (pp. 1–42). Hershey, PA: IGI Global. doi:10.4018/978-1-5225-0843-4.ch001

Wang, C., Schofield, M., Li, X., & Ou, X. (2017). Do Chinese Students in Public and Private Higher Education Institutes Perform at Different Level in One of the Leadership Skills: Critical Thinking?: An Exploratory Comparison. In V. Wang (Ed.), *Encyclopedia of Strategic Leadership and Management* (pp. 160–181). Hershey, PA: IGI Global. doi:10.4018/978-1-5225-1049-9.ch013

Wang, J. (2017). Multi-Agent based Production Management Decision System Modelling for the Textile Enterprise. *Journal of Global Information Management, 25*(4), 1–15. doi:10.4018/JGIM.2017100101

Wiedemann, A., & Gewald, H. (2017). Examining Cross-Domain Alignment: The Correlation of Business Strategy, IT Management, and IT Business Value. *International Journal of IT/Business Alignment and Governance, 8*(1), 17–31. doi:10.4018/IJITBAG.2017010102

Wolf, R., & Thiel, M. (2018). Advancing Global Business Ethics in China: Reducing Poverty Through Human and Social Welfare. In S. Hipsher (Ed.), *Examining the Private Sector's Role in Wealth Creation and Poverty Reduction* (pp. 67–84). Hershey, PA: IGI Global. doi:10.4018/978-1-5225-3117-3.ch004

Yablonsky, S. (2018). Innovation Platforms: Data and Analytics Platforms. In *Multi-Sided Platforms (MSPs) and Sharing Strategies in the Digital Economy: Emerging Research and Opportunities* (pp. 72–95). Hershey, PA: IGI Global. doi:10.4018/978-1-5225-5457-8.ch003

Yaşar, B. (2021). The Impact of COVID-19 on Volatility of Tourism Stocks: Evidence From BIST Tourism Index. In M. Demir, A. Dalgıç, & F. Ergen (Eds.), *Handbook of Research on the Impacts and Implications of COVID-19 on the Tourism Industry* (pp. 23–44). IGI Global. https://doi.org/10.4018/978-1-7998-8231-2.ch002

Yusoff, A., Ahmad, N. H., & Halim, H. A. (2017). Agropreneurship among Gen Y in Malaysia: The Role of Academic Institutions. In N. Ahmad, T. Ramayah, H. Halim, & S. Rahman (Eds.), *Handbook of Research on Small and Medium Enterprises in Developing Countries* (pp. 23–47). Hershey, PA: IGI Global. doi:10.4018/978-1-5225-2165-5.ch002

Zacher, D., & Pechlaner, H. (2021). Resilience as an Opportunity Approach: Challenges and Perspectives for Private Sector Participation on a Community Level. In A. Zehrer, G. Glowka, K. Schwaiger, & V. Ranacher-Lackner (Eds.), *Resiliency Models and Addressing Future Risks for Family Firms in the Tourism Industry* (pp. 75–102). IGI Global. https://doi.org/10.4018/978-1-7998-7352-5.ch004

Zanin, F., Comuzzi, E., & Costantini, A. (2018). The Effect of Business Strategy and Stock Market Listing on the Use of Risk Assessment Tools. In *Management Control Systems in Complex Settings: Emerging Research and Opportunities* (pp. 145–168). Hershey, PA: IGI Global. doi:10.4018/978-1-5225-3987-2.ch007

Zgheib, P. W. (2017). Corporate Innovation and Intrapreneurship in the Middle East. In P. Zgheib (Ed.), *Entrepreneurship and Business Innovation in the Middle East* (pp. 37–56). Hershey, PA: IGI Global. doi:10.4018/978-1-5225-2066-5.ch003

About the Contributors

Agustinus Borgy Waluyo holds a PhD in Computer Science. He has worked in various research projects funded by Research Organization, University and Government, and actively reported his research outcomes in peer-reviewed journals, book chapters, as well as conferences on mobile, wireless and pervasive computing. He also actively serves in the organizing and technical program committee of international conferences including IEEE International Conference on Communications and IEEE Global Communications Conference, and is currently the chief editor of the International Journal of Mobile Computing and Multimedia Communications.

Ling Tan is a Research Fellow at the Australian Council for Educational Research (ACER). His current research interests include psychometric analysis, computerised adaptive testing algorithms, and applying statistical learning and data mining technology to educational measurement.

* * *

Aysu Altaş, Ph.D., has been working as an Associate Professor at Aksaray University, Faculty of Tourism, Department of Tourism Guidance. Dr. Altaş has published numerous papers in several journals, conference proceedings, and chapters in edited books since she got her MA. Her major is gastronomy and gastronomy tourism. She also has books on coffee tourism, tea tourism, beer tourism, and wine tourism.

Nedra Bahri-Ammari is an associate professor of marketing and responsible of the Research Master of Marketing at IHEC of Carthage, Tunisia. Dr. Bahri's researches include: customer relationships management (CRM), social-CRM, Digital Marketing, Artificial Intelligence, Augmented reality, Web 4.0, tourism marketing and tourism behavior. She has published in reputed peer-reviewed academic journals (indexed JCR, ABDC, CNRS, Scopus…) as: Journal of Retailing and Consumer Services; International Journal of Contemporary Hospitality Management; International Journal of Mobile Communications, Journal of Hospitality and Tourism

Technology; Journal of Research in Marketing and Entrepreneurship; Management Research Review; Journal of Marketing & Case studies; International Journal of Customer Relationship Marketing and Management. She has published a book on CRM and loyalty strategy, also participated in the writing of books on tourism and social CRM, Competitive Social Media & marketing Strategies IGI Global; Encyclopedia of Information Science and Technologies, 4th Edition of the IGI Global Scholarly reference book. She has published in international and national academic conferences proceedings (EUROMED, AFM, World Research Summit for Hospitality and Tourism, Academy of Marketing Science). And she serves as an editorial board member and reviewer for some of international journals.

Yulita Hanum P. Iskandar is a Senior Lecturer in Graduate School of Business, Universiti Sains Malaysia. She currently teaches 'Management Information Systems' and 'Technology Management' for postgraduates and specializes in research related to technological and innovation.

Mehmet Altuğ Şahin works as a Research Assistant at the Department of Tourism Management, Istanbul University Faculty of Economics. He holds a Ph.D. and a Master's Degree in Tourism Management. His research interests include big data and smart tourism, social media analysis, revenue management, and hospitality accounting..

Siti Salina Saidin is a senior lecturer at the Faculty of Hospitality, Tourism and Wellness, Universiti Malaysia Kelantan (UMK). She obtained a bachelor's degree in Computer Science from Universiti Sains Malaysia (USM). She holds a Master's degree in Business Administration from Universiti Teknologi MARA (UiTM). Subsequently, she received a Doctor of Philosophy in Business Information Technology from Universiti Sains Malaysia (USM) in 2019. Before joining the academic field, she was working as a computer engineer for two multinational companies in Kuala Lumpur and Cyberjaya. She also has experience as a lecturer at the Asian Metropolitan University and has been teaching information technology subject for almost five years. Her research area includes business information technology, cybersecurity, internet research, e-learning and digital tourism.

Phoebe Yueng Hee Sia is a solution manager responsible for software compliance solutions at SAP R/3 and ERP/ECC 6.0 include modules: Financials and Controlling (FICO), Human Resources (HR), Materials Management (MM), Sales & Distribution (SD), Production Planning (PP) and Plant Maintenance (PM), covering the typical functions in place in an organization listed on the Main Board of Bursa Malaysia Securities Berhad. Has over 20 years of experience working in different roles in information development and product and solution management. Comprise roles of IT Application Assistant Manager, Senior Application Specialist and IT Developer responsible to establish project plan/timelines, project implementation/execution, application supports and development across multiple industries. Holds a Master Degree in Business Administration from the Universiti Sains Malaysia, graduated from the Campbell University, North Carolina USA with Bachelor of Science in Business Information Technology and awarded an Advance Diploma cum Diploma of Business Information System and Certification of Computing with Accounting from Tunku Abdul Rahman University College.

Anusha Thakur has done B.Tech in "Electronics and Instrumentation Technology" from Tumkur, Karnataka, and Master's in Business Administration (MBA) in "International Business" from University of Petroleum & Energy Studies, (UPES) Dehradun. In addition to this, Anusha has also completed a Post-Graduation certification in "Market Research and Data Analytics" from MICA, Ahmedabad. Anusha has also been writing few papers and chapters for the journals/books in India and abroad. One of her published works includes chapter on "Market for Plant-Based Meat Alternatives" for the book "Environmental, Health, and Business Opportunities in the New Meat Alternatives Market" which has won accolades across the world. The book was considered #1 in the WORLD in the category of Vegetarian Writing. Further, the book was also announced as the national winner for Australia in the Vegetarian Writing Category. Further, she has a work experience of approximately 4 years in the Market Research industry. Her work primarily focuses, on the secondary research, market analysis, market sizing, interpretation, forecasting and report writing.

Zafer Türkmendağ is a Ph.D. at the Faculty of Tourism, Atatürk University, Turkey. He holds his Ph.D. in tourism management in 2019 from the Gazi University, Turkey. His research interests include tourism technology, tourism marketing and management, accessible tourism, and tourism guidance. He has certificates in information technology (Middle East Technical University) and hotel management systems (Oracle Opera 5.0). His researches have been published in the Journal of Organizational Change Management.

Ali Yuce is born and grew up in Turkey. After working independently for his business, he changed his life dramatically and radically to pursue his passion for academia. After receiving his Bachelor's degree from DePaul University, he obtained his Master's and Doctoral Degrees from Eastern Mediterranean University. Prior to graduation from the Ph.D. program, he is the author of several articles and book chapters in international journals along with numerous congress papers about the impact of technology on the landscape of tourism, education, and marketing. On the other side, he is a lifelong learner who loves conveying every bit of knowledge that he learns for others. He enjoys swimming, cooking, jogging, reading about history, technology, tourism; he also loves animals and nature.

Yiğit Yurder graduated from Istanbul University, Faculty of Business Administration in 2009. He completed his doctoral dissertation in the field of business administration in 2017, titled "Factors Affecting the Consumer Factor in the Adoption of Technology". Since 2019, he has been working as an Assistant Professor at Istanbul University, Faculty of Economics, Department of Marketing. Yurder's field of work includes marketing, brand management and digital marketing.

Index

V

www.ingramcontent.com/pod-product-compliance
Lightning Source LLC
LaVergne TN
LVHW061540290926
849162LV00004B/858